Designed Molecular Space in Material Science and Catalysis

Seiji Shirakawa
Editor

Designed Molecular Space in Material Science and Catalysis

Springer

Editor
Seiji Shirakawa
Department of Environmental Science
Graduate School of Fisheries and
Environmental Sciences
Nagasaki University
Nagasaki, Japan

ISBN 978-981-13-1255-7 ISBN 978-981-13-1256-4 (eBook)
https://doi.org/10.1007/978-981-13-1256-4

Library of Congress Control Number: 2018956411

This Springer imprint is published by the registered company Springer Nature Singapore Pte Ltd.
The registered company address is: 152 Beach Road, #21-01/04 Gateway East, Singapore 189721, Singapore

Preface

Design of molecular space is an important theme for the development of novel functional materials and catalysts. This theme connects a range of diverse research areas such as supramolecular chemistry, material science, chemical biology, organic synthesis, and catalysis. Based on this important theme, we started the Chemical Society of Japan (CSJ) research group "Creation of Molecular Space toward Innovative Transformations" (website: http://molspace.chemistry.or.jp/index.html) in 2014 to advance molecular space chemistry. Chapters of this book were written by the research group members. Professors Maeda and Haketa, Ritsumeikan University, warmly accepted our invitation to join in this book as additional authors.

A wide variety of topics and ideas included in this book are based on the concept of molecular space chemistry. The book showcases recent representative examples of molecular space design to create functional materials and catalysts possessing unique properties. This unique volume will be of great interest to chemists in a wide variety of research fields, including organic, inorganic, biological, polymer, and supramolecular chemistry. Readers will obtain new ideas and directions to create novel functional molecules, and those ideas will lead to innovative views of science.

Finally, I would like to warmly thank all the authors for their contributions to create this book. I also would like to thank the team at Springer Japan, especially Shinichi Koizumi, Asami Komada, and Taeko Sato, for their helpful assistance during the preparation of this book.

Nagasaki, Japan
April 2018

Seiji Shirakawa

Contents

Chapter 1
π-Electronic Ion-Pairing Supramolecular Assemblies

Yohei Haketa and Hiromitsu Maeda

Abstract Appropriately designed anion-responsive π-electronic molecules provide tunable π-electronic ion pairs comprising receptor–anion complexes and countercations. Such π-electronic ion pairs form highly organized dimension-controlled assemblies such as liquid crystals, gels, and other functional soft materials through electrostatic interactions with the support of other weak noncovalent interactions. Well-designed pyrrole-based π-electronic molecules can effectively bind anions, thus creating planar receptor–anion complexes and resulting assembled structures. Therefore, precise modifications of oligopyrrole π-electronic systems are important for the preparation of receptor–anion-complexing ion pairs as components of ion-pairing dimension-controlled assemblies. This chapter describes recent progresses in the area of ion-pairing dimension-controlled assemblies, especially those built with pyrrole-based anion-responsive π-electronic molecules, focusing on molecular design.

Keywords π-Electronic anions · π-Electronic cations · Anion-responsive molecules · Pyrrole derivatives · Ion-pairing assemblies · Dimension-controlled assemblies · Liquid crystals

1.1 Introduction

Assemblies of ion pairs can provide fascinating functional materials including solids (single crystals), soft materials (gels, liquid crystals, etc.) [1–4], and liquid states (ionic liquids) [5, 6], depending on the geometries and electronic states of the constituent charged species. Among noncovalent interactions, electrostatic attractive and repulsive interactions between charged species are the main driving forces for ion-pairing assemblies, thus affording various assembling states in combination with

Y. Haketa · H. Maeda (✉)
Department of Applied Chemistry, College of Life Sciences, Ritsumeikan University, Kusatsu, Japan
e-mail: maedahir@ph.ritsumei.ac.jp

S. Shirakawa (ed.), *Designed Molecular Space in Material Science and Catalysis*,
https://doi.org/10.1007/978-981-13-1256-4_1

Fig. 1.1 Three representative examples of π-electronic ion pairs: (**a**) a pair of electronically neutral π-electronic systems possessing a charged part(s) and counterion, (**b**) a pair of charged π-electronic systems, and (**c**) a pair of π-electronic receptor–ion complex and counterion

other weak noncovalent interactions (e.g., hydrogen bonding, van der Waals, π–π, dipole–dipole, etc.) [7–13]. Such noncovalent interactions between designed charged species can provide highly organized assembling structures, namely, an ionic self-assembly [14, 15], based on the anisotropic orientation of π-electronic ions, thus affording various unique and functional dimension-controlled ion-pairing assemblies [16–21].

π-Electronic molecules have several advantages as building units of molecular assemblies: various substituents including aliphatic chains can be introduced into π-electronic molecules, whose characteristic electronic properties are applicable for electrical and optical materials. Incorporation of ionic unit(s) into π-electronic molecules provides various organized assemblies using electrostatic interactions. There are three representative methods to prepare π-electronic ion pairs forming dimension-controlled assemblies (Fig. 1.1): (a) a pair of electronically neutral π-electronic systems connecting charged part(s) and corresponding counterion(s), (b) a pair consisting of a π-electronic cation and a π-electronic anion (genuine π-electronic ion pair), and (c) a pair consisting of a π-electronic receptor–ion complex (pseudo π-electronic ion) and its counterion(s). Based on these ion-pair formation strategies, various ion-pairing assemblies have been reported to date, focusing on characteristic assembling behaviors, morphologies, and functions. Among the three strategies for the preparation of π-electronic ionic species shown above, π-electronic receptor–ion-complexing ion pairs (Fig. 1.1c) have a great advantage for preparation of fine-tuned components, because synthesis of electronically neutral π-electronic molecules is much easier than that of charged species. Most importantly, the appropriate combinations of π-electronic receptors, modified anions, and countercations can provide diverse ion pairs. Receptor–ion complexes, which are in equilibrium in solution state, should be stable enough for the formation of ion-pairing assemblies in bulk materials. The design of anion-responsive π-electronic molecules that form assemblies is important in this method.

Appropriately designed π-electronic ion pairs with suitable geometries and electronic states result in an ordered arrangement of the ionic components and the formation of dimension-controlled ion-pairing assemblies with diverse modes. A charge-by-charge assembly is defined as an assembling mode comprising alternately stacking positively and negatively charged species (Fig. 1.2 left), whereas a charge-segregated assembly results from the stacking of identically charged species by

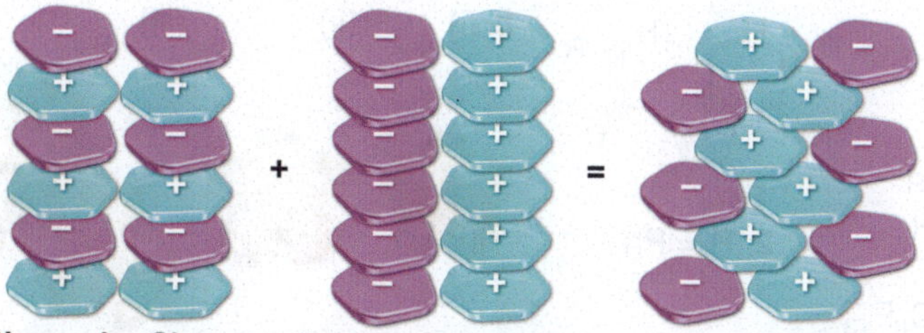

Fig. 1.2 Conceptual diagram of π-electronic ion-pairing assemblies comprising the ion pairs in Fig. 1.1b

overcoming electrostatic repulsion (Fig. 1.2 center). A charge-by-charge mode is effective for constructing columnar assemblies because of the attractive force between opposite charges supporting π–π stacking. In most cases, partial contributions of charge-by-charge and charge-segregated assemblies would provide intermediate assembling modes (Fig. 1.2 right).

This chapter presents an overview of the design and synthesis of anion-responsive π-electronic molecules that form dimension-controlled assemblies, especially pyrrole-based receptor–anion-complexing ion-pairing systems reported by our group. Furthermore, various ion-pairing assemblies, including crystals, liquid crystals, and supramolecular gels, are shown with their characteristic assembling modes and properties.

1.2 π-Electronic Anionic Species Based on Anion-Responsive π-Electronic Molecules

1.2.1 Pyrrole-Based π-Electronic Anion-Responsive Molecules

Design of anion-responsive π-electronic molecules that form planar receptor–anion complexes requires the incorporation of appropriate anion-binding sites. To date, various anion receptors comprising π-electronic building units have been investigated [22–27]. Anion binding is mainly induced by hydrogen bonding of polarized hydrogens in aromatic units [28]. For example, the NH moieties of pyrroles and indoles can act as effective hydrogen-bonding donor units. Their π-electronic planes are also suitable for stacking assemblies with the support of peripheral substituents. Covalently linked appropriate arrangement of such anion-interacting subunits affords planar receptor–anion complexes. Among anion-interacting π-electronic units, pyrrole has been widely investigated because it is found in various naturally occurring anion-binding molecules such as tripyrrolic prodigiosin derivatives [27]. While the anion-binding behaviors of cyclic oligopyrrole derivatives have

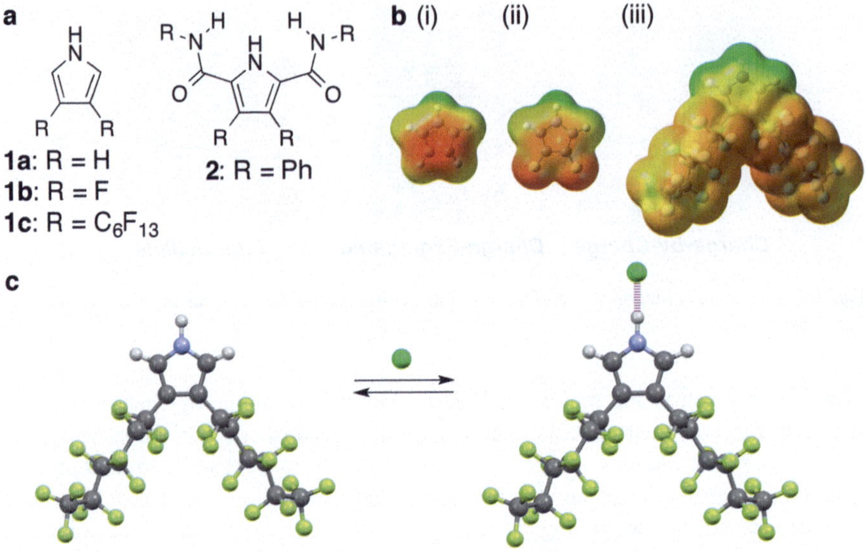

Fig. 1.3 (**a**) Anion-responsive monopyrroles with electron-withdrawing units **1a–c** and **2** (a preorganized structure), (**b**) ESP maps of (*i*) **1a**, (*ii*) **1b**, and (*iii*) **1c**, and (**c**) anion-binding mode of **1c**

been extensively studied, acyclic pyrrole-based receptors have been less investigated, due to the low binding ability of less preorganized structures. However, appropriate modifications of pyrrole-based anion-responsive molecules would enhance the binding abilities. Importantly, appropriate acyclic oligopyrroles show planar geometries and the facile introduction of aliphatic chains induces the formation of molecular assemblies.

The hydrogen-bonding abilities of pyrrole NH sites can be tuned by modification of the substituents. Introduction of electron-withdrawing moieties at the pyrrole α- and β-positions as seen in **1b**,**c** and **2** (Fig. 1.3a) enhances their anion-binding behaviors using the pyrrole NH site. In fact, 3,4-difluoro-substituted **1b** and 3,4-diperfluorohexyl-substituted **1c** were found to be electron-deficient π-systems as evaluated by density functional theory (DFT) calculations [29]. The electrostatic potential (ESP) of **1c** was more electron-deficient, suggesting that the pyrrole NH was more polarized than **1a**,**b** (Fig. 1.3b). Anion binding through single-point hydrogen bonding via pyrrole NH (Fig. 1.3c) was revealed by ^{1}H NMR. The anion-binding constants (K_a) followed the order of **1c** > **1b** > **1a**: the K_a value of **1c** is 1400 M^{-1} for Cl^- in CD_2Cl_2, which is about ten times greater than that of **1b**. Introduction of electron-withdrawing units at pyrrole α-positions is also effective. For example, the introduction of phenylamide units at pyrrole α-positions as seen in **2** shows enhanced binding ability compared to **1a** (Fig. 1.3a) [30]. In this case, the electron-withdrawing carbonyl units increase the stability of hydrogen bonding of pyrrole NH, and the amide NH also interacts supportively with the anion. The electron-withdrawing moieties showed the contributions of effective hydrogen-

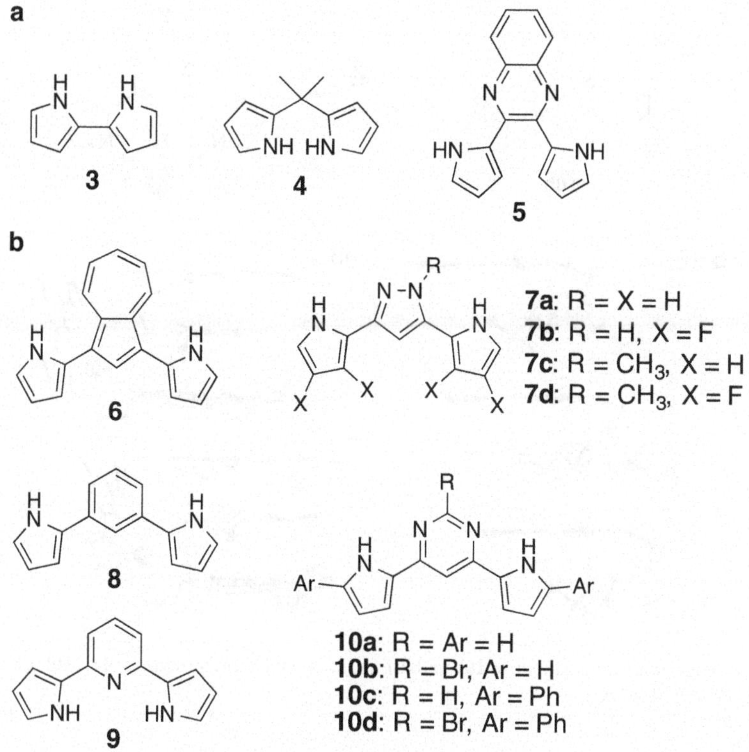

Fig. 1.4 (**a**) C_0–C_2-bridged dipyrrole receptors **3–5** and (**b**) C_3-bridged dipyrrole receptors **6–10**

bonding and ion–dipole interactions. Modification of pyrrole derivatives as the building units of π-electronic anion receptors is essential for tunable anion-binding abilities and the resulting formation of π-electronic anions.

As the arrangement of pyrrole NH sites is important for effective anion binding, various oligopyrrole-based anion-responsive molecules were developed (e.g., **3–5**, Fig. 1.4a). The number of atoms bridging two pyrrole rings is important. For example, C_0-bridged bipyrrole **3** [31] and C_1-bridged dipyrromethane **4** [32] show effective anion-binding behaviors compared to pyrrole **1a**. Although C_2-bridged dipyrrolylquinoxaline **5** [33] also binds anions in [1+1]-type mode, the close location of two pyrrole NH units is interfered with, thus forming non-planar anion complexes. Planar receptor–anion complexes can be obtained by bridging multiple pyrroles via an appropriate aromatic moiety. Based on theoretical studies of molecular geometries, five- and six-membered rings are more suitable as C_3-bridged units for pyrroles to provide planar anion-responsive molecules. As examples of π-electronic anion receptors consisting of two pyrroles bridged with a five-membered ring (e.g., **6** and **7a–d**, Fig. 1.4b), dipyrrolylazulene **6** [34] binds anion by hydrogen bonding of two pyrrole NH sites as well as the bridging CH unit; the K_a value for Cl^- added as a tetrabutylammonium (TBA) salt is 110 M^{-1} in CH_2Cl_2.

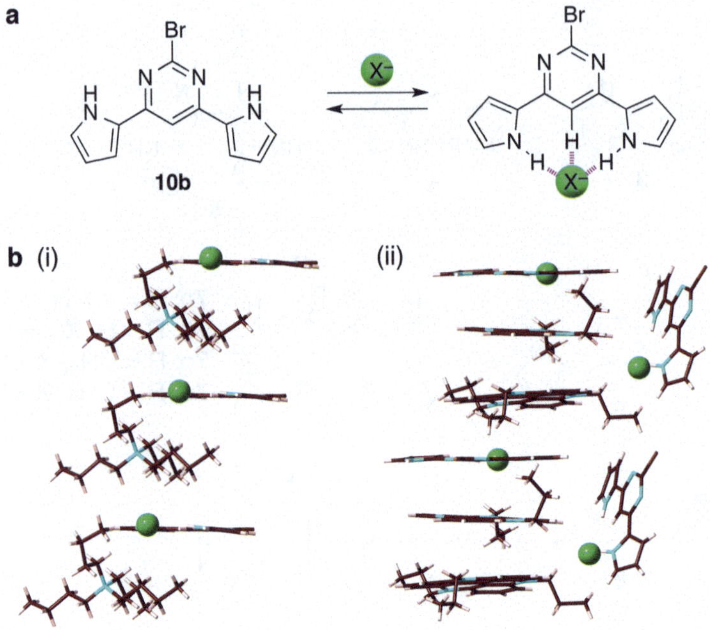

Fig. 1.5 (**a**) Anion-binding mode of **10b** and (**b**) single-crystal X-ray structures of (*i*) **10b**·Cl^--TBA^+ and (*ii*) **10b**·Cl^--$TATA^+$

Furthermore, dipyrrolylpyrazoles (e.g., **7a**, Fig. 1.4b) bind anions through hydrogen bonding of the pyrrole NH and bridging pyrazole CH with the inversion of the pyrrole rings [35–37]. In particular, *N*-methyl-substituted dipyrroles **7c,d** with β-hydrogen and β-fluorine substituents showed K_a values of 1600 and 28,000 M^{-1}, respectively, indicating that the electron-withdrawing β-fluorine moieties enhance the anion-binding ability [35].

Other anion-responsive π-electronic molecules include those consisting of two pyrroles bridged with a C_3-bridged six-membered ring. For example, 1,3-dipyrrolylbenzene **8** (Fig. 1.4b), which was first reported by Sessler et al. as a building subunit of larger macrocycles, showed effective anion binding through hydrogen bonding of pyrrole NH [32]. Introduction of electron-withdrawing units at the pyrrole α-positions of dipyrrolylbenzene enhances the K_a values [38]. In contrast, 2,6-dipyrrolylpyridine **9** showed poor anion-binding behavior due to repulsion between the electron pair of the pyridyl nitrogen and anions along with intramolecular pyrrole-N–H···pyridyl-N hydrogen bonding [38, 39]. On the other hand, 4,6-dipyrrolylpyrimidines **10b,d** (Fig. 1.4b) were prepared using Suzuki coupling reaction starting from bromo-substituted pyrimidine and pyrroles with boronic acid or its ester, and subsequent debromination gave **10a,c** (Fig. 1.4b) [40]. The dipyrrolylpyrimidines showed anion binding through hydrogen bonding by the inversion of two pyrrole rings (Fig. 1.5a). DFT studies suggested that the conformation with both pyrrole-NH groups pointing toward the pyrimidine-N side is more

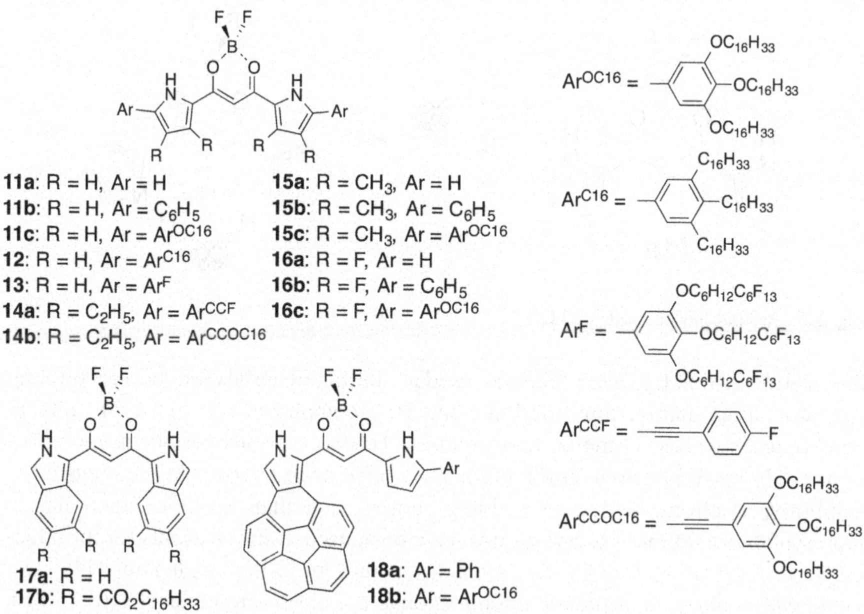

Fig. 1.6 Representatives of dipyrrolyldiketone BF_2 complexes **11–18**

stable at 9.84 kcal/mol than that with the opposite orientations of pyrimidine and pyrrole dipoles. In fact, the anion-binding modes of the dipyrrolylpyrimidines (Fig. 1.5a) were revealed by ^{1}H NMR upon the addition of Cl^- as a TBA salt in CD_2Cl_2. At –50 °C, the signals of the pyrrole NH and pyrimidine 5-CH of **10b** at 9.72 and 7.49 ppm gradually disappeared, and those at 11.84 and 9.40 ppm concurrently appeared with Cl^- binding. Such independent signal changes due to the anion-free and anion-binding states were originally observed for the [1+1]-type anion binding of dipyrrolyldiketone BF_2 complexes with anions (*vide infra*). The [1+1]-type pyrrole-inverted receptor–Cl^- complex of **10b** was also revealed by single-crystal X-ray analysis, providing a one-by-one charge-by-charge assembling mode with counter TBA^+ in **10b**·Cl^--TBA^+ (Fig. 1.5b(i)), whereas one **10b**·Cl^- and two propyl-substituted triazatriangulenium cations ($TATA^+$) [41, 42] formed a columnar structure in **10b**·Cl^--$TATA^+$ (Fig. 1.5b(ii)) [40]. It should be noticed that **10b**·Cl^- forms planar geometries (planar anions) in combination with both countercations in the crystal structures (Fig. 1.5b).

As dipyrrole-based π-electronic anion receptors bridged with a six-membered ring, we have been investigating dipyrrolyldiketone boron complexes since 2005. A series of dipyrrolyldiketone BF_2 complexes have been investigated (**11–18** as examples, Fig. 1.6) as effective anion receptors that provide various π-electronic anionic species [16–21, 43–84].

Dipyrrolyldiketone derivatives were synthesized mainly by two methods: (i) starting pyrrole derivatives are reacted with malonyl chloride in CH_2Cl_2, followed by addition of $BF_3·OEt_2$ and (ii) α-iodinated derivatives of β-modified

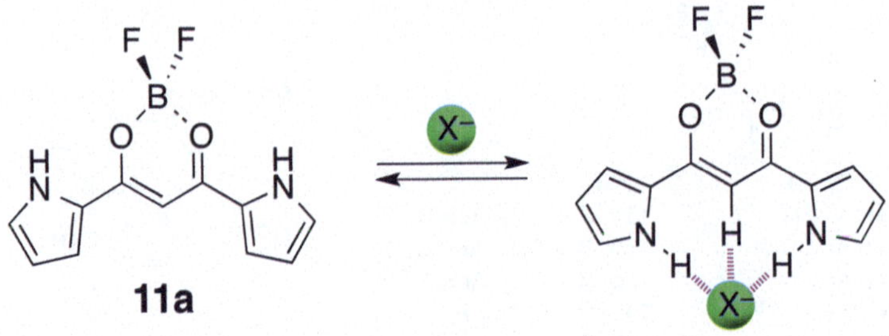

Fig. 1.7 Anion-binding mode of **11a**

dipyrrolyldiketone BF_2 complexes are used as starting materials for various coupling reactions. Importantly, dipyrrolyldiketone BF_2 complexes are stable in ambient conditions as well as at higher temperatures. To date, various peripheral modifications of dipyrrolyldiketone boron complexes have been investigated to control the solubility, electronic states, assembling modes, and their specific functions. In representative examples, α-aryl groups have been used as the scaffolds for introducing various substituents to the core π-electronic unit for further applications [44–53], as shown in alkoxy-substituted anion-responsive π-electronic molecules (e.g., **11c**, Fig. 1.6) that form anion-responsive supramolecular gels [44, 49, 52, 53] and thermotropic liquid crystals [47, 51, 53]. The coupling reaction of pyrrole β-modified receptors [54–72] with iodination at the pyrrole α-positions results in the introduction of various aryl moieties that exhibit diverse electronic properties [63], substituent-dependent assemblies (e.g., **15c** and **16c**, Fig. 1.6) [67], covalently linked oligomers for high anion-binding affinities and ion-pairing chirality induction [65, 68], and cooperatively interlocked [2+1]-type receptor–anion complexes (e.g., **14a**, Fig. 1.6) [69–72]. Furthermore, the boron unit can also be modified [73–76], as seen in catechol–boron and 1,1′-bi-2-naphthol–boron complexes, exhibiting anion-driven circularly polarized luminescence (CPL) enhancement for the latter [75].

As two pyrrole rings in the most stable conformations are not located at preorganized positions for anion binding, they should be inverted for anion binding (Fig. 1.7). Efficient [1+1]-type anion-binding abilities were elucidated by ^{1}H NMR spectral changes, wherein anion-free and anion-complexing signals were independently observed due to the slow anion-binding processes that correlate with pyrrole inversion. K_a values have been evaluated by UV/Vis absorption spectral changes upon the addition of anions as TBA salts; the K_a value of parent **11a** for Cl^- is 15,000 M^{-1} in CH_2Cl_2 at r.t. [43]. α-Aryl-substituted receptors also form [1+1]-type receptor–anion complexes with the help of *o*-CH hydrogen bonding to the anion, thus showing a higher K_a, for example, for **11b** of 30,000 M^{-1}.

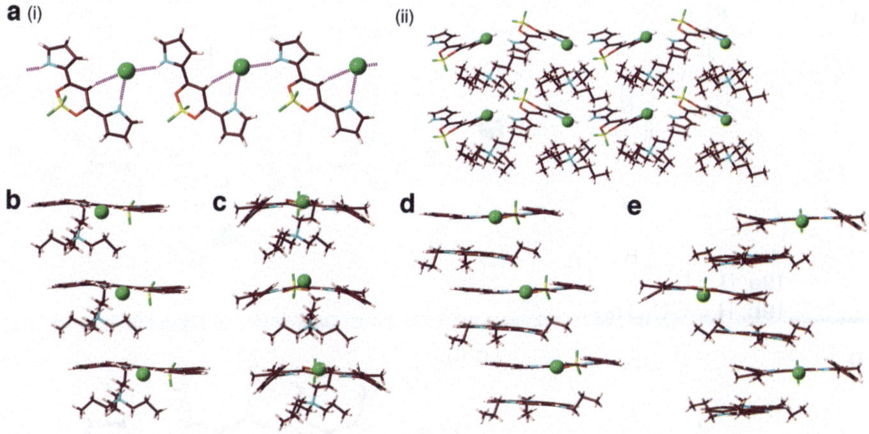

Fig. 1.8 Single-crystal X-ray structures of (**a**) **11a**·Cl⁻-TBA⁺ ((*i*) hydrogen-bonding chain and (*ii*) packing structures)), (**b**) **11b**·Cl⁻-TPA⁺, (**c**) **15b**·Cl⁻-TPA⁺, (**d**) **11b**·Cl⁻-TATA⁺, and (**e**) **15b**·Cl⁻-TATA⁺

1.2.2 Crystal-State Ion-Pairing Assemblies of Receptor–Anion-Complexing Ion Pairs

In the solid state, receptor–anion complexes exhibit various assembling modes in combination with countercations depending on the geometries and substituents of the components. For example, **11a** with tetrabutylammonium chloride (TBACl) forms a hydrogen-bonding chain structure based on one pyrrole inversion (Fig. 1.8a) [43]. The pyrrole β-fluorine-substituted **16a** also formed a hydrogen-bonding chain structure without pyrrole inversions [54]. In both cases, layers of hydrogen-bonding chain consisting of receptor–Cl^- complex and those of counter TBA^+ alternate to form the charge-by-charge assembly. On the other hand, α-aryl-substituted **11b**, **15b**, and **16b** formed planar [1+1]-type Cl^- complexes as revealed by solution-state NMR studies as well as single-crystal X-ray analysis: **11b**·Cl^- (Fig. 1.8b) [44] and **15b**·Cl^- (Fig. 1.8c) [67] with tetrapropylammonium cation (TPA^+) and **16b**·Cl^- with TBA^+. The receptor–anion complexes and cations formed charge-by-charge assemblies via one-by-one stacking of the receptor–Cl^- complexes and cations for **11b**·Cl^- and **15b**·Cl^-, whereas **16b**·Cl^- showed two-by-two stacking. Meanwhile, the π-electronic $TATA^+$ cation afforded an ion pair **11b**·Cl^--$TATA^+$, forming a charge-by-charge assembly with a smaller distance of 6.85 Å between the **11b**·Cl^- units (Fig. 1.8d), in contrast to 7.29 Å for the ion pair with TPA^+

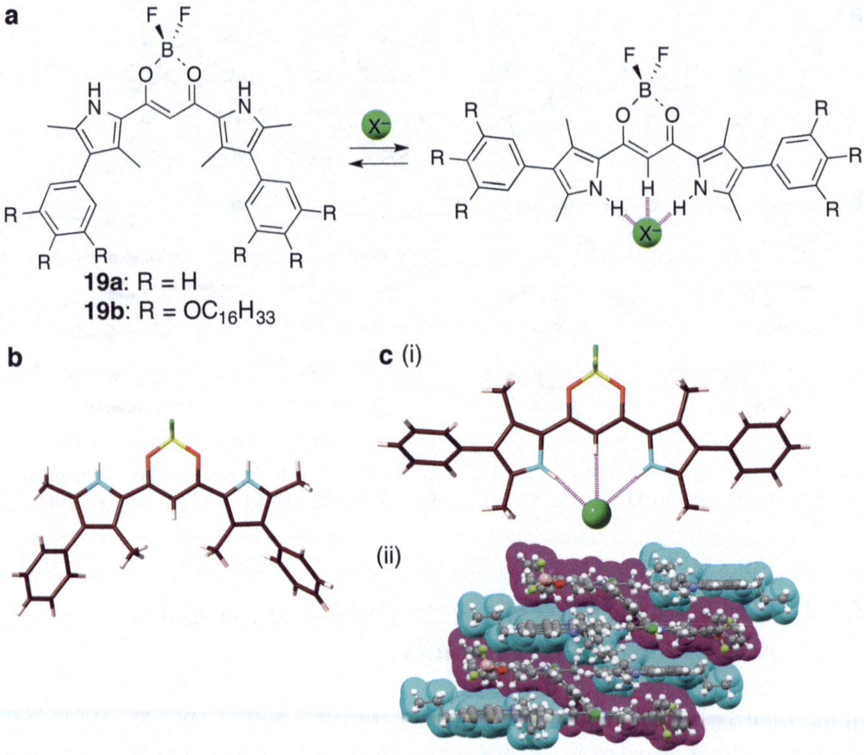

Fig. 1.9 (**a**) Structures of β-aryl **19a**,**b** and anion-binding mode and single-crystal X-ray structures of (**b**) **19a** and (**c**) **19a**·Cl^--$TATA^+$ as (*i*) [1+1]-type complex and (*ii*) packing diagram (magenta and cyan represent **19a**·Cl^- and $TATA^+$, respectively)

[77]. Furthermore, both β-methyl **15b**·Cl^--$TATA^+$ and β-fluorinated **16b**·Cl^--$TATA^+$ also formed charge-by-charge assemblies (Fig. 1.8e), wherein the larger $Cl^-\cdots Cl^-$ distances along the columns were estimated as 12.43 and 10.57 Å, respectively [81]. The smaller overlap of the receptor–Cl^- complexes and $TATA^+$ in **15b**·Cl^--$TATA^+$ and **16b**·Cl^--$TATA^+$ indicates that the β-substituents of the anion-responsive π-electronic molecules may interfere with the formation of stable charge-by-charge columnar structures.

One of the important properties of dipyrrolyldiketone BF_2 complexes is their pyrrole inversion behavior with respect to anion binding. By using such conformation changes, rod-shaped molecules can be converted to fan-shaped geometries. In fact, α-aryl-substituted receptors exhibit rod-shaped geometries as their most stable structures, which are converted to fan-shaped geometries by anion binding. The control of molecular geometries by anion binding can also be achieved in the derivatives possessing pyrrole β-aryl substituents. β-Aryl derivatives **19a**,**b** (Fig. 1.9a) were synthesized by the Suzuki coupling reaction of β-iodo-substituted receptor and corresponding boronic acid or ester. In fact, **19a** showed a fan-shaped

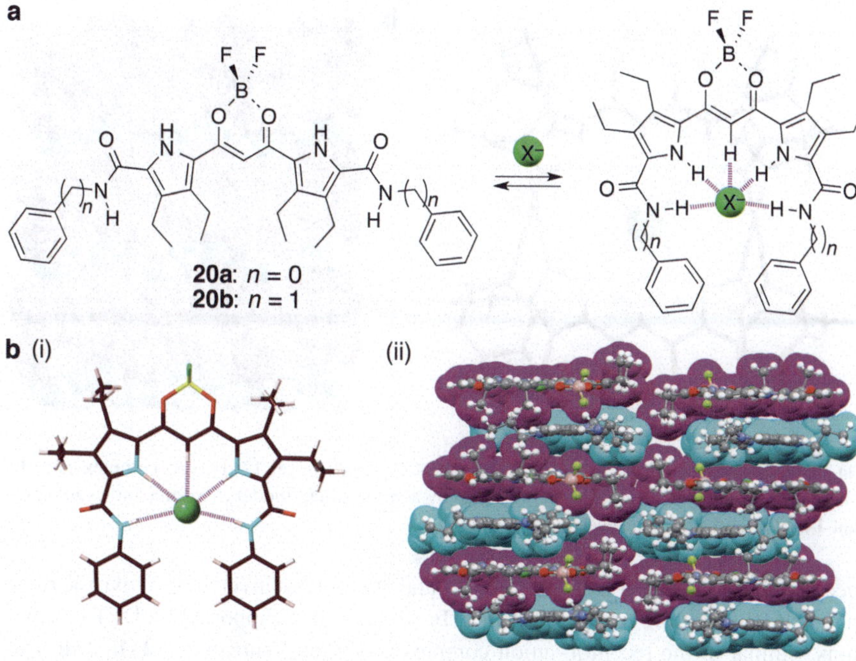

Fig. 1.10 (**a**) Structures of **20a**,**b** and anion-binding mode and (**b**) single-crystal X-ray structure of **20a**·Cl^--$TATA^+$ as (*i*) [1+1]-type complex and (*ii*) packing structure (magenta and cyan represent **20a**·Cl^- and $TATA^+$, respectively)

geometry (Fig. 1.9b), whereas the Cl^- complex of **19a** formed a rod shape (Fig. 1.9c (i)) as revealed by single-crystal X-ray analysis [60]. Similar to the charge-by-charge stacking structures of α-aryl-substituted receptors such as **11b**, **15b**, and **16b**, β-aryl **19a** formed a charge-by-charge assembly with a rod-shaped planar [1+1]-type anion complex and $TATA^+$ (Fig. 1.9c(ii)). Such a structural change enables the control of organized structures in ion-pairing dimension-controlled assemblies.

[1+1]-Type receptor–anion complexes are also constructed with the receptors possessing multiple hydrogen-bonding sites. The introduction of electron-withdrawing and additional hydrogen-bonding donor moieties, as shown in phenylamide-substituted pyrrole **2**, enhances anion-binding ability. Recently, phenyl- and benzylamide-substituted anion-responsive molecules **20a**,**b** (Fig. 1.10a) were synthesized from carboxy-substituted dipyrrolyldiketone derivative and corresponding amines using Mukaiyama reagent [61]. Phenylamide **20a** showed a K_a value of $>10^8$ M^{-1} for Cl^- in CH_2Cl_2, suggesting an extremely high binding affinity supported by multiple interaction sites. ^{1}H NMR signals of **20a**,**b** for pyrrole NH, amide NH, and bridging CH gradually shifted downfield upon the addition of Cl^- as a TBA salt in CD_2Cl_2, suggesting the formation of [1+1]-type Cl^- complexes using the amide NH units as well. The gradual shifts of ^{1}H NMR signals by anion binding in contrast to the signal shift of **11b** suggested a fast equilibrium between the

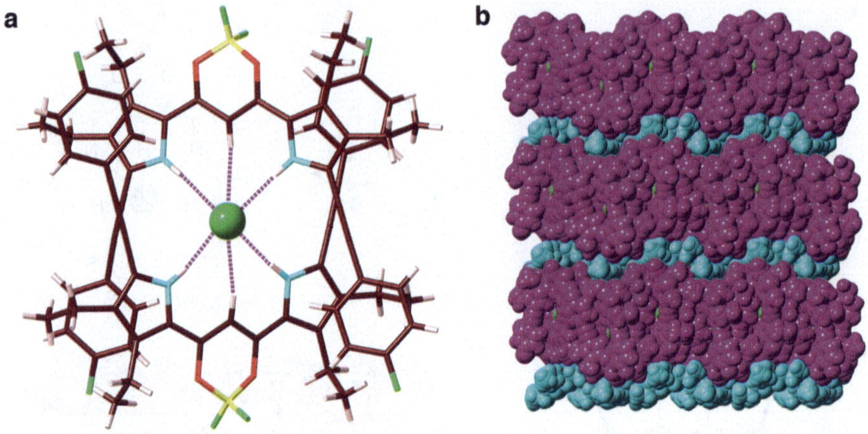

Fig. 1.11 Single-crystal X-ray structure of $\mathbf{14a}_2{\cdot}Cl^-$-$TBA^+$ as (**a**) [2+1]-type complex and (**b**) totally charge-segregated assembly represented by a space-filling model (magenta and cyan represent $\mathbf{14a}_2{\cdot}Cl^-$ and TBA^+, respectively)

receptors **20a**,**b** and their anion complexes and also the fast inversions of pyrrole rings in the NMR timescale compared with **11b**, which is also supported by DFT calculations. Similar to the receptor–anion complexes of α-aryl-substituted **11b**, **15b**, and **16b**, [1+1]-type receptor–anion complexes were observed for $\mathbf{20a,b}{\cdot}Cl^-$-$TATA^+$ in the crystal state (Fig. 1.10b), wherein the π electronic anions **20a**,**b** Cl^- and $TATA^+$ formed charge-by-charge stacking assemblies [61].

In general, acyclic anion receptors can form [2+1]-type interlocked receptor–anion complexes, which are not easily formed compared to [1+1]-type planar complexes. For example, α-phenyl-substituted **11b** (Fig. 1.6) forms a [2+1]-type Cl^- complex at low temperatures with a small amount of the guest anion [44]. In contrast, ethynyl-substituted **14a** (Fig. 1.6), as a π-extended derivative, shows the cooperative formation of [2+1]-type receptor–anion complexes in solution and also in the crystal state. The stabilities of [2+1]-type complexes are greater when cooperative binding occurs with receptors with more electron-withdrawing substituents, such as CF_3 moieties, at the terminal *para* positions. Cooperatively formed [2+1]-type complexes have also been examined by theoretical studies, suggesting effective intermolecular (intracomplex) interactions between ethynyl moieties and pyrrole units [69, 72]. As for ion-pairing assemblies in the solid state, the ion pair with TBA^+ formed totally charge-segregated assemblies comprising the charged layers of a [2+1]-type Cl^- complex $\mathbf{14a}_2{\cdot}Cl^-$ and those of the cation in the crystal state (Fig. 1.11) [69]. The guest anion is surrounded by the two interlocked receptor moieties, resulting in delocalization of the negative charge of the anion into the receptors and the formation of charged layers due to the small electrostatic repulsion between the anion complexes. Interestingly, the ion pair $\mathbf{14a}_2{\cdot}Cl^-$-$TBA^+$ provided two other more stable crystal polymorphs with the contribution of charge-by-charge assemblies [71]. Furthermore, $\mathbf{14a}_2{\cdot}Cl^-$ formed diverse solid-state ion-pairing assemblies with various cations such as TPA^+, $(C_8H_{17})_3MeN^+$, $TATA^+$ [69], and an azobenzene-appended ammonium cation [70].

1.3 Dimension-Controlled Ion-Pairing Assemblies

1.3.1 Supramolecular Gels Based on Modified Anion-Responsive Molecules

The combination of appropriately modified π-electronic receptors, guest anions, and countercations is important for the formation of receptor–anion-complexing ion pairs (Fig. 1.12). In general, there are three ways to functionalize the building components of dimension-controlled assemblies: the modification of π-electronic receptors, guest anions, and countercations. Synthetic modification of electronically neutral molecules is much easier than those for charged species, as can be used to achieve multiple variations in structural and electronic tunings. On the other hand, modification of the guest anions and the countercations is also possible but rather difficult. Introduction of aliphatic chains to the building units is a practical way to induce the formation of dimension-controlled assemblies. Therefore, through modifications, the receptor–anion-complexing ion-pairing assemblies exhibit unique assembling modes as well as electronic properties based on the combinations of various charged species.

In the preparation protocol of receptor–anion-complexing ion pairs, π-electronic anion-responsive molecules and guest anions accompanied by countercations are mixed in appropriate solvents and the precipitated. The solvent conditions for the precipitation depend on the solubility of π-electronic anion-responsive molecules, added guest salts, and produced ion pairs. In the case of dipyrrolyldiketone BF_2 complexes, hydrocarbons (such as octane) and ethers (including 1,4-dioxane) are often the first choices for the precipitation. Characterization of receptor–ion-complexing ion pairs is mainly conducted by ^{1}H NMR and elemental analysis. Differential scanning calorimetry (DSC) and polarized optical microscopy (POM) also support the identification of target ion pairs in the bulk state.

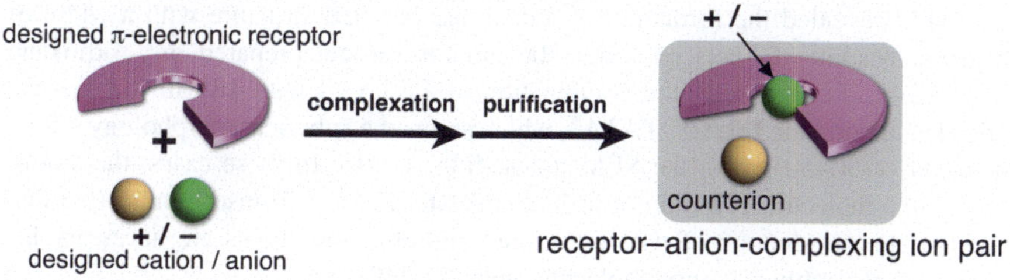

Fig. 1.12 General preparation process of receptor–anion-complexing ion pairs

π-Electronic anion-responsive molecules with aliphatic chains, such as hexadecyloxy-substituted **11c** and hexadecyl-substituted **12** (Fig. 1.6), formed supramolecular gels with less polar hydrocarbon solvents, such as octane (10 mg mL^{-1}) [44, 53]. The octane gel of **12** showed a higher melting point (43.0 °C) than that of **11c** (27.5 °C) due to alkyl chains being more rigid than alkoxy chains. Supramolecular gels are formed via noncovalent interactions between the rigid core π-electronic moieties as well as interactions between their peripheral substituents, as suggested by atomic force microscopy (AFM), scanning electron microscopy (SEM), and X-ray diffraction (XRD) analyses. The addition of Cl^- as a solid TBA salt to the gel of **11c** resulted in a gradual transition to the solution state, which can be attributed to the formation of the soluble ion pair **11c**·Cl^--TBA^+ in the less polar solvent [44, 53].

In sharp contrast to gel decomposition by the addition of the aliphatic bulky TBA salt, the addition of π-electronic cation salts afforded dimension-controlled assemblies based on charge-by-charge stacking structures (Fig. 1.2 left) [77]. Ion pair **11c**·Cl^--$TATA^+$ (Fig. 1.13a) afforded an opaque octane gel with a gel-to-solution transition temperature of 35 °C (Fig. 1.13b inset). Optical microscopy (OM), AFM, and SEM analyses of the xerogel of **11c**·Cl^--$TATA^+$ showed the formation of entangled fibers with diameters of 0.1–2 μm and lengths of >100 μm (Fig. 1.13b). The detailed structure of the fibrous xerogel of **11c**·Cl^--$TATA^+$ was investigated by synchrotron XRD analysis at SPring-8 and was shown to exhibit a hexagonal columnar (Col_h) arrangement with $a = 4.25$ nm and $c = 0.73$ nm based on a trimeric assembly ($Z = 3$ for $\rho = 1$) (Fig. 1.13c). The c value of 0.73 nm was derived from the two alternately stacking planar components, **11c**·Cl^- and $TATA^+$ in this case, indicating the formation of a charge-by-charge assembly as seen in the crystal of **11b**·Cl^--$TATA^+$ (Fig. 1.8d). The disk-shaped components include three sets of **11c**·Cl^--$TATA^+$, as BF_2 units and aliphatic chains are arranged on the center and outside of the disk, respectively [77].

Another modification of anion-responsive π-electronic molecules is the extension of the π-conjugation of building units, providing organized assemblies applicable for electric conductive materials. π-Extended derivatives **17a,b** [57, 58] and **18a,b** [59] possessing benzo and corannulene moieties, respectively, at pyrrole β-positions (Fig. 1.6) showed interesting assembling behaviors. For example, **18b** formed supramolecular gels in octane, 1,4-dioxane (Fig. 1.14a inset), and toluene (20 mg mL^{-1}) with gel–sol transition temperatures of 17, 22, and 28 °C, respectively. SEM and AFM revealed the formation of a distorted bundled structure with a width of 8 μm consisting of fibers of width 300 nm for xerogel prepared in 1,4-dioxane (Fig. 1.14a). Furthermore, the combination of TATA·Cl with **18b** in 1,4-dioxane afforded an ion pair **18b**·Cl^--$TATA^+$, which showed a spherical morphology with a diameter of 30–50 μm in the SEM image (Fig. 1.14b). In these cases, the anion-responsive molecule **17b** did not form a gel state (Fig. 1.14b inset), indicating that appropriate introduction of π-extended unit and aliphatic chains are necessary for organized assemblies as supramolecular gels [57–59].

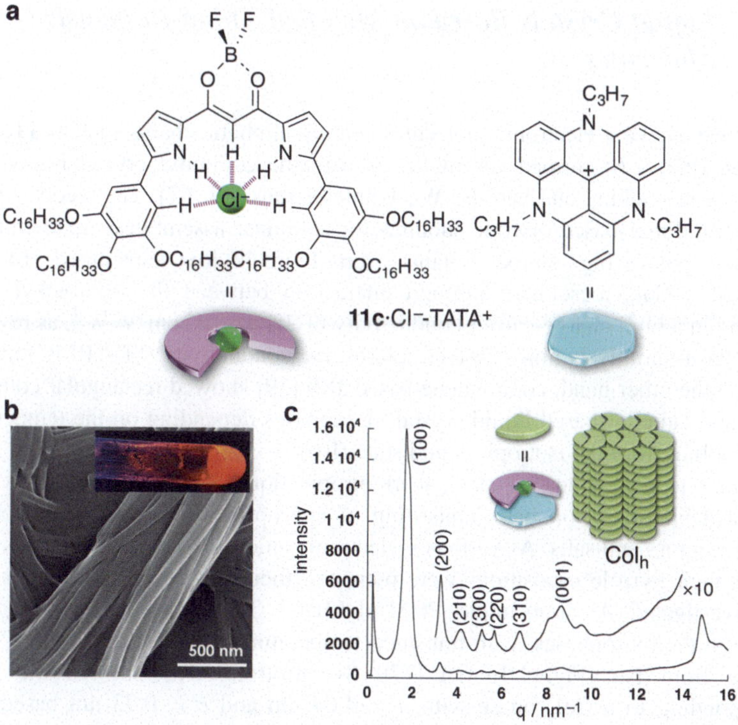

Fig. 1.13 (**a**) Structure of **11c**·Cl^--$TATA^+$, (**b**) SEM image of **11c**·Cl^--$TATA^+$ as the xerogel at 20 °C (inset: photograph of the gel under UV light (365 nm)), and (**c**) the corresponding XRD and proposed assembled model at 25 °C

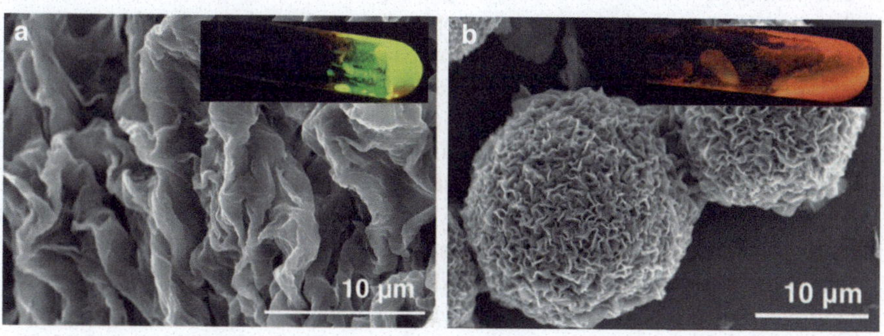

Fig. 1.14 SEM images and photographs under UV light (365 nm) of (**a**) xerogel of **18b** in 1,4-dioxane (20 mg mL^{-1}) and (**b**) dried precipitate of **18b**·Cl^--$TATA^+$ prepared in 1,4-dioxane

1.3.2 Liquid Crystals Based on Modified Anion-Responsive Molecules

Anion-responsive π-electronic molecules bearing aliphatic chains such as **11c**, **15c**, and **16c** (Fig. 1.6) formed assemblies as Col_h-based liquid crystal mesophases through π–π stacking and van der Waals interactions [47, 67]. The receptor molecules form dimer-based discotic subunits for columnar assemblies. Apart from the receptors possessing alkoxy chains, semifluoroalkyloxy-substituted **13** also exhibited a Col_h mesophase through interaction between the fluoroalkyl units, resulting in a high charge-carrier conductivity of 2.4×10^{-5} $cm^2V^{-1}s^{-1}$, as revealed by the flash-photolysis time-resolved microwave conductivity (FP-TRMC) method [51]. On the other hand, corannulene-based **18b** [59] showed rectangular columnar (Col_r) and lamellar-based liquid crystal mesophases depending on the temperature upon cooling from the isotropic liquid state (Iso).

Similar to the anion-free state, thermotropic liquid crystalline materials were fabricated based on ion pairs consisting of appropriately combined π-electronic anion receptors and salts. As various π-electronic anion-responsive molecules modified at their pyrrole α-positions were prepared, their ion-pairing assemblies were also investigated. As an example, POM of **11c**·Cl^--$TATA^+$ revealed a focal conic texture in the mesophases, with transition temperatures of 94, 88, and 42 °C upon cooling from Iso (Fig. 1.15a(i)) [77]. Synchrotron XRD showed the peaks corresponding to a Col_h phase with $a = 4.64$ nm and $c = 0.73$ nm based on a tetrameric component ($Z = 4$ for $\rho = 1.12$) (Fig. 1.15a(ii)). The c value of 0.73 nm corresponds to the stacking distance between the ion pairs of **11c**·Cl^- and $TATA^+$, indicating the formation of charge-by-charge assemblies as observed in the crystal structure of **11b**·Cl^--$TATA^+$ (6.85 Å) (Fig. 1.8d) [77].

Ion-pairing assemblies with charge-by-charge modes were observed even with bulky aliphatic cations. In fact, Col_h mesophases based on charge-by-charge assemblies were observed for the receptor–anion complexes with bulky cations, such as **11c**·Cl^--TBA^+ [79]. In this case, the XRD showed a broad diffraction peak identified as the repeating distance of the charge-by-charge stacking **11c**·Cl^- and bulky TBA^+, indicating a more weakly stacking structure compared to those comprising planar $TATA^+$. Furthermore, a combination of **11c**·Cl^- and various tetraalkylammonium cations afforded Col_h structures with very low mesophase temperatures due to the less efficient stacking of the bulky cations [79].

Tuning of assemblies has been achieved by the modification of aliphatic chains at the pyrrole α-positions of anion receptors. In fact, the ion-pairing assembly of **12**·Cl^- and $TATA^+$ afforded a Col_r mesophase at 50 °C upon cooling, with $a = 3.75$ nm, $b = 2.14$ nm, and $c = 0.39$ nm ($Z = 1$ for $\rho = 1.17$) (Fig. 1.15b) [53]. The broad peak at 0.39 nm was derived from the partial contribution of a charge-segregated assembly with stacking of identically charged species, as seen in the cases of **15c**·Cl^--$TATA^+$ and **16c**·Cl^--$TATA^+$. Interestingly, the diffraction at 0.76 nm observed at 20 °C upon cooling suggested a small contribution of the charge-segregated assembly at lower temperatures [53].

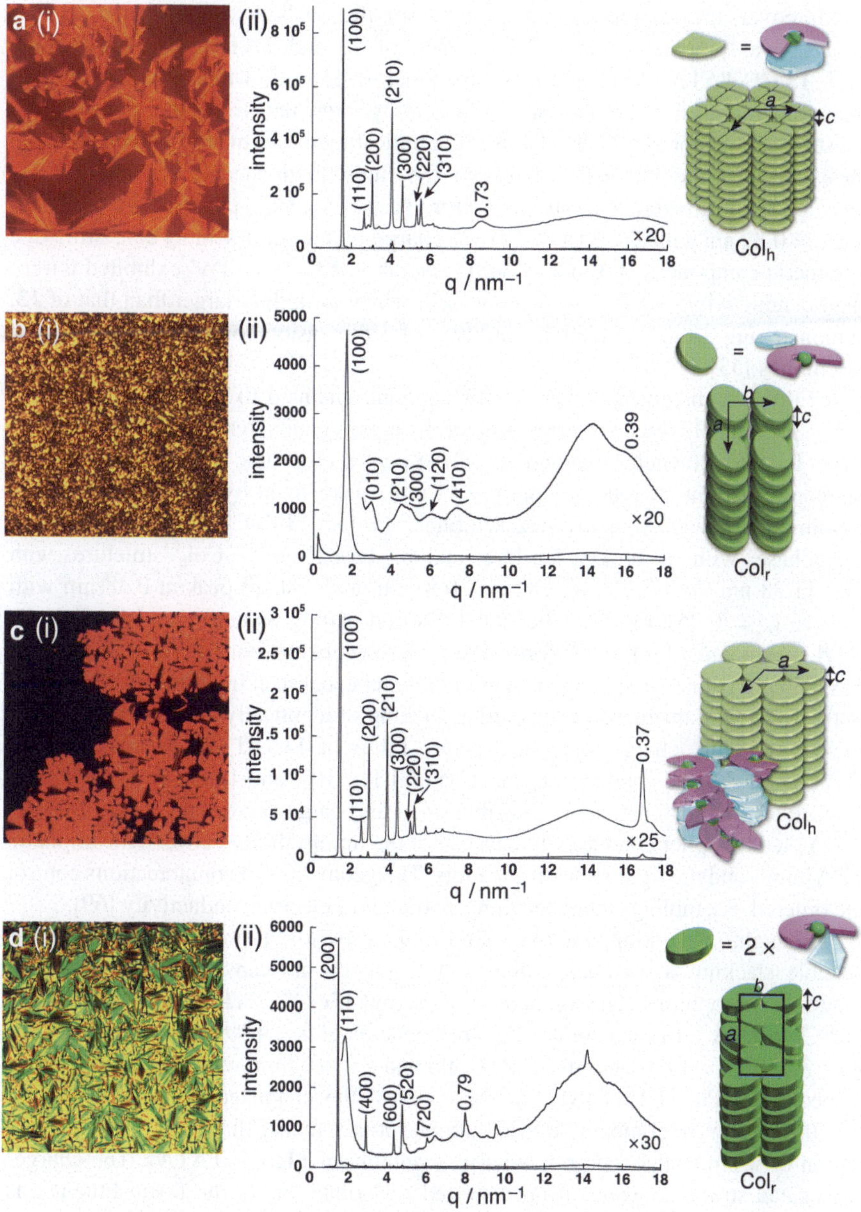

Fig. 1.15 (*i*) POM images and (*ii*) XRD and proposed assembled models of (**a**) **11c**·Cl⁻-TATA⁺ at (*i*) 70 °C and (*ii*) 101 °C, (**b**) **12**·Cl⁻-TATA⁺ at 62 °C, (**c**) **15c**·Cl⁻-TATA⁺ at (*i*) 150 °C and (*ii*) 101 °C, and (**d**) **19b**·Cl⁻-TBA⁺ at 63 °C upon cooling from Iso

Moreover, the ion-pairing assembly of semifluoroalkyloxy-substituted **13**·Cl^--$TATA^+$ was prepared from a 1:1 solution of **13** and TATA·Cl in 1,4-dioxane [51]. **13**·Cl^--$TATA^+$ formed a mesophase below 153 °C, which was higher than the temperature for the ion-pairing assembly prepared from the dodecyloxy-substituted receptor and TATA·Cl (80 °C), indicating the formation of a more stable mesophase by the interaction between the fluoroalkyl chains. POM and XRD revealed the formation of a fan-like texture based on a Col_h phase of $a = 4.20$ nm and $c = 0.87$ nm with $Z = 4$ ($\rho = 1.7$) as a charge-by-charge assembly consisting of a tetrameric component of **13**·Cl^--$TATA^+$. Notably, **13**·Cl^--$TATA^+$ exhibited a transient conductivity of 2.7×10^{-5} $cm^2V^{-1}s^{-1}$, which is slightly larger than that of **13**, demonstrating the potential utility of dimension-controlled ion-pairing assemblies [53].

α-Ethynyl-substituted receptor with aliphatic chains **14b** showed less ordered states, presumably due to rotation around the ethynyl units, resulting in less suitable states for intermolecular interactions of the core π-electronic units [69]. However, the conformations of rotatable peripheral units were fixed by anion binding, thus forming organized ion-pairing assemblies. In fact, **14b**·Cl^--TBA^+ exhibited a mesophase with a broken-fan-like texture based on a Col_r structure with $a = 11.28$ nm, $b = 7.95$ nm, and $c = 0.81$ nm and a small peak at 0.38 nm with $Z = 16$ ($\rho = 0.85$). On the other hand, the ion-pairing assembly of **14b**·Cl^- with octyl-substituted $(TATA^{C8})^+$ formed a Col_h mesophase with $a = 4.34$ nm and $c = 0.36$ nm with $Z = 2$ ($\rho = 1.5$). The smaller repeated stacking distance c observed in the XRD was attributed to the partial stacking of identically charged π-electronic species (partially charge-segregated state). Films of **14b**·Cl^--TBA^+ and **14b**·Cl^--$(TATA^{C8})^+$ showed conductivity maxima of 4.1×10^{-5} and 1.5×10^{-5} $cm^2V^{-1}s^{-1}$, respectively, indicating that the contribution of the charge-segregated state in **14b**·Cl^--TBA^+ led to higher conductivity because of the immiscibility between the aliphatic TBA^+ units and the rigid π-electronic units. The geometries of countercations control the ordered assembling states, resulting in tuning of electric conductivity [69].

Although, as mentioned above, ethyl units at pyrrole β-positions interfere with suitable stacking assemblies, rather small substituents can change the stacking modes in mesophases. The ion pairs of β-methyl **15c**·Cl^--$TATA^+$ and β-fluorinated **16c**·Cl^--$TATA^+$ afforded similar Col_h mesophases of $a = 4.99$ nm and $c = 0.37$ nm with $Z = 2$ ($\rho = 1.03$) and of $a = 4.92$ nm and $c = 0.37$ nm with $Z = 2$ ($\rho = 1.01$), respectively (Fig. 1.15c) [81]. In these cases, the characteristic XRD peaks of $c = 0.37$ nm were comparable to the regular π–π stacking distances, in contrast to that at 0.73 nm in the charge-by-charge assembly of **11c**·Cl^--$TATA^+$. The charge-segregated structures result from distorted aryl rings due to the β-substituents, as observed in the single-crystal X-ray structures (Fig. 1.8e). Furthermore, time-of-flight (TOF) electrical conductivity measurements of **11c**·Cl^--$TATA^+$, **15c**·Cl^--$TATA^+$, and **16c**·Cl^--$TATA^+$ indicated ambipolar charge-carrier transporting behaviors, with well-balanced values at high mobility (10^{-3}–10^{-2} $cm^2V^{-1}s^{-1}$) for both holes and electrons [81].

π-Electronic units at pyrrole β-positions are also applicable for the introduction of aliphatic chains, as seen in **17b** [58]. Benzo-fused **17b** formed a Col_h phase with $a = 4.2$ nm and $c = 0.45$ nm based on a dimeric component ($Z = 2$ for $\rho = 0.74$). On the other hand, ion-pairing **17b**·Cl^--$TATA^+$ afforded a tetragonal columnar (Col_t) mesophase with $a = 4.21$ nm and $c = 0.73$ nm based on a tetrameric component ($Z = 4$ for $\rho = 1.0$). In the assembling disk consisting of four sets of **17b**·Cl^--$TATA^+$, the receptors face the center of the circular unit, presumably because the anions are arranged on the inside and the aliphatic chains at the benzo units are located at the periphery of an assembled disk. It is noteworthy that dipyrrolyldiketone BF_2 complexes with alkoxy and aryl substituents at pyrrole β-positions formed no ordered assemblies as mesophases, suggesting that more planar receptor–anion complexes are appropriate for the formation of ion-pairing assemblies. Despite the formation of the ordered mesophase structure in **18b**, ion-pairing **18b**·Cl^--$TATA^+$ showed no clear mesophase by DSC and POM [59]. However, XRD of **18b**·Cl^--$TATA^+$ at 80 °C upon cooling indicated the formation of a lamellar structure with broad diffraction peaks in the small-angle region, suggesting a less-ordered assembly comprising various components, such as planar and bowl-shaped π-electronic receptor units, aliphatic chains, and $TATA^+$ cations [58, 59].

Conformational changes by anion binding from rod-shaped receptors to fan-shaped anion complexes enable changes in the assembling structures. For example, the derivative substituted with β-aryl moieties with aliphatic chains **19b** forms a Col_h mesophase with $a = 3.84$ nm and $c = 0.45$ nm based on a dimeric component ($Z = 2$ for $\rho = 1.10$) [60]. On the other hand, the ion pair **19b**·Cl^--TBA^+ formed a Col_r (*P*2/*a*) phase of $a = 8.66$ nm, $b = 3.79$ nm, and $c = 0.79$ nm (Fig. 1.15d), indicating the formation of a charge-by-charge assembly. The average number of ion pairs in the unit cell was estimated as $Z = 8$ ($\rho = 1.12$), suggesting that two ion pairs form each circular disk [60].

1.3.3 Receptor–Anion-Complexing Ion Pairs with Modified Anions

Combination of chemically modified anions with π-electronic anion receptors provides tunable π-electronic anionic species as the building components of ion-pairing assemblies. For example, the combination of **11a,b** with 3,4,5-trialkoxy-substituted benzoates **21a**$^-$–**c**$^-$ (Fig. 1.16a) as TBA^+ salts formed lamellar mesophases based on charge-by-charge assemblies (Fig. 1.16b) [78]. These ion-pairing materials showed moderately high charge-carrier transporting properties, as seen in the values of 0.02 and 0.05 $cm^2V^{-1}s^{-1}$ for **11a**·**21a**$^-$-TBA^+ and **11b**·**21a**$^-$-TBA^+, respectively [78].

Furthermore, carboxylate-appended azobenzene derivatives with TBA^+ (**22a**$^-$,**b**$^-$·TBA^+) were prepared for photo-responsive ion-pairing assemblies (Fig. 1.17a) [85]. Alkyl chains are connected to the azobenzene unit avoiding the electronic effects of the electron-donating substituents on the photo-isomerization properties. As an example, DSC of **22b**$^-$·TBA^+ elucidated the mesophase only upon heating. XRD revealed the formation of a lamellar structure with a repeating distance of 3.8 nm, corresponding to the summed length of **22b**$^-$ and TBA^+. The photo-

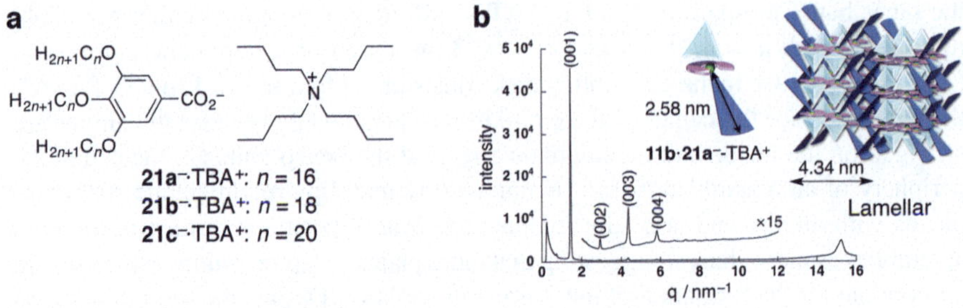

Fig. 1.16 (**a**) Modified carboxylates **21a**⁻–**c**⁻ as TBA⁺ salts and (**b**) XRD and a packing model of **11b**·**21a**⁻-TBA⁺ at r.t

responsive property of ionic crystal assemblies of **22b**⁻·TBA⁺ was investigated by UV (365 nm) irradiation. Without the irradiation, the XRD of **22b**⁻·TBA⁺ at 40 °C after heating from 10 °C showed a lamellar structure with *d* spacing of 3.8 and 1.9 nm (Fig. 1.17b(i)), which can be indexed to (001) and (002), respectively. UV irradiation (365 nm, 100 mWcm^{-2}) induced the gradual disappearance of these XRD peaks and concurrent appearance of the peaks at 3.4 and 1.7 nm (Fig. 1.17b(ii, iii)). As the sizes of the ion pairs as *trans* and *cis* isomers were estimated as 3.9 and 3.3 nm, respectively, the difference between these phases can be rationalized by the *trans*-to-*cis* photo-isomerization. Further irradiation of visible light (436 nm, 60 mWcm^{-2}) induced reversion to the initial state (Fig. 1.17b(iv)). In contrast to the photo-responsive crystal–crystal switching of **22b**⁻·TBA⁺, the ion pair **22a**⁻·TBA⁺ underwent the conversion to optically isotropic states by UV irradiation. This result suggests that the interactions between long aliphatic chains are important for crystal–crystal phase transitions [85]. Furthermore, the ion pair **11a**·**22b**⁻-TBA⁺ (Fig. 1.17c), also forming a lamellar structure, exhibited no photo-responsive properties, presumably due to strong packing around the receptor–carboxylate complex in **11a**·**22b**⁻-TBA⁺ [86].

1.3.4 Receptor–Anion-Complexing Ion Pairs with Modified Cations

Control of the geometries and electronic states of counter species of π-electronic anions (receptor–anion complexes) is also important for achieving functional ion-pairing assemblies. Based on the modification of diverse π-electronic cationic species, which is easier than those of anions, more specific cations could afford fascinating ion-pairing assemblies. Metal complexes of π-electronic ligands also work as π-electronic cations. For example, Pt^{II}Cl complexes of 2,2′6′,2″-terpyridine (trpy) (**23a**⁺,**b**⁺, Fig. 1.18a) behave as planar monocationic species [84]. The ion-pairing assembly of 4′-hexadecyloxy-substituted cation **23b**⁺ and **11c**·Cl⁻ was prepared from a 1:1 mixture of **11c** and a **23b**⁺·Cl⁻ salt in 1,4-dioxane. The

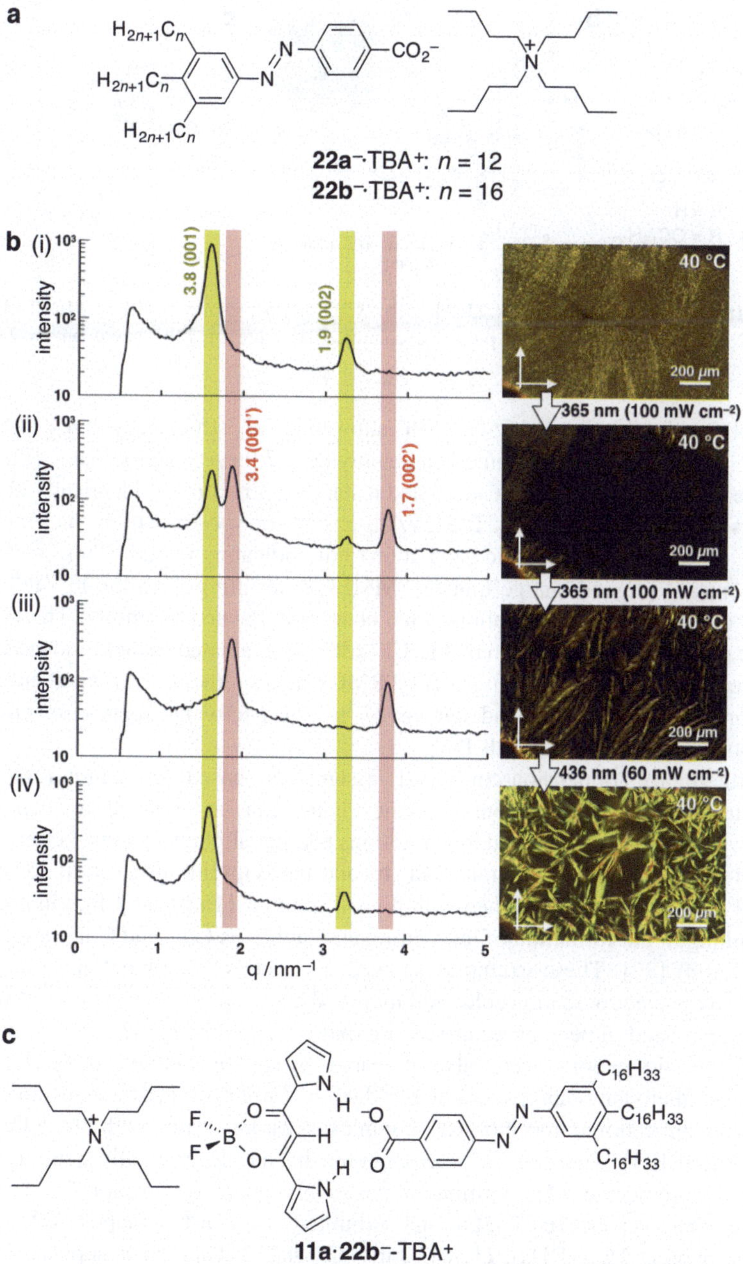

Fig. 1.17 (**a**) Azobenzene carboxylate ion pair **22a**,**b**$^-$·TBA$^+$; (**b**) changes in XRD patterns (small-angle region) and corresponding POM images of **22b**$^-$·TBA$^+$ in a sandwiched cell (8 μm spacer) at variable temperatures at (*i*) 40 °C, (*ii, iii*) 40 °C under UV (365 nm, 100 mWcm^{-2}) irradiation with an interval of approximately 0.5 h, and (*iv*) 40 °C under visible light (436 nm, 60 mWcm^{-2}); and (**c**) receptor–anion-complexing ion pair **11a**·**22b**$^-$-TBA$^+$

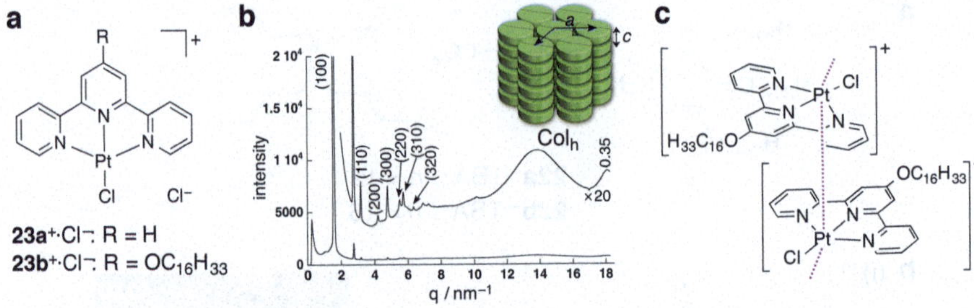

Fig. 1.18 (**a**) trpy–Pt^{II} complexes **23a**$^+$,**b**$^+$, (**b**) XRD and a packing model of **11c**·Cl$^-$-**23b**$^+$ at 120 °C upon cooling, and (**c**) a proposed columnar structure of **23b**$^+$ based on the Pt^{II}–Pt^{II} interaction

mesophase of the ion pair **11c**·Cl$^-$-**23b**$^+$ showed a Col$_h$ structure with $a = 4.57$ nm and $c = 0.35$ nm based on a dimeric assembly ($Z = 2$ for $\rho = 1.36$) (Fig. 1.18b). The intense XRD peak at 0.35 nm suggested a stacking structure of identically charged species in the Col$_h$ mesophase. The UV/Vis absorption spectrum of the mesophase at 100 °C showed a characteristic band at 667 nm, indicating the dσ*(Pt$_2$) → π*(trpy) metal–metal-to-ligand charge transfer (MMLCT) resulting from the Pt^{II}–Pt^{II} interaction (Fig. 1.18c), which stabilizes the charge-segregated assembly. The charge-carrier mobility of the thin film of **11c**·Cl$^-$-**23b**$^+$, as evaluated by field-induced time-resolved microwave conductivity (FI-TRMC) measurements, exhibited hole and electron mobilities of 0.7 and 0.6 cm^{2}V^{-1}s^{-1}, respectively, suggesting effective ambipolar conductive materials [84].

Furthermore, dimension-controlled assemblies based on anion-responsive π-electronic molecules without aliphatic chains can be obtained by combining them with anions accompanied by the cations bearing aliphatic chains. For example, benzyl-trialkylammonium cations **24a**$^+$,**b**$^+$ and benzylpyridinium cation **25**$^+$ as Cl$^-$ salts (Fig. 1.19a) were combined with β-unsubstituted **11a**,**b** and β-fluorinated **16a**, **b**, resulting in the formation of ion-pairing assemblies as Col$_h$ and Col$_r$ mesophases (Fig. 1.19b) [82]. These examples showed a variety of combinations for anion-responsive π-electronic molecules and modified anions as well as cations, affording highly organized dimension-controlled ion-pairing assemblies [82].

π-Electronic cations were also prepared by anion binding of π-electronic dications. Phenylene- and pyrimidine-bridged bis(imidazolium)dications **26a**$^{2+}$,**b**$^{2+}$ formed monocationic receptor–Cl$^-$ complexes as ion pairs with the other Cl$^-$ (Fig. 1.20a) [83]. The free Cl$^-$ was captured by an electronically neutral anion-responsive molecule **11c**, forming a negatively charged receptor–Cl$^-$ complex **11c**·Cl$^-$. The ion pair **11c**·Cl$^-$-**26b**$^{2+}$·Cl$^-$ exhibited a broken-fan-shaped POM texture as a mesophase. The XRD of **11c**·Cl$^-$-**26b**$^{2+}$·Cl$^-$ at 115 °C in the mesophase showed a Col$_h$ structure with $a = 4.34$ nm, as well as a broad and weak peak at ca. 0.75 nm, indicating the contribution of a charge-by-charge assembly (Fig. 1.20b). A Col$_h$ structure was also observed at 20 °C upon cooling with a lattice parameter of $a = 4.73$ nm along with a clearer peak at 0.75 nm based on a trimeric assembly ($Z = 3$ for $\rho = 1$). The similarity in the sizes and shapes of two ionic components is

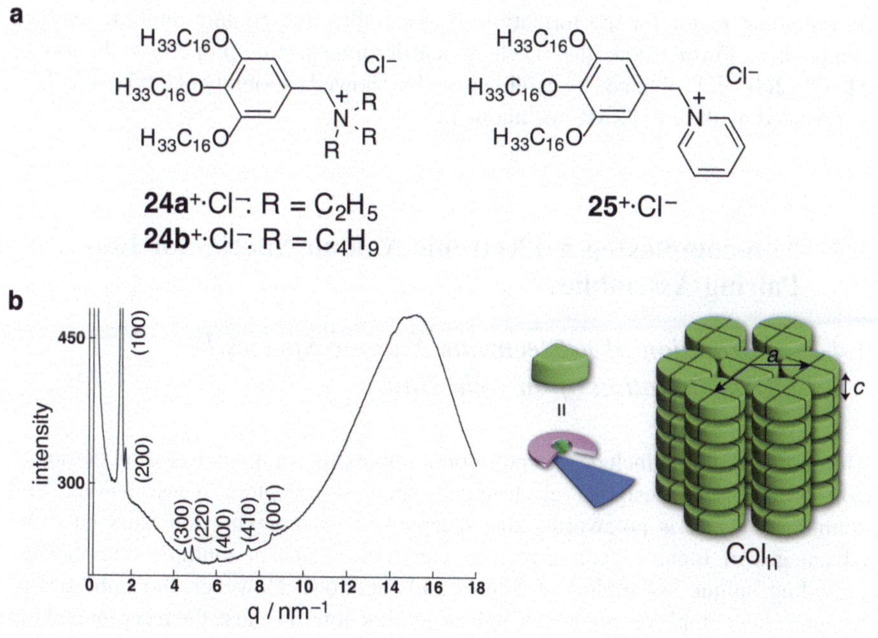

Fig. 1.19 (**a**) Modified cation Cl^- salts **24a**$^+$,**b**$^+$·Cl^- and **25**$^+$·Cl^- and (**b**) XRD and a packing model of **11a**·Cl^--**24b**$^+$ at 90 °C upon heating

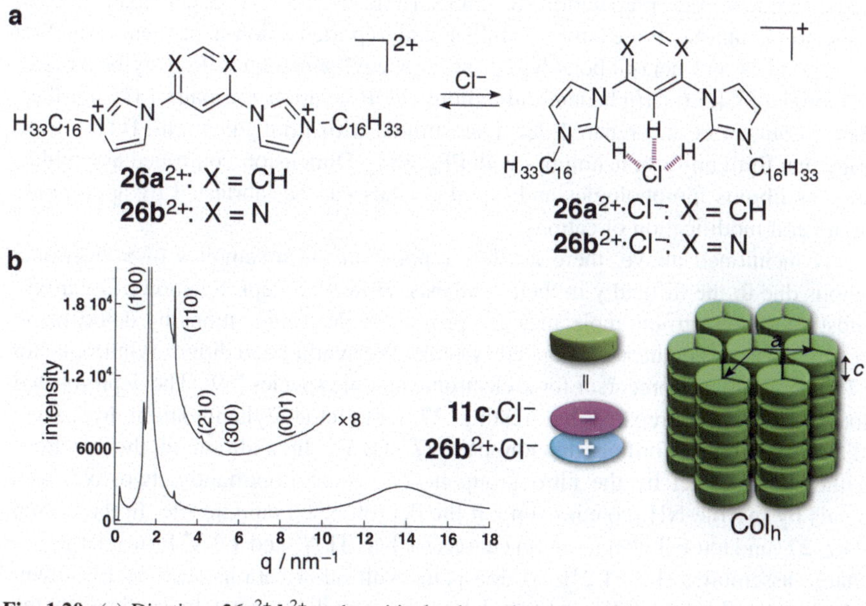

Fig. 1.20 (**a**) Dications **26a**$^{2+}$,**b**$^{2+}$ and positively charged receptor–Cl^- complexes and (**b**) XRD and a packing model of **11c**·Cl^--**26b**$^{2+}$·Cl^- at 115 °C upon cooling

an important factor for the formation of assemblies that contain multiple charged components. Furthermore, the 1D charge-carrier transporting property of the film of **11c**·Cl^--**26b**$^{2+}$·Cl^- obtained by cooling from Iso showed a mobility of 0.05 $cm^2V^{-1}s^{-1}$, as revealed by the FP-TRMC technique [83].

1.4 Non-complexing π-Electronic Anionic Species for Ion-Pairing Assemblies

1.4.1 Formation of π-Electronic Anionic Species by Deprotonation of an Acid Unit

The former sections include receptor–ion-complexing ion pairs based on combination of precisely designed electronically neutral receptors, guest anions, and countercations. It is noteworthy that receptor–ion-complexing ion pairs have the advantages of tunable geometries and electronic states of multiple components, providing unique assembling structures and functions. However, the stabilities of receptor–ion complexes are not as high as genuine ions, because the receptor and ion are noncovalently connected. As shown in Fig. 1.1b, the pairing of a π-electronic cation and a π-electronic anion (genuine π-electronic ion pair) is another candidate method for making components of ion-pairing dimension-controlled assemblies. There are a few reports on ion-pairing assemblies based on π-electronic cation ion pairs, because the preparation of π-electronic cations is easier than that of π-electronic anions. For example, Müllen et al. reported anion-dependent formation of nanotubes and nanoribbons based on tetradecyl-substituted 9-phenylbenzo[1,2]quinolizino[3,4,5,6-*fed*]phenanthridinylium (PQP^+) cation ion pairs [87]. Furthermore, Laursen et al. prepared decyl-substituted trioxatriangulenium ($TOTA^+$) ion pairs that form uniform nanotubes with PF_6^- [88]. Dimension-controlled assemblies such as fibrous morphologies and liquid crystals can be fabricated by appropriate peripheral modification of cations.

As mentioned above, there are few reports on the assemblies of π-electronic anions due to the difficulty in their synthesis. However, deprotonation of hydroxy-substituted π-electronic molecules can provide π-electronic anions by delocalizing negative charges in the π-conjugated systems. We synthesized dipyrrolylnitrophenol **27** (Fig. 1.21a) as a precursor for π-electronic anionic species [89]. The deprotonated species **27**$^-$ was prepared by treating **27** with tetrabutylammonium hydroxide (TBAOH), thus affording the ion pair **27**$^-$·TBA^+. In addition to the electron-withdrawing effect by the nitro group in **27**$^-$, more importantly, two hydrogen bonds by pyrrole-NH moieties support the deprotonated anionic site. In the crystal state, **27**$^-$ and tetraalkylammonium cations such as TPA^+ and TBA^+ form charge-by-charge assemblies (Fig. 1.21b(i)). Ion pairs with other cations such as 15-crown-5·Na^+ and 18-crown-6·K^+, prepared by corresponding metal hydroxides in the presence of crown ethers, afforded solid-state charge-by-charge assemblies (Fig. 1.21b(ii)) [89].

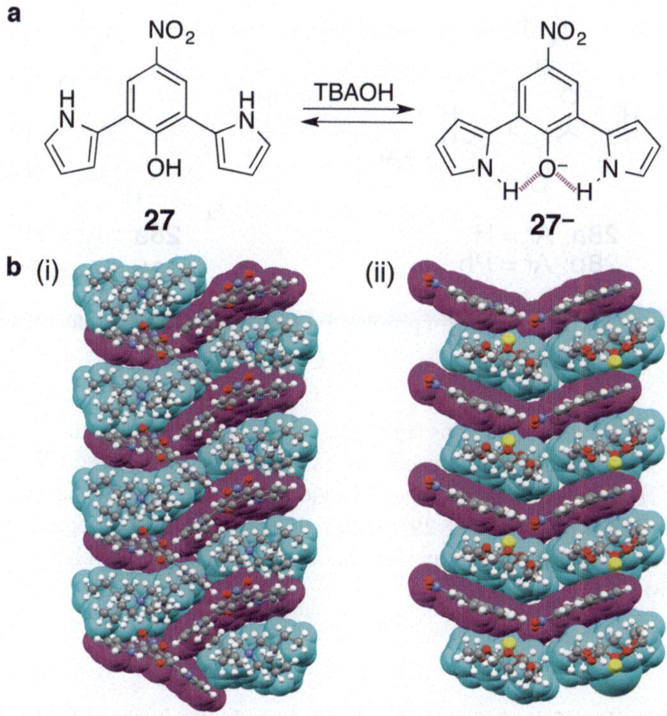

Fig. 1.21 (**a**) Dipyrrolylnitrophenol **27** and the formation of deprotonated species **27**$^-$ by TBAOH and (**b**) single-crystal X-ray structures of (*i*) **27**$^-$·TBA$^+$ and (*ii*) **27**$^-$·15-crown-5·Na$^+$ (magenta and cyan represent **27**$^-$ and cations, respectively)

As shown in the deprotonation of dipyrrolylnitrophenol, the stabilization of the anionic part by neighboring hydrogen-bonding donors is crucial in the preparation of diverse π-electronic anionic species. Similar to dipyrrolylnitrophenol, containing five-, six-, and five-membered rings, as the precursor of π-electronic anionic species, *meso*-hydroxy-substituted dipyrrolyldiketone BF_2 complexes **28a**,**b** were also synthesized via the hydrolysis of *meso*-acetoxy-substituted derivatives (Fig. 1.22) [90]. Upon the addition of TBAOH to CH_2Cl_2 solutions of **28a**,**b**, the solution colors changed upon deprotonation (Fig. 1.22), as also shown by UV/Vis absorption spectra. The deprotonation behaviors, investigated by ^{1}H NMR and DFT studies, suggested deprotonation pathways including characteristic pyrrole inversions. DFT-based ESP of **28a**$^-$,**b**$^-$ showed delocalization of the negative charge in the core dipyrrolyldiketone unit. However, the anionic species were less stable and easily converted to fragmented species [90]. Furthermore, *meso*-thiol-substituted derivatives were also synthesized, spontaneously affording disulfide-bridged dimers with intramolecularly H-aggregated conformations [91].

Large π-electronic molecules with an acid unit could provide stable anionic species owing to the delocalization of negative charges in the π-electronic core unit. In fact, we synthesized appropriately designed porphyrin **29** with a hydroxy

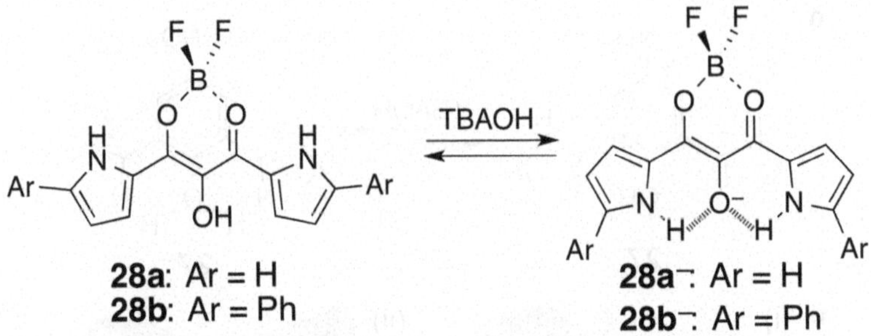

Fig. 1.22 *meso*-Hydroxy-substituted dipyrrolyldiketones **28a**,**b** and the formation of deprotonated species **28a**$^-$,**b**$^-$ by TBAOH

unit at one of the *meso* positions (Fig. 1.23a) [92]. Deprotonation of **29** by treating with TBAOH provided deprotonated species **29**$^-$ as indicated by the UV/Vis absorption and ^{1}H NMR spectral changes. Single-crystal X-ray analysis of **29**$^-$·TBA$^+$ revealed the exact structure of **29**$^-$, with delocalization of the negative charge in the porphyrin core, and the formation of a charge-by-charge assembly (Fig. 1.23b) [92].

1.4.2 Ion-Pairing Assemblies Based on Genuine π-Electronic Anions

Stable genuine π-electronic anions are rare [93–96], because π-electronic anions could suffer oxidation and electrophilic attacks. For this reason, limited numbers of stable genuine π-electronic anions and their ion pairs have been known to date. In particular, assemblies of π-electronic ion pairs consisting of π-electronic anions have mainly been studied in crystal states. For example, White et al. prepared an ion pair consisting of penta(methoxycarbonyl)-substituted cyclopentadienide anion (PMCp$^-$) and tetramethylammonium cation **30a** (Fig. 1.24a), whose solid-state charge-by-charge assembly was revealed by single-crystal X-ray analysis [97]. Furthermore, a π-electronic ion pair consisting of PMCp$^-$ with cycloheptatrienyl (tropylium) cation (Ch$^+$) **30b** (Fig. 1.24a) as a π-electronic cation afforded a charge-by-charge assembly in the crystal state [98]. On the other hand, cyano substituents can also stabilize the negative charge as seen in pentacyanocyclopentadienide (PCCp$^-$) [94, 99]. Based on the ion-exchange strategy, a Na$^+$ salt of PCCp$^-$ and Cl$^-$ salts of desired cations were combined in appropriate conditions, affording ion pairs of PCCp$^-$ **31a–h** (Fig. 1.24b) [100]. For example, the ion pair comprising PCCp$^-$ and Ch$^+$ was prepared as a genuine π-electronic ion pair, showing the formation of a charge-by-charge assembly in the crystal state **31a** (Fig. 1.25a) [100]. The stacking distances between oppositely charged π-electronic ions are 3.45 and 3.41 Å, suggesting an effective π–π stacking assembly. Furthermore, PCCp$^-$-based ion pairs with crystal violet **31b** and TATA$^+$

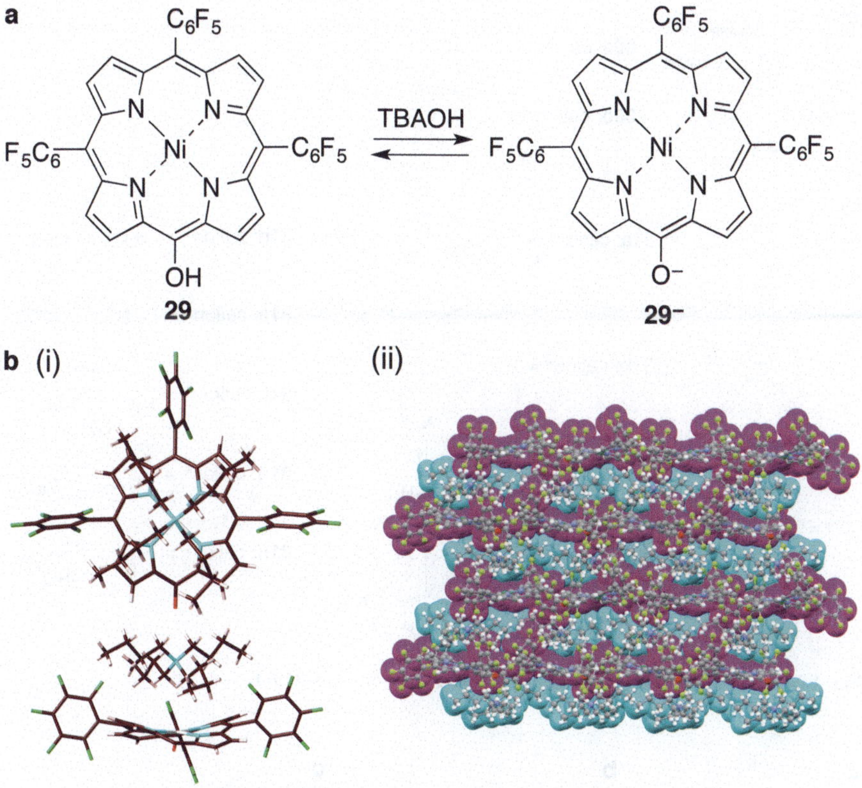

Fig. 1.23 (**a**) *meso*-Hydroxy-substituted porphyrin **29** and the formation of deprotonated species **29**$^-$ and (**b**) single-crystal X-ray structure of **29**$^-$·TBA$^+$ as (*i*) top and side views and (*ii*) packing diagram (magenta and cyan represent **29**$^-$ and cations, respectively)

31c were prepared by similar ion-exchange process. Single-crystal X-ray analysis revealed the formation of charge-by-charge assemblies (Fig. 1.25b,c). Interestingly, **31c** exhibited a two-by-two stacking columnar structure of oppositely charged π-electronic units (TATA$^+$ and PCCp$^-$) [100].

Ion pairs consisting of π-electronic anions with relatively planar or small cations provided charge-by-charge assemblies. On the other hand, modification of the geometries of cations afforded charge-segregated assemblies. For example, the ion pairs of PCCp$^-$ with triethylammonium cation **31d** (Fig. 1.24b) by Wood and Wright et al. [101], butyltrimethylammonium **31e**, and tetrapropylammonium **31f** (Fig. 1.24b) provided charge-segregated assemblies based on stacking columnar structures of PCCp$^-$ in the crystal state (Fig 1.26a) [100]. Charge-segregated dimension-controlled assemblies are fascinating, because the effective electronic properties can be achieved. Accordingly, ion pairs consisting of PCCp$^-$ with appropriate aliphatic cations such as dimethyldioctadecylammonium **31g** and tridodecylmethylammonium **31h** (Fig. 1.24b) were prepared. The ion pairs **31g,h**

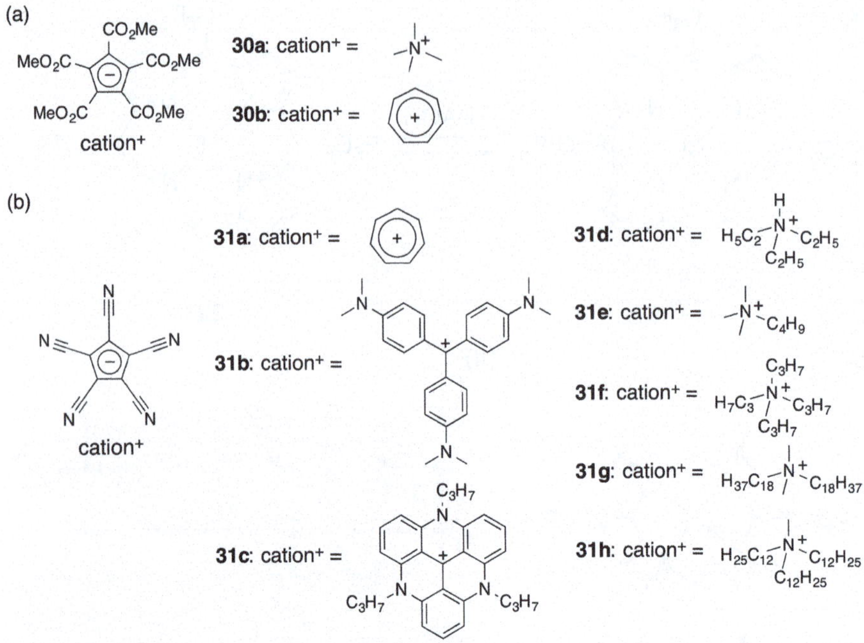

Fig. 1.24 (**a**) PMCp⁻ ion pairs **30a,b** and (**b**) PCCp⁻ ion pairs **31a–h**

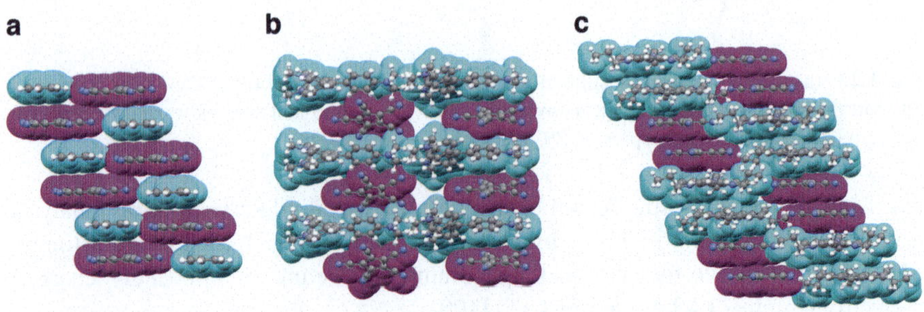

Fig. 1.25 Single-crystal X-ray structures of (**a**) **31a**, (**b**) **31b**, and (**c**) **31c** (magenta and cyan represent PCCp⁻ and cations, respectively)

formed smectic B (SmB) and Col_h phases, respectively, in their mesophases (Fig. 1.26b). Characteristic diffraction peaks at 0.60 and 0.36 nm observed in the XRD corresponded to the repeating distances of the ammonium cations and $PCCp^-$, respectively. These observations strongly suggested the formation of charge-segregated assemblies in the mesophases. Furthermore, the stacking columnar assembly of **31h** showed a hole-transporting property (0.4 $cm^2V^{-1}s^{-1}$) in the film prepared by drop-casting of a $CHCl_3$ solution, as revealed by FI-TRMC method [100].

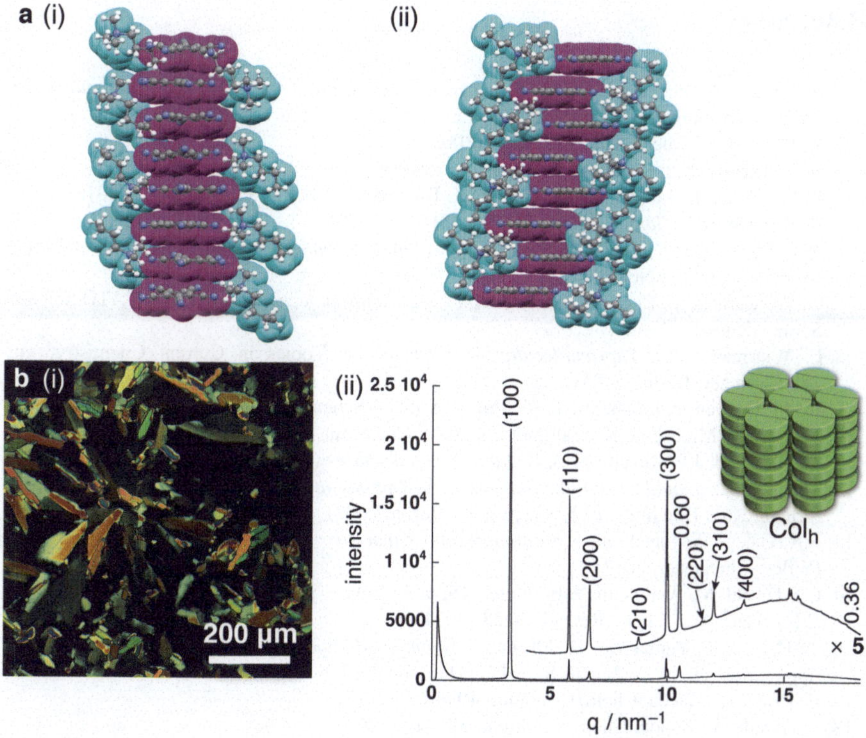

Fig. 1.26 (**a**) Single-crystal X-ray structures of (*i*) **31e** and (*ii*) **31f** (magenta and cyan represent $PCCp^-$ and cations, respectively) and (**b**) (*i*) POM and (*ii*) XRD of **31h** at 70 °C upon cooling from Iso

1.5 Conclusion

It is highly important to control the geometries and electronic states of ionic π-electronic components in order to fabricate highly organized ion-pairing assemblies. We have shown various pyrrole-based anion-responsive π-electronic molecules that effectively form receptor–anion complexes as π-electronic anionic species. The obtained π-electronic anion-based ion pairs provide unique assembling behaviors based on charge-by-charge and charge-segregated assemblies. A great advantage of receptor–anion-complexing ion-pairing assemblies is the preparation of numerous ion pairs by combining designed π-electronic receptors, guest anions, and countercations. Based on the characteristic assembling behaviors of receptor–anion-complexing ion pairs, we have demonstrated dimension-controlled assemblies comprising genuine π-electronic anions. π-Electronic ion pairs afforded various organized structures in crystal states as well as liquid crystal mesophases depending on the countercations. Further designed and modified π-electronic ion pairs [102–105] would exhibit intriguing optical and electronic properties originating from their specific organized structures.

References

1. A. Ciferri, A. Perico (eds.), *Ionic Interactions in Natural and Synthetic Macromolecules* (Wiley, New Jersey, 2012)
2. K. Binnemans, Chem. Rev. **105**, 4148 (2005)
3. K.V. Axenov, S. Laschat, Materials **4**, 206 (2011)
4. K. Goossens, K. Lava, C.W. Bielawski, K. Binnemans, Chem. Rev. **116**, 4643 (2016)
5. B. Kirchner (ed.), *Ionic Liquids* (Springer, Berlin, 2009)
6. N.V. Plechkova, K.R. Seddon (eds.), *Ionic Liquids Completely UnCOILed: Critical Expert Overviews* (Wiley, New Jersey, 2015)
7. G. Tsoucaris (ed.), *Current Challenges on Large Supramolecular Assemblies,* NATO Science Series, (Kluwer, Dordrecht, 1999)
8. F. Würthner (ed.), *Supramolecular Dye Chemistry*, Topics in Current Chemistry, vol 258 (Springer, Berlin, 2005)
9. F.J.M. Hoeben, P. Jonkheijm, E.W. Meijer, A.P.H.J. Schenning, Chem. Rev. **105**, 1491 (2005)
10. T. Kato, N. Mizoshita, K. Kishimoto, Angew. Chem. Int. Ed. **45**, 38 (2006)
11. J.L. Atwood, J.W. Steed (eds.), *Organic Nanostructures* (Wiley, Weinheim, 2007)
12. T. Kato (ed.), *Liquid Crystalline Functional Assemblies and Their Supramolecular Structures*, Structure and Bonding, vol 128 (Springer, Berlin, 2008)
13. P.A. Gale, J.W. Steed (eds.), *Supramolecular Chemistry: From Molecules to Nanomaterials* (Wiley, Chichester, 2012)
14. C.F.J. Faul, M. Antonietti, Adv. Mater. **15**, 673 (2003)
15. C.F.J. Faul, Acc. Chem. Res. **47**, 3428 (2014)
16. Y. Haketa, R. Yamakado, H. Maeda, in *Conjugated Objects: Developments, Synthesis, and Application*, ed. by A. Nagai, K. Takagi (Pan Stanford, Singapore, 2017), Ch.13, p. 349
17. B. Dong, H. Maeda, Chem. Commun. **49**, 4085 (2013)
18. H. Maeda, Y. Bando, Chem. Commun. **49**, 4100 (2013)
19. H. Maeda, Bull. Chem. Soc. Jpn. **86**, 1359 (2013)
20. Y. Haketa, H. Maeda, Chem. Commun. **53**, 2894 (2017)
21. Y. Haketa, H. Maeda, Bull. Chem. Soc. Jpn. **91**, 420 (2018)
22. A. Bianchi, K. Bowman-James, E. García-España (eds.), *Supramolecular Chemistry of Anions* (Wiley, New York, 1997)
23. R.P. Singh, B.A. Moyer (eds.), *Fundamentals and Applications of Anion Separations* (Kluwer, New York, 2004)
24. I. Stibor (ed.), *Anion Sensing*, Topics in Current Chemistry, vol 255 (Springer, Berlin, 2005)
25. J.L. Sessler, P.A. Gale, W.S. Cho, *Anion Receptor Chemistry* (RSC, Cambridge, 2006)
26. R. Vilar (ed.), *Recognition of Anions*, Structure and Bonding, vol 129 (Springer, Berlin, 2008)
27. P.A. Gale, W. Dehaen (eds.), *Anion Recognition in Supramolecular Chemistry*, Topics in Heterocyclic Chemistry, vol 24 (Springer, Berlin, 2010)
28. Y. Hua, A.H. Flood, Chem. Soc. Rev. **39**, 1262 (2010)
29. Y. Haketa, R. Takasago, H. Maeda, Chem. Commun. **52**, 7364 (2016)
30. P.A. Gale, S. Camiolo, C.P. Chapman, M.E. Light, M.B. Hursthouse, Tetrahedron Lett. **42**, 5095 (2001)
31. J.L. Sessler, D. An, W.S. Cho, V. Lynch, J. Am. Chem. Soc. **125**, 13646 (2003)
32. J.L. Sessler, D. An, W.S. Cho, V. Lynch, M. Marquez, Chem. Eur. J. **11**, 2001 (2005)
33. C.B. Black, B. Andrioletti, A.C. Try, C. Ruiperez, J.L. Sessler, J. Am. Chem. Soc. **121**, 10438 (1999)
34. H. Salman, Y. Abraham, S. Tal, S. Meltzman, M. Kapon, N. Tessler, S. Speiser, Y. Eichen, Eur. J. Org. Chem. 2207 (2005)
35. H. Maeda, Y. Ito, Y. Kusunose, T. Nakanishi, Chem. Commun. **28**, 1136 (2007)
36. H. Maeda, K. Chigusa, T. Sakurai, K. Ohta, S. Uemura, S. Seki, Chem. Eur. J. **19**, 9224 (2013)
37. H. Maeda, K. Chigusa, R. Yamakado, T. Sakurai, S. Seki, Chem. Eur. J. **21**, 9520 (2015)

38. W.V. Rossom, T.G. Terentyeva, K. Sodeyama, Y. Matsushita, Y. Tateyama, K. Ariga, J.P. Hill, Org. Biomol. Chem. **12**, 5492 (2014)
39. B.A. Trofimov, A.M. Vasil'tsov, E.Y. Schmidt, N.V. Zorina, A.V. Afonin, A.I. Mikhaleva, K.B. Petrushenko, I.A. Ushakov, L.B. Krivdin, V.K. Belsky, L.I. Bryukvina, Eur. J. Org. Chem. **4338** (2005)
40. Y. Haketa, Y. Tamura, N. Yasuda, H. Maeda, Org. Biomol. Chem. **14**, 8035 (2016)
41. B.W. Laursen, F.C. Krebs, Angew. Chem. Int. Ed. **39**, 3432 (2000)
42. B.W. Laursen, F.C. Krebs, Chem. Eur. J. **7**, 1773 (2001)
43. H. Maeda, Y. Kusunose, Chem. Eur. J. **11**, 5661 (2005)
44. H. Maeda, Y. Haketa, T. Nakanishi, J. Am. Chem. Soc. **129**, 13661 (2007)
45. H. Maeda, N. Eifuku, Chem. Lett. **38**, 208 (2009)
46. H. Maeda, Y. Ito, Y. Haketa, N. Eifuku, E. Lee, M. Lee, T. Hashishin, K. Kaneko, Chem. Eur. J. **15**, 3706 (2009)
47. H. Maeda, Y. Terashima, Y. Haketa, A. Asano, Y. Honsho, S. Seki, M. Shimizu, H. Mukai, K. Ohta, Chem. Commun. **46**, 4559 (2010)
48. H. Maeda, N. Eifuku, Y. Haketa, Y. Ito, E. Lee, M. Lee, Phys. Chem. Chem. Phys. **13**, 3843 (2011)
49. H. Maeda, Y. Terashima, Chem. Commun. **47**, 7620 (2011)
50. H. Maeda, K. Kinoshita, K. Naritani, Y. Bando, Chem. Commun. **47**, 8241 (2011)
51. Y. Terashima, T. Sakurai, Y. Bando, S. Seki, H. Maeda, Chem. Mater. **25**, 2656 (2013)
52. H. Maeda, W. Hane, Y. Bando, Y. Terashima, Y. Haketa, H. Shibaguchi, T. Kawai, M. Naito, K. Takaishi, M. Uchiyama, A. Muranaka, Chem. Eur. J. **19**, 16263 (2013)
53. Y. Haketa, D. Katayama, S. Fukunaga, Y. Bando, T. Sakurai, W. Matsuda, S. Seki, H. Maeda, Chem. Asian J. **11**, 2025 (2016)
54. H. Maeda, Y. Ito, Inorg. Chem. **45**, 8205 (2006)
55. H. Maeda, Y. Kusunose, Y. Mihashi, T. Mizoguchi, J. Org. Chem. **72**, 2612 (2007)
56. Y. Maeda, Y. Haketa, S. Bando, S. Sakamoto, Synth. Met. **159**, 792 (2009)
57. H. Maeda, Y. Bando, Y. Haketa, Y. Honsho, S. Seki, H. Nakajima, N. Tohnai, Chem. Eur. J. **16**, 10994 (2010)
58. Y. Bando, S. Sakamoto, I. Yamada, Y. Haketa, H. Maeda, Chem. Commun. **48**, 2301 (2012)
59. Y. Bando, T. Sakurai, S. Seki, H. Maeda, Chem. Asian J. **8**, 2088 (2013)
60. V. Lakshmi, Y. Haketa, R. Yamakado, N. Yasuda, H. Maeda, Chem. Commun. **53**, 3834 (2017)
61. A. Kuno, N. Tohnai, N. Yasuda, H. Maeda, Chem. Eur. J. **23**, 11357 (2017)
62. H. Maeda, Y. Haketa, Org. Biomol. Chem. **6**, 3091 (2008)
63. H. Maeda, Y. Mihashi, Y. Haketa, Org. Lett. **10**, 3179 (2008)
64. H. Maeda, R. Fujii, Y. Haketa, Eur. J. Org. Chem. 1469 (2010)
65. Y. Haketa, H. Maeda, Chem. Eur. J. **17**, 1485 (2011)
66. H. Maeda, K. Kitaguchi, Y. Haketa, Chem. Commun. **47**, 9342 (2011)
67. Y. Haketa, S. Sakamoto, K. Chigusa, T. Nakanishi, H. Maeda, J. Org. Chem. **76**, 5177 (2011)
68. Y. Haketa, Y. Bando, K. Takaishi, M. Uchiyama, A. Muranaka, M. Naito, H. Shibaguchi, T. Kawai, H. Maeda, Angew. Chem. Int. Ed. **51**, 7967 (2012)
69. R. Yamakado, T. Sakurai, W. Matsuda, S. Seki, N. Yasuda, S. Akine, H. Maeda, Chem. Eur. J. **22**, 626 (2016)
70. R. Yamakado, H. Maeda, J. Photochem. Photobiol. A **331**, 215 (2016)
71. R. Yamakado, R. Sato, Y. Shigeta, H. Maeda, J. Org. Chem. **81**, 8530 (2016)
72. R. Yamakado, Y. Ashida, R. Sato, Y. Shigeta, N. Yasuda, H. Maeda, Chem. Eur. J. **23**, 4160 (2017)
73. H. Maeda, Y. Fujii, Y. Mihashi, Chem. Commun. 4285 (2008)
74. H. Maeda, M. Takayama, K. Kobayashi, H. Shinmori, Org. Biomol. Chem. **8**, 4308 (2010)
75. H. Maeda, Y. Bando, K. Shimomura, I. Yamada, M. Naito, K. Nobusawa, H. Tsumatori, T. Kawai, J. Am. Chem. Soc. **133**, 9266 (2011)
76. Y. Terashima, M. Takayama, K. Isozaki, H. Maeda, Chem. Commun. **49**, 2506 (2013)

77. Y. Haketa, S. Sasaki, N. Ohta, H. Masunaga, H. Ogawa, N. Mizuno, F. Araoka, H. Takezoe, H. Maeda, Angew. Chem. Int. Ed. **49**, 10079 (2010)
78. H. Maeda, K. Naritani, Y. Honsho, S. Seki, J. Am. Chem. Soc. **133**, 8896 (2011)
79. B. Dong, Y. Terashima, Y. Haketa, H. Maeda, Chem. Eur. J. **18**, 3460 (2012)
80. Y. Haketa, M. Takayama, H. Maeda, Org. Biomol. Chem. **10**, 2603 (2012)
81. Y. Haketa, Y. Honsho, S. Seki, H. Maeda, Chem. Eur. J. **18**, 7016 (2012)
82. B. Dong, T. Sakurai, Y. Honsho, S. Seki, H. Maeda, J. Am. Chem. Soc. **135**, 1284 (2013)
83. B. Dong, T. Sakurai, Y. Bando, S. Seki, K. Takaishi, M. Uchiyama, A. Muranaka, H. Maeda, J. Am. Chem. Soc. **135**, 14797 (2013)
84. R. Sekiya, Y. Tsutsui, W. Choi, T. Sakurai, S. Seki, Y. Bando, H. Maeda, Chem. Commun. **50**, 10615 (2014)
85. R. Yamakado, M. Hara, S. Nagano, T. Seki, H. Maeda, Chem. Eur. J. **23**, 9244 (2017)
86. R. Yamakado, M. Hara, S. Nagano, T. Seki, H. Maeda, Chem. Lett. **47**, 404 (2018)
87. D. Wu, L. Zhi, C.J. Bodwell, G. Cui, N. Tsao, K. Müllen, Angew. Chem. Int. Ed. **46**, 5417 (2007)
88. D. Shi, C. Schwall, G. Sfintes, E. Thyrhaug, P. Hammershøj, M. Cárdenas, J.B. Simonsen, B.W. Laursen, Chem. Eur. J. **20**, 6853 (2014)
89. H. Maeda, A. Fukui, R. Yamakado, N. Yasuda, Chem. Commun. **51**, 17572 (2015)
90. Y. Sasano, N. Yasuda, H. Maeda, Chem. Asian J. **11**, 3423 (2016)
91. Y. Sasano, R. Sato, Y. Shigeta, N. Yasuda, H. Maeda, J. Org. Chem. **82**, 11166 (2017)
92. Y. Sasano, N. Yasuda, H. Maeda, Dalton Trans. **46**, 8924 (2017)
93. E. Le Goff, R.B. LaCount, J. Am. Chem. Soc. **85**, 1354 (1963)
94. O.W. Webster, J. Am. Chem. Soc. **87**, 1820 (1965)
95. R. Kuhn, D. Rewicki, Angew. Chem. Int. Ed. Eng. **6**, 635 (1967)
96. K. Okamoto, T. Kitagawa, K. Takeuchi, K. Komatsu, T. Kinoshita, S. Aonuma, M. Nagai, A. Miyabo, J. Org. Chem. **55**, 996 (1990)
97. M.I. Bruce, P.A. Humphrey, B.W. Skelton, A.H. White, Aust. J. Chem. **37**, 2441 (1984)
98. M.I. Bruce, P.A. Humphrey, B.W. Skelton, A.H. White, Aust. J. Chem. **39**, 165 (1986)
99. T. Sakai, S. Seo, J. Matsuoka, Y. Mori, J. Org. Chem. **78**, 10978 (2013)
100. Y. Bando, Y. Haketa, T. Sakurai, W. Matsuda, S. Seki, H. Takaya, H. Maeda, Chem. Eur. J. **22**, 7843 (2016)
101. R.J. Less, M.J. McPartlin, M. Rawson, P.T. Wood, D.S. Wright, Chem. Eur. J. **17**, 13723 (2010)
102. H. Maeda, Y. Takeda, Y. Haketa, Y. Morimoto, N. Yasuda, Chem. Eur. J. **24**, 8910 (2018)
103. G. Hirata, H. Maeda, Org. Lett. **20**, 2853 (2018)
104. S. Kaname, Y. Haketa, N. Yasuda, H. Maeda, Org. Lett. **20**, 3268 (2018)
105. A. Kuno, M. Fujiwara, Y. Haketa, H. Maeda, Chem. Asian J. **14**, in press (2019). https://doi.org/10.1002/asia.201801040

Chapter 2
Molecular Space Chemistry Based on Pillar[n]arenes

Tomoki Ogoshi, Takahiro Kakuta, and Tada-aki Yamagishi

Abstract In 2008, our group reported on pillar[n]arenes as novel-shaped macrocyclic compounds. Owing to the para-bridge connection between 1,4-dialkoxybenzene units, pillar[n]arenes adopt a highly symmetrical pillar and polygonal-shaped structure, which is different from typical host molecules. Because pillar[n]arenes exhibit these highly symmetrical structures with high functionality, many chemists have used pillar[n]arenes as building blocks to construct supramolecular assemblies. Therefore, pillar[n]arenes have applications in various fields such as material science and biochemistry. We discuss the host–guest ability of pillar[n]arenes, application of pillar[5]arenes as catalysts and reductants, and assembly of pillar[n]arenes on surfaces based on their pillar-shaped structures. The highly ordered assembled structures, based on their highly symmetrical polygonal prism shape, are also discussed.

Keywords Pillar[n]arenes · Molecular recognition · Supramolecular catalysts and reductants · Surface chemistry · Assembly · Porous materials

2.1 Introduction

Macrocyclic compounds have an intrinsic molecular space owing to their cyclic structures. The intrinsic space is very useful to capture guest molecules that are able to fit into the space. Using the host–guest ability of macrocyclic compounds, various supramolecular assemblies have been developed. As macrocyclic compounds, cyclodextrins, [1–6] calix[n]arenes, [7–12] cucurbit[n]urils [13–15], and crown ethers [16, 17] are well-known and have been widely used. However, these macrocyclic compounds had been thoroughly investigated by 2000, which meant that a

T. Ogoshi (✉)
Graduate School of Natural Science and Technology, Kanazawa University, Kanazawa, Japan

WPI Nano Life Science Institute, Kanazawa University, Kanazawa, Japan
e-mail: ogoshi@se.kanazawa-u.ac.jp

T. Kakuta · T.-a. Yamagishi
Graduate School of Natural Science and Technology, Kanazawa University, Kanazawa, Japan

S. Shirakawa (ed.), *Designed Molecular Space in Material Science and Catalysis*,
https://doi.org/10.1007/978-981-13-1256-4_2

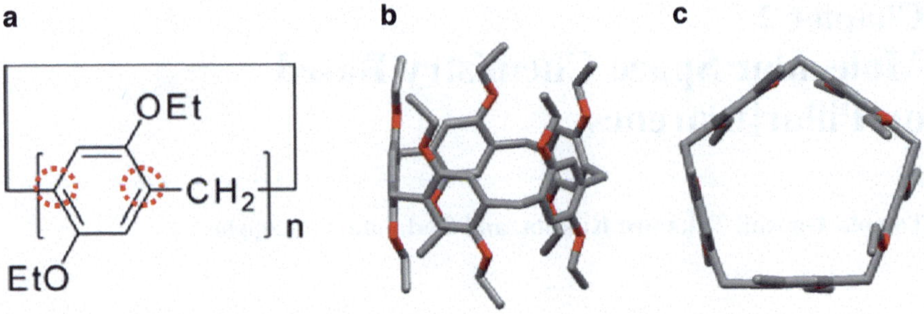

Fig. 2.1 (**a**) Chemical and (**b** and **c**) single crystal X-ray structures of pillar[n]arenes

quest for new host molecules was required. Against this historical background, in 2008, our group reported on pillar[n]arenes as novel-shaped macrocyclic compounds (Fig. 2.1) [18]. Owing to the para-bridge connection between 1,4-dialkoxybenzene units, they are able to adopt highly symmetrical pillar and polygonal-shaped structures, which are different from typical host molecules. Because pillar[n]arenes exhibit these new highly symmetrical structures with high functionality, many chemists have used pillar[n]arenes as building blocks to construct supramolecular assemblies, and pillar[n]arenes have applications in various fields such as material science and biochemistry [19–22]. In this chapter, we discuss the host–guest ability of pillar[n]arenes, application as catalysts and reductants for organic reactions, and assembly of pillar[n]arenes on surfaces based on their pillar-shape structures. The highly ordered assembled structures based on their highly symmetrical polygonal prism shape are also discussed.

2.2 Host Ability of Pillar[n]arenes

The cavity size of pillar[5]arenes is approximately 4.7 Å (Fig. 2.2), which is almost the same as those of α-cyclodextrin (ca. 4.7 Å) and cucurbit[6]uril (ca. 5.8 Å).

Similar to α-cyclodextrin and cucurbit[6]uril, linear alkanes are good guest molecules for pillar[5]arenes owing to the size matching between the sizes of these guests and the cavity size of pillar[5]arenes. The cavity size of pillar[6]arenes is approximately 6.7 Å, which is almost equal to those of β-cyclodextrin (ca. 6.0 Å) and cucurbit[7]urils (ca. 7.3 Å). Thus, as with β-cyclodextrin and cucurbit[7]urils, the cavity size of pillar[6]arenes can fit bulky hydrocarbons such as branched, cyclic alkanes and substituted benzenes. Pillar[n]arenes are composed of electron-donating 1,4-dialkoxybenzene units. Therefore, the cavity is an electron-rich space. Owing to this electron-rich space, molecules with cationic moieties and with electron-withdrawing groups are good guest molecules [23]. For pillar[5]arenes, linear alkanes with cationic groups, such as pyridinium and trimethyl ammonium cations, and electron-withdrawing groups, such as cyano, halogens, and triazole groups, are

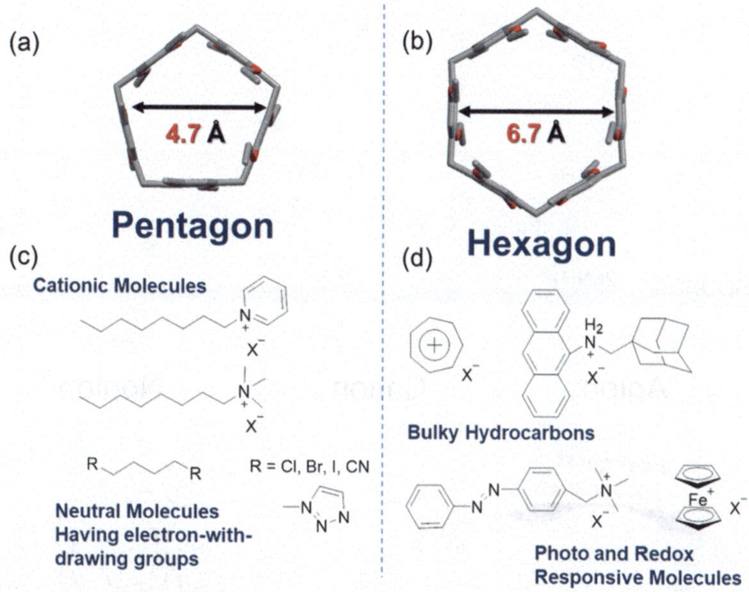

Fig. 2.2 Cavity size of (**a**) pillar[5]arenes and (**b**) pillar[6]arenes and good guest molecules for (**c**) pillar[5]arenes and (**d**) pillar[6]arenes

good guest molecules. C4 alkanes possessing these electron-withdrawing groups at both ends form very stable host–guest complexes because the C4 length fits into the height of pillar[5]arene. For pillar[6]arenes, bulky hydrocarbons with cationic groups are good guest molecules. One of the advantages of pillar[n]arene chemistry, compared with other host molecules, is its versatile functionality. The host–guest complexation ability of pillar[n]arenes can be tailored by introducing functional groups. A straightforward approach to the modification with functional groups is rim functionalization. Mono-, di-, tri-, tetra-, and per-rim-functionalized and rim-different pillar[n]arenes can be produced by various organic methods [21, 24]. Water-soluble pillar[n]arenes can be produced by introducing anionic, cationic, and nonionic groups on the both rims of pillar[n]arenes (Fig. 2.3a).

The complexation ability is enhanced in aqueous media because a hydrophilic–hydrophobic interaction exists in aqueous media. The introduction of interaction points into the rims also contributes to stabilize the host–guest complexation. For example, stable host–guest complexes between diamine and pillar[5]arenes can be obtained by introducing two carboxylic acid groups at the same unit (Fig. 2.3b) [25].

Furthermore, Yang and co-workers recently reported the change of cavity electron density by π-surface metalation of pillar[5]arene (Fig. 2.4) [26].

One, two, and three charged transition metal cations can directly bond to the outer π-surface of pillar[5]arene. The transition metal cation coordination contributes to change the cavity from being electron rich to electron deficient, which allows for the formation of complexes with electron-rich guests containing a $CF_3SO_3^-$ anion

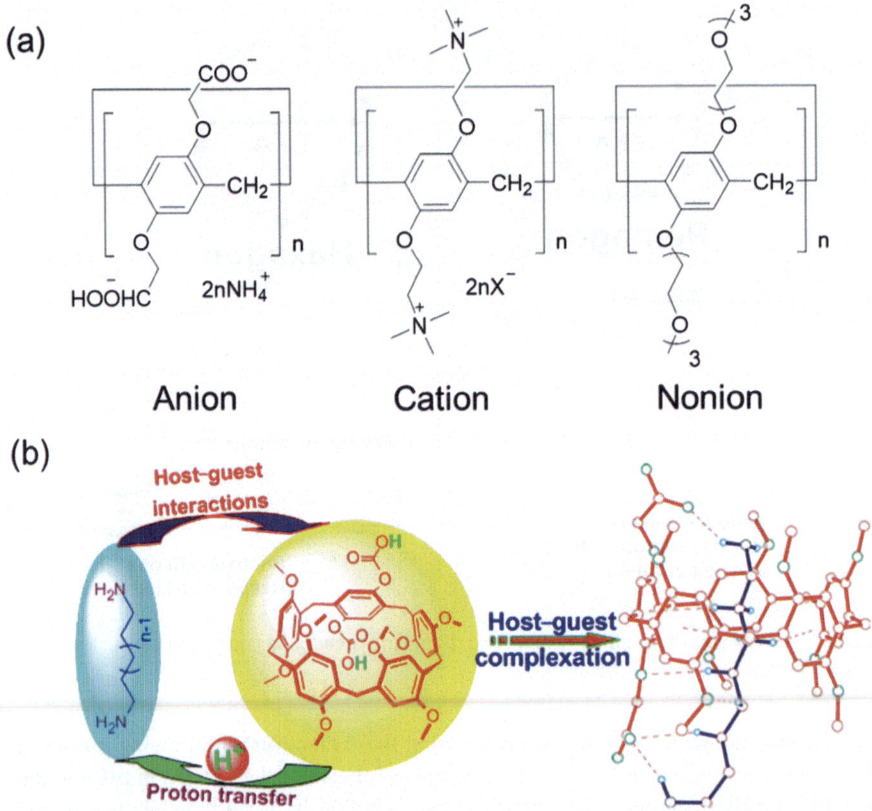

Fig. 2.3 (**a**) Water-soluble pillar[n]arenes by introducing anionic, cationic, and nonionic moieties at both rims of pillar[n]arenes. (**b**) Proton transfer in host–guest complexation between a pillar[5] arene with two carboxylic acid groups at the same unit and alkyldiamines. (Reproduced with permission from Ref. [25]. Copyright 2014 American Chemical Society)

through anion-π interactions. Introducing one benzoquinone unit contributes to the construction of a redox responsive host–guest complexation system (Fig. 2.5) [27]. A pillar[5]arene with one benzoquinone unit forms a weak host–guest complex with a C4 alkane possessing triazole groups at both ends. In contrast, the reduced pillar[5]arene with one hydroquinone unit forms a stable host–guest complex with the same guest.

2.3 Supramolecular Catalysts and Oxidants

Pillar[n]arenes have designable functionality; thus functional groups can be installed which work as catalysts. Our group synthesized a new phase-transfer catalyst consisting of ten tetra-alkyl phosphonium bromide groups and a pillar[5]arene core (Fig. 2.6) [28].

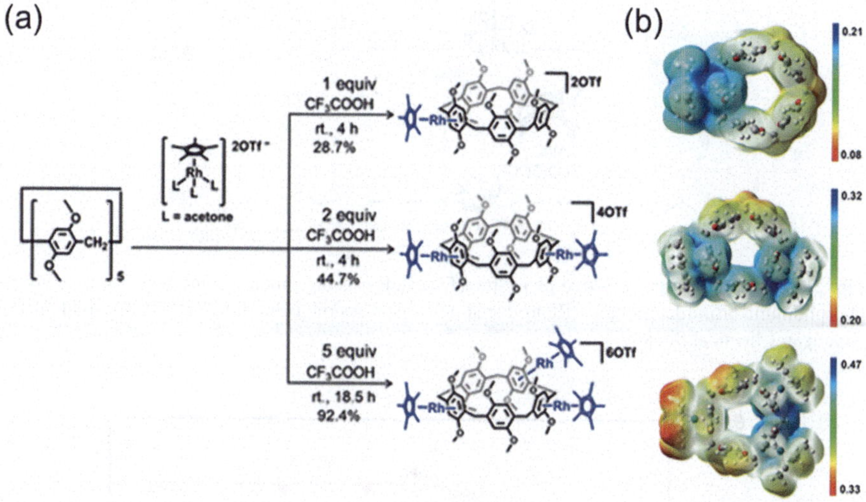

Fig. 2.4 (**a**) Synthesis and (**b**) electron potential mapping of mono-, di-, and tri-nuclear pillar[5] arene complexes. (Reproduced with permission from Ref. [26]. Copyright 2017 Wiley-VCH Verlag GmbH & Co. KGaA)

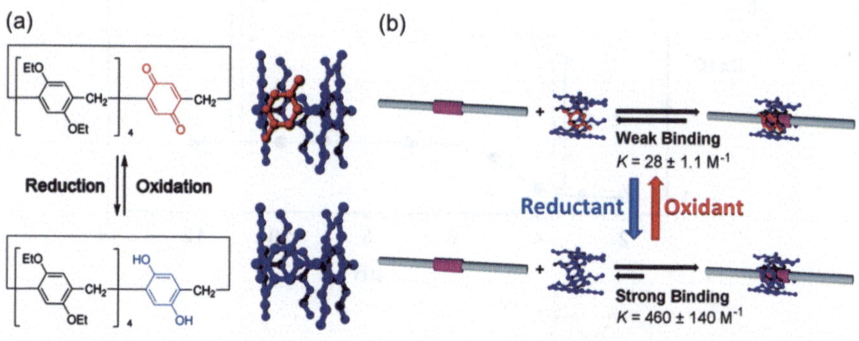

Fig. 2.5 (**a**) Chemical structures of redox-responsive pillar[5]arene and (**b**) cartoon representation of the formation of a redox-responsive inclusion complex. (Reproduced with permission from Ref. [27]. Copyright 2016 The Royal Society of Chemistry)

Owing to the selective complexation of linear-shaped molecules by pillar[5] arenes, the amphiphilic pillar[5]arene selectively works as a phase-transfer catalyst for linear-shaped alkenes. The conversion from alkene to aldehyde by $KMnO_4$ oxidation using linear-shaped alkenes was over 99%, while that using branched-shaped alkenes was quite low (31%). Complexation between the linear-shaped alkene and the amphiphilic pillar[5]arene and increasing the local concentration of oxidant MnO_4^- around the rims contribute to the high oxidation efficiency.

Garcia-Rio, Nome, and co-workers reported phosphate monoester hydrolysis using a cationic pillar[5]arene (Fig 2.7) [29].

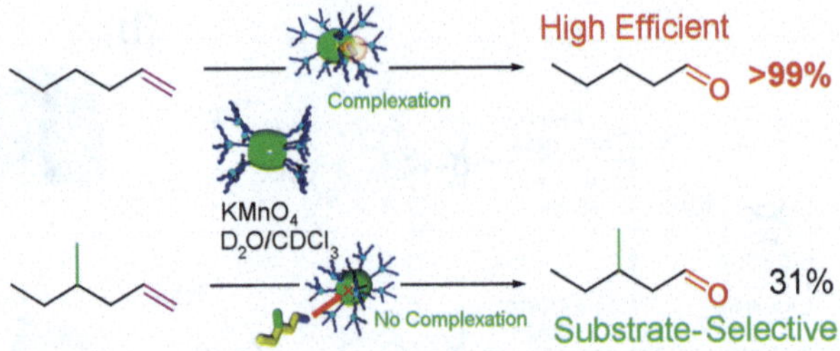

Fig. 2.6 Substrate-shape selective phase-transfer catalyst using pillar[5]arene with ten tetra-alkyl phosphonium bromide groups. (Reproduced with permission from Ref. [28]. Copyright 2013 American Chemical Society)

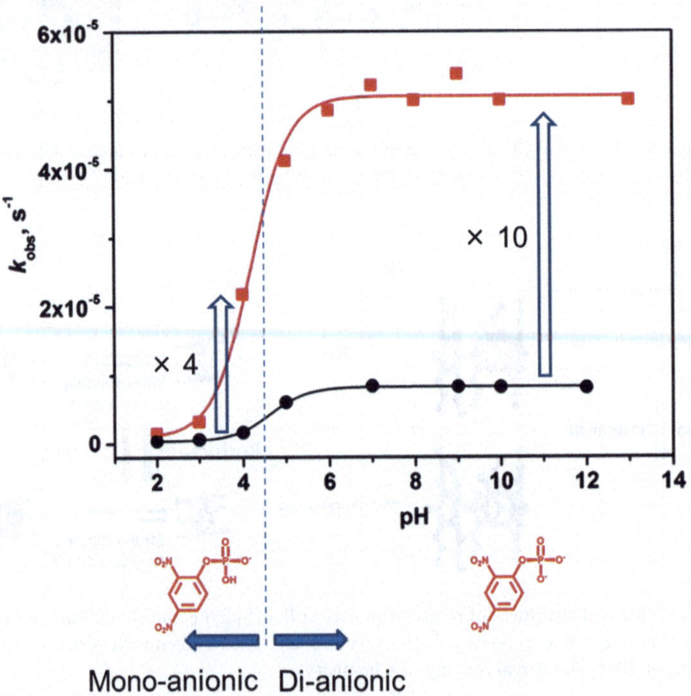

Fig. 2.7 pH-rate profiles for the hydrolysis of dinitrophenylphosphate in the presence of pillar[5] arene (red squares, 0.002 M) and in aqueous solutions in the absence of pillar[5]arene (black circles). (Reproduced with permission from Ref. [29]. Copyright 2016 The Royal Society of Chemistry)

The rate constant for dinitrophenylphosphate monoester hydrolysis increased in the presence of cationic pillar[5]arene (Fig. 2.3) owing to the formation of a host–guest complex between the monoester and cationic pillar[5]arene. The rate constant in the monoanionic state increased approximately fourfold and that in the dianionic state increased almost tenfold. The enhancement of the rate constants resulted from

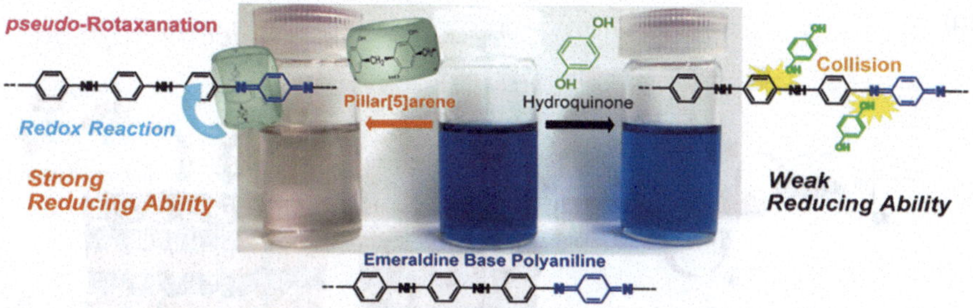

Fig. 2.8 Color change of the EB form of PANI upon addition of pillar[5]arene and hydroquinone and proposed structures in the reduction mechanism using hydroquinone and pillar[5]arene as reducing agents. (Reproduced with permission from Ref. [30]. Copyright 2011 American Chemical Society)

reducing the negative electron density on the PO_3^- oxygens by the formation of cationic groups of pillar[5]arene and the steric effects from the wall of the pillar[5] arene cavity.

Hydroquinone is easily oxidized and converts to benzoquinone, which means hydroquinone can be used as a reducing agent. Therefore, pillar[n]arenes consisting of hydroquinone units can be used as macrocyclic reducing agents (Fig. 2.8) [30].

Polyaniline (PANI) has four forms. The fully reduced form is named the leucoemeraldine base (LB). The fully oxidized form is called the pernigraniline base (PB). The half-oxidized and protonated forms are named the emeraldine base (EB) and the emeraldine salt (ES), respectively. Reversible interconversion between these different forms of PANI can be electrically and chemically controlled and induces color changes. When pillar[5]arene, consisting of five hydroquinone units, was mixed with the EB form of PANI, the blue solution from the EB form became completely colorless as the LB form. However, in the unit model of hydroquinone, the blue color remained, and conversion to the LB form was 40%, which indicated that the cyclic structure of pillar[5]arene exhibited a stronger reducing ability than the unit model of hydroquinone. The formation of the polypseudorotaxane structure between PANI and pillar[5]arene contributed to the strong reducing ability of pillar [5]arene compared with the unit model of hydroquinone.

2.4 Surface Chemistry

One-, two-, and three-dimensional molecular spaces are able to be produced by connecting macrocyclic compounds in various dimensions. For example, by connecting macrocyclic compounds in a lengthwise direction, tubular assemblies can be constructed. One of the characteristic points of pillar[n]arenes is their original shape—pillar-shaped structures. The pillar-shaped structures are very useful to modify various surfaces. Our group reported multilayer formation by alternating

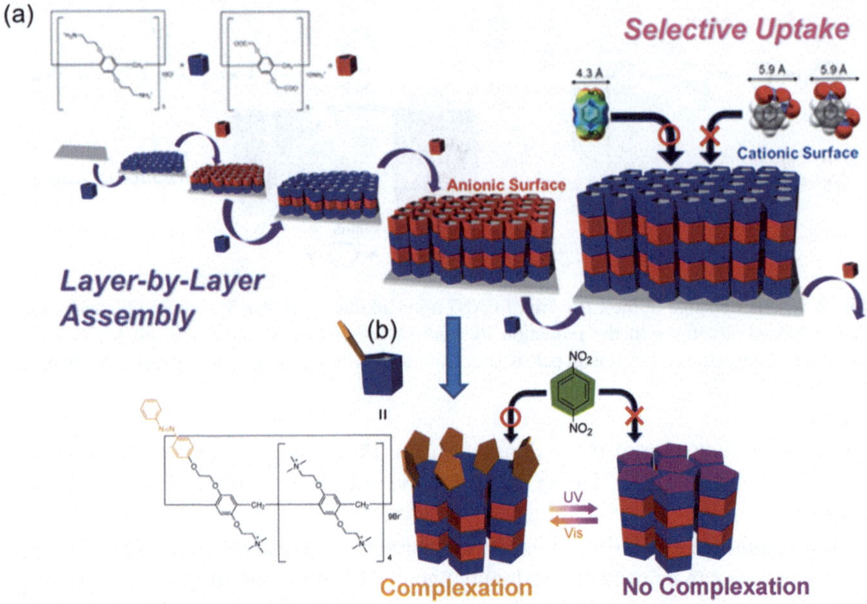

Fig. 2.9 (**a**) Layer-by-layer assembly by the consecutive adsorption of cationic pillar[5]arene and anionic pillar[5]arene. The multilayered film showed size-selective molecular recognition properties. (**b**) Modification of azobenzene valves and schematic representation of photoresponsive guest uptake. (Reproduced with permission from Refs. [31, 32]. Copyright 2015 and 2017 American Chemical Society)

the adsorption of cationic and anionic pillar[5]arenes on an inorganic surface (Fig 2.9a) [31].

An inorganic surface usually has a negative charge; thus, dipping the inorganic substrate in the cationic pillar[5]arene solution resulted in the surface modification with cationic pillar[5]arenes. At this stage, the surface has a cationic charge; thus dipping the substrate in the anionic pillar[5]arene solution resulted in forming a second anionic layer on the first cationic pillar[5]arene layer. By repeating this process, multilayers were able to be constructed on the substrate. The multilayer films could take up *p*-dinitrobenzene which could fit into the cavity size of pillar[5]arenes but were unable to take up *o*- and *m*-dinitrobenzenes, which are larger than the cavity size. By increasing the number of the layers, the amount of the guest uptake increased, which meant that the uptake amount of the guest was tunable.

Access of the guest molecule can be regulated by introducing a photoresponsive azobenzene group as a valve (Fig. 2.9b) [32]. First, multilayer films were produced by repeating the adsorption of cationic and anionic pillar[5]arenes. The surface was anionic with four layers. Thus, to attach azobenzene valves onto the micropore outlets, a cationic pillar[5]arene carrying one azobenzene group was adsorbed on

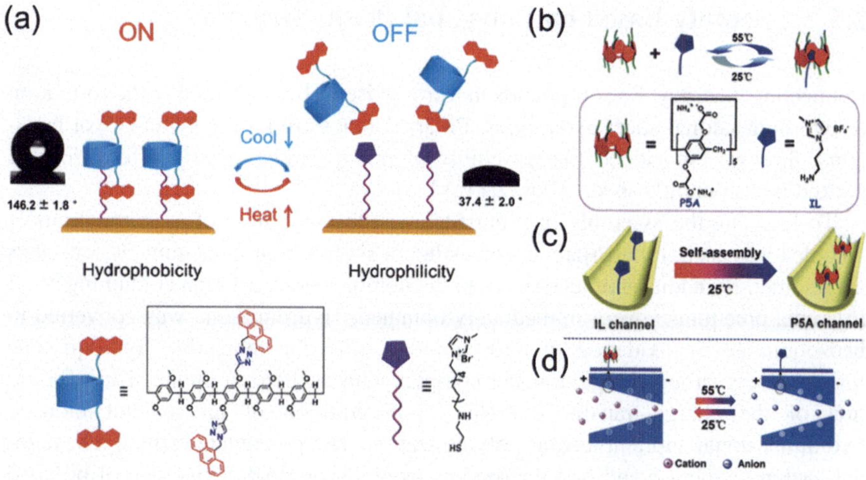

Fig. 2.10 (**a**) Schematic diagram of a temperature-responsive hydrophobic/hydrophilic switch constructed using an anthracene-functionalized pillar[5]arene-based host–guest interaction. (Reproduced with permission from Ref. [33]). Copyright 2016 American Chemical Society. (**b**) The host–guest interaction of anionic pillar[5]arene and an ionic liquid moiety. (**c**) The modification process in the nanochannel. (**d**) The mechanism of the temperature-triggered charge-reversal nanochannel constructed by pillar[5]arene-based host–guest interactions. (Reproduced with permission from Ref. [34]. Copyright 2017 Wiley-VCH Verlag GmbH & Co. KGaA)

the surface instead of a cationic pillar[5]arene. The azobenzene groups worked as photo-valves. In the trans form, the guest uptake took place as in the absence of the azobenzene valves. Upon UV irradiation, the trans form changed to the cis form. The cis form azobenzene blocked the access of the guest.

Li and co-workers reported temperature-responsive hydrophobic/hydrophilic surface switching based on the thermal-responsive host–guest interaction between pillar[5]arene and an ionic liquid modified on the surface (Fig. 2.10a) [33].

When pillar[5]arenes form complexes with ionic liquid groups modified on the surface, the surface is hydrophobic owing to the hydrophobic nature of pillar[5]arenes. Upon heating, the surface becomes hydrophilic because the pillar[5]arenes are released and the surface has hydrophilic ionic liquid groups. The thermal-responsive system has been applied as temperature-sensitive artificial channels by the same research group (Fig. 2.10b–d) [34]. Anionic pillar[5]arene and a cationic ionic liquid group were used as the host and guest, respectively. Ionic liquid groups were modified on the surface of the nanochannels. At 25 °C, the surface was anionic because of the formation of host–guest complexes between anionic pillar[5]arenes and ionic liquid groups, which resulted in cation transport at 25 °C. At high temperatures, the surface was cationic owing to the dissociation of the host–guest complexes, and anion transport was preferential.

2.5 Assembly Based on Polygonal Prism Structure

Connecting macrocyclic compounds in a crosswise fashion resulted in the formation of two-dimensional sheet assemblies. Pillar[6]arenes are highly symmetrical hexagons; thus we envisioned that assembly of pillar[6]arenes would afford densely packed hexagonal 2D sheets (Fig. 2.11a) [35].

To facilitate the assembly of pillar[6]arenes, the formation of a charge–transfer complex was used. Pillar[6]arene, consisting of six hydroquinone units, was used as a monomer. By addition of an oxidant to the homogeneous solution containing pillar[6]arene, precipitates were immediately obtained. Hydroquinone was converted to benzoquinone by oxidation of hydroquinone; then the remaining hydroquinone formed a very strong charge–transfer complex with the benzoquinone. Thus, formation of the charge–transfer complex upon addition of an oxidant induced two-dimensional supramolecular polymerization. The precipitate formed a hexagonal packing structure and had micropores from the original cavity size of pillar[6]arene. In contrast, when pillar[5]arene, consisting of five hydroquinone units, was used as a monomer, the assembled structure was an amorphous state. The small difference between the pentamer and hexamer resulted in the assembled structures.

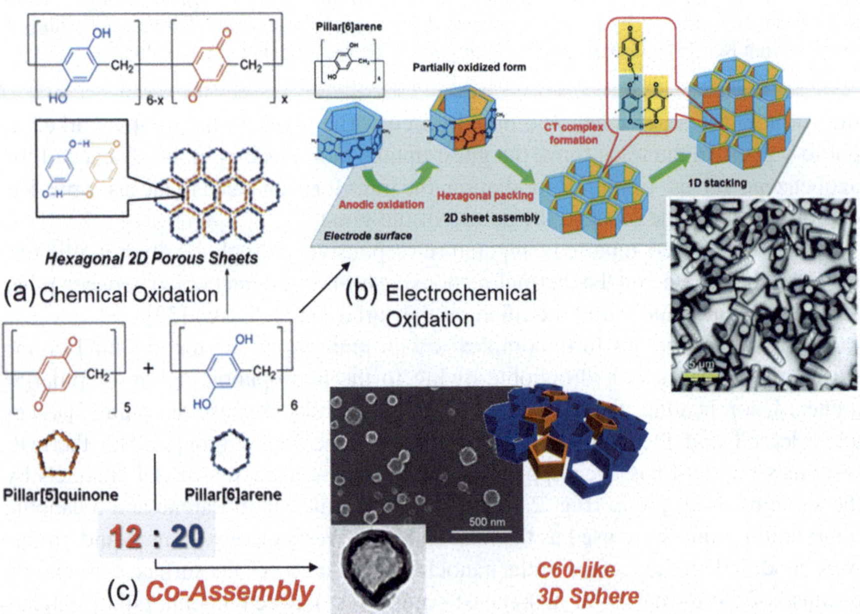

Fig. 2.11 Formation of 2D porous sheets by oxidation of the pillar[6]arene hydroquinone units by (**a**) chemical oxidation to obtain bulk 2D sheets and (**b**) electrochemical oxidation to modify the 2D sheets on the surface. (Reproduced with permission from Refs. [35, 36]). Copyright 2015 Wiley-VCH Verlag GmbH & Co. KGaA and 2017 The Royal Society of Chemistry. (**c**) Co-assembly of hexagonal pillar[6]arene and pentagonal pillar[5]quinone to form spherical vesicular assemblies. (Reproduced with permission from Ref. [37]. Copyright 2016 American Chemical Society)

Bulk two-dimensional sheets were obtained by assemblies of pillar[6]arene with a chemical oxidant. Using electrochemical oxidation, electrode surface modification of the hexagonal assemblies could be achieved [36]. Fig. 2.11b shows the cyclic voltammogram of pillar[6]arene. Peaks were observed at 1.2 V versus SCE, which corresponded to the oxidation of hydroquinone units of pillar[6]arene. By repeating the cycles, the anodic current gradually decreased, which indicated that insulating molecular layers were formed on the surface of the working electrode. Indeed, a white color thin film was deposited on the ITO electrode after the CV cycles. Figure 2.11b shows a 3D laser microscope image of the films. Uniform hexagonal assemblies were observed on the ITO surface. Electrochemical oxidation enabled modification of the 2D sheet assembly on the ITO surface and controlled the density and length of the 2D assembled structures.

C_{60} consists of 12 pentamers and 20 hexamers. With inspiration from the C_{60} structure, a co-assembly of pentagonal pillar[5]arenes and hexagonal pillar[6]arenes was investigated (Fig. 2.11c) [37]. Hexagons are easy-to-assemble compared with pentagons because hexagons have a higher symmetrical structure than pentagons. Thus, to incorporate the pentagons into the hexagon assembly, pillar[5]quinone, consisting of five benzoquinones, was first prepared. Then, the co-assembly of pillar [5]quinone and pillar[6]arene was investigated. In the case of an excess feed ratio of pillar[5]quinone, the assembled structures mainly adopted a tubular structure. In contrast, hexagonal disk-shaped assemblies were observed with an excess feed ratio of pillar[6]arene. A spherical assembly was observed by co-assembly of pillar[5] quinone and pillar[6]arene in 5:5 and 3:7 molar feed ratios. However, their shapes were inhomogeneous, and aggregation of the spheres was observed. Uniform spherical aggregates with diameters of 100–300 nm were observed in a 12:20 pillar[5] quinone/pillar[6]arene molar feed ratio. From the TEM measurements, spherical assemblies were observed, which indicated vesicle assemblies. Generally, vesicle assemblies are obtained by amphiphilic molecules in aqueous media. However, in this case, vesicle assemblies were produced by co-assembly of pentagons and hexagons. Thus, this method is a unique approach to obtain vesicle assemblies based on geometric design.

Li and co-workers prepared 2D sheets and 3D spheres by connecting pillar[5] arene at lateral positions (Fig. 2.12) [38]. Laterally functionalized pillar[5]arenes were synthesized for the first time by bromination at the methylene bridge of pillar [5]arene. 2D sheets and 3D spheres were obtained by connecting pillar[5]arenes at lateral positions with diamine linkers.

2.6 Porous Materials

Pillar[n]arenes are highly symmetrical polygonal prisms; thus, assembled structures of pillar[n]arenes in the crystal state afford well-ordered assembled structures in most cases. Activated crystals of pillar[n]arenes, produced by removing the solvent upon heating, are able to take up gases and organic vapors depending on the

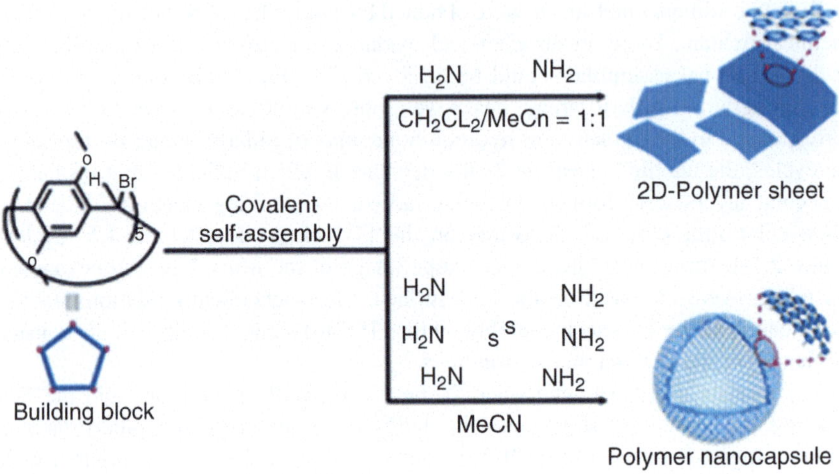

Fig. 2.12 The mechanism of the generation of 2D-polymer nanocapsules and 2D-polymer films by reaction of laterally functionalized pillar[5]arene and diamines. (Reproduced with permission from Ref. [38]. Copyright 2017 The Royal Society of Chemistry)

substituents of the pillar[n]arenes. In the case of pillar[n]arenes possessing phenolic groups, their crystal structures are stabilized by intra-/intermolecular hydrogen bonds; thus, they retain the original X-ray structures even after the removal of solvate-containing crystals. In the case of pillar[5]arene with ten phenolic groups, the crystals had intrinsic pores from the cavity of pillar[5]arene and another space between the pillar[5]arene molecules. Yang and co-workers reported that the activated crystals highly selectively took up CO_2 (375/1, 339/1 over CH_4 and N_2, respectively) owing to the porous nanostructure of the compounds and the dipole interactions between CO_2 and phenolic groups (Fig. 2.13) [39].

In the case of pillar[6]arene, the assembled structures are highly symmetrical 1D channel structures, owing to the highly symmetrical structure of hexagonal pillar[6]arene (Fig. 2.14). Our group reported that the activated crystals not only took up gases, such as CO_2 and N_2, but also alkane vapors, such as *n*-butane, *n*-hexane, and cyclohexane, because the cavity size of pillar[6]arene (6.7 Å) is larger than that of pillar[5]arene (4.7 Å) [40]. Another reason for this could be the highly ordered 1D channel structure. Owing to the 1D channel, pore blocking, which is observed in nitrogen adsorption at low temperatures, was not observed.

In the case of ethyl substituents, the activation of pillar[5,6]arenes resulted in the formation of nonporous crystals. However, the nonporous activated crystals exhibited shape-selective alkane uptake behavior. In the case of the pillar[5]arene, the activated crystals quickly took up vapors of linear-shaped alkanes such as *n*-hexane (Fig. 2.15a) [41]. Crystal transformation was the driving force to uptake linear alkanes. No uptake of alkane vapors was observed when branched- and cyclic-shaped alkane vapors were used. This is because the cavity size of pillar[5]arene could fit linear-shaped alkanes but was too small to capture branched and cyclic

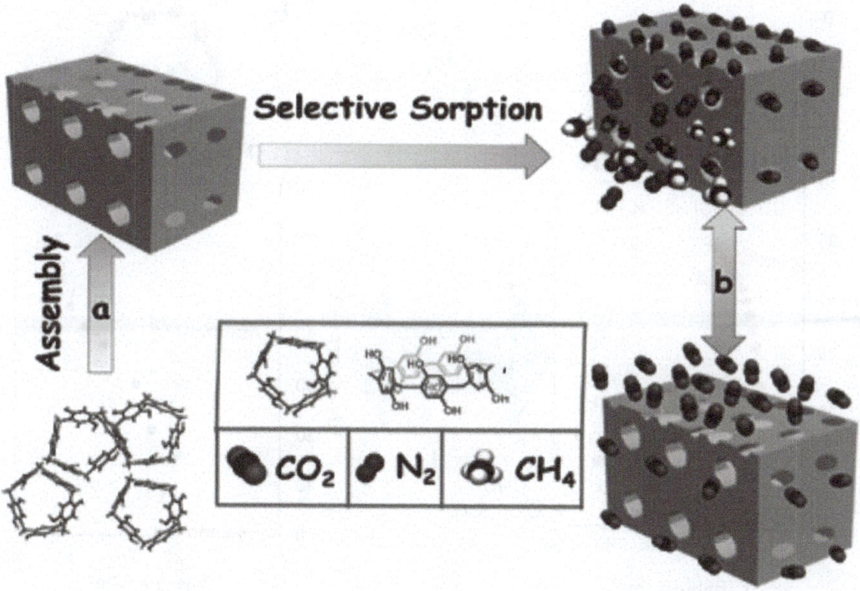

Fig. 2.13 Schematic representation of the supramolecular organic framework constructed from per-hydroxylated pillar[5]arene and its selective uptake of CO_2. The supramolecular organic framework can be reused for CO_2-selective adoption. (Reproduced with permission from Ref. [39]. Copyright 2014 Wiley-VCH Verlag GmbH & Co. KGaA)

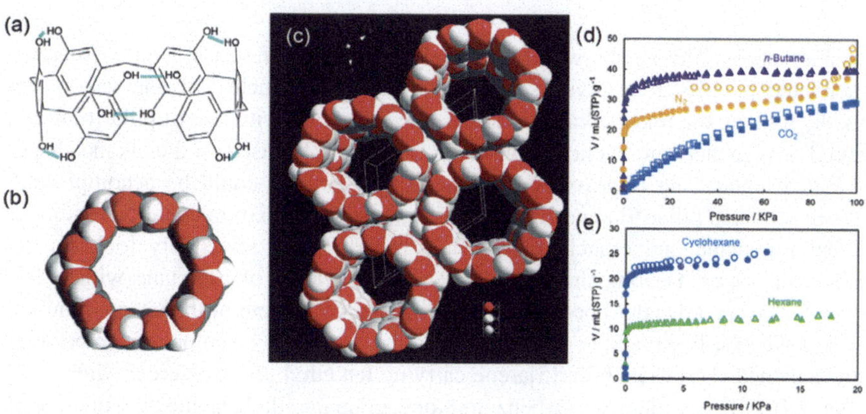

Fig. 2.14 (**a**) Chemical and (**b, c**) X-ray crystal structures of the per-hydroxylated pillar[6]arene. (**c**) View along the c axis. (**d**) Nitrogen (77 K, yellow circles), CO_2 (298 K, blue squares), and *n*-butane (298 K, purple triangles) sorption isotherms of powdered pillar[6]arene. (**e**) Sorption isotherms of powdered pillar[6]arene with pores for the vapors of *n*-hexane (green triangles) and cyclohexane (blue circles) at 298 K. Solid symbols = adsorption; open symbols = desorption. (Reproduced with permission from Ref. [40]. Copyright 2014 The Royal Society of Chemistry)

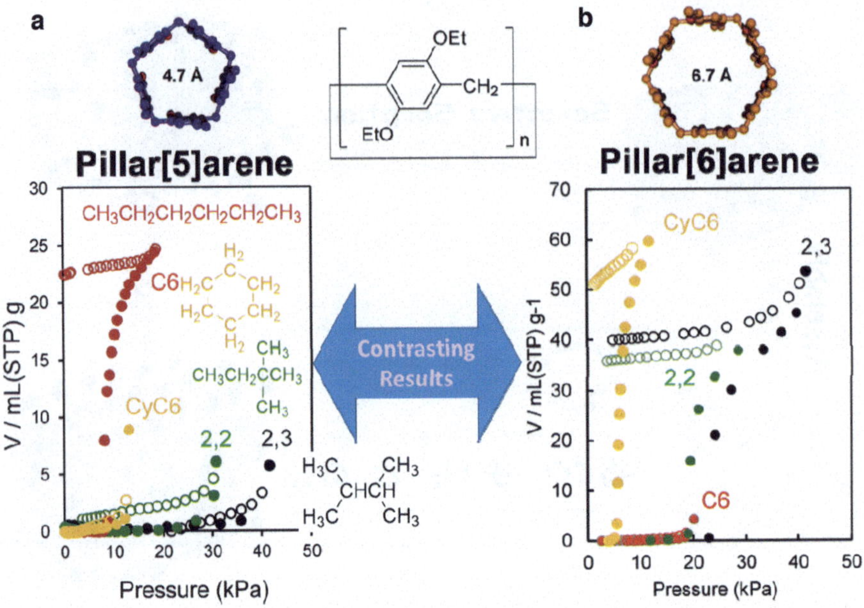

Fig. 2.15 Contrast sorption isotherms of (**a**) pillar[5]arene with 10 ethyl groups and (**b**) pillar[6] arene with 12 ethyl groups toward linear (*n*-hexane), branched (2,2-dimethylbutane and 2,3-dimethylbutane), and cyclic (cyclohexane) alkane vapors. Solid symbols = adsorption; open symbols = desorption

alkane vapors. The results were completely opposite to the case of pillar[6]arene. Activated pillar[6]arene crystals took up cyclic and branched alkane vapors but hardly took up the linear alkane, *n*-hexane, because the cavity size of pillar[6]arenes could fit branched and cyclic alkanes (Fig. 2.15b) [42]. Based on the alkane shape-selective uptake, an improvement of the octane number could be accomplished. When activated pillar[6]arene crystals were exposed to a mixture vapor consisting of linear *n*-heptane and branched isooctane, pillar[6]arene selectively took up the isooctane vapor. Heating the crystals resulted in release of isooctane with >99% purity. The crystals could be used again for the improvement of the octane number.

Uptake of *n*-linear alkane vapor could not be detected by a color change because the activated crystals of pillar[5]arene carrying ten ethyl groups were a white solid (Fig. 2.16). By introducing one benzoquinone group in pillar[5]arene by oxidation of one dialkoxybenzene unit, the color of the product was brown, owing to the formation of charge–transfer complexes between 1,4-diethoxybenzene and benzoquinone [43]. Interestingly, a color change occurred from brown to red by exposing the crystals to *n*-hexane vapor. The *n*-hexane vapor uptake triggered the crystal transformation, which was proposed to trigger a change in the stacking of the charge–transfer complexes and induce the color change. In contrast, no color change was observed when the brown crystals were exposed to branched and cyclic alkane vapors because the activated crystals of pillar[5]arene could take up linear alkane vapors but not branched and cyclic alkane vapors owing to the cavity size-dependent

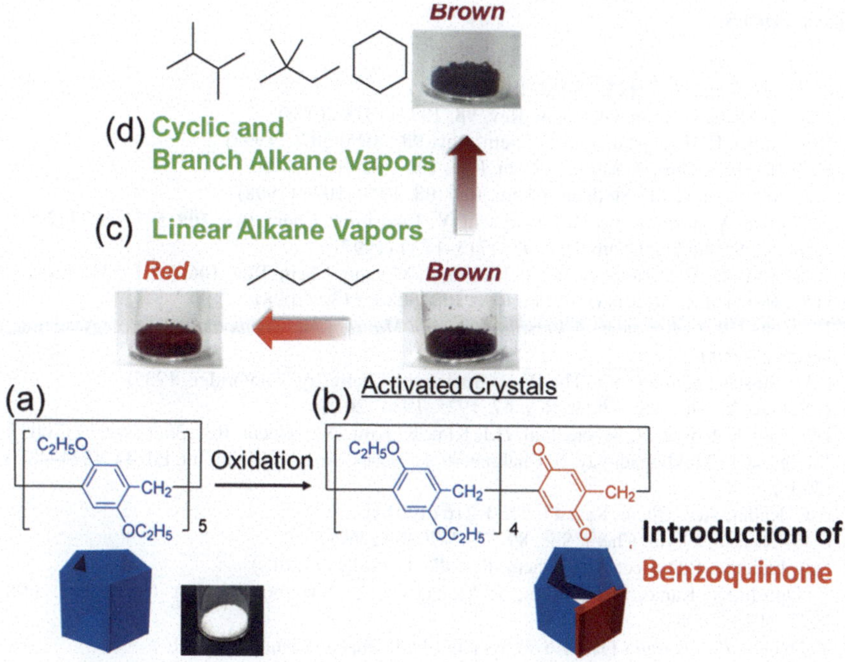

Fig. 2.16 Chemical structures of pillar[5]arenes (**a**) bearing ten ethyl groups and (**b**) containing one benzoquinone unit. The activated crystals of the monoquinone showed that a color change occurred from brown to red upon exposure to *n*-hexane vapor (**c**), while no color change was observed by exposing the crystals to branched and cyclic alkane vapors (**d**)

alkane uptake. Overall, we could detect the shape of alkane vapors by the color change of the pillar[5]arene crystals.

2.7 Conclusions

In this chapter, we describe molecular space chemistry using pillar[n]arenes. Owing to the high functionality, easy fabrication, redox property, and cavity-dependent host–guest ability of pillar[n]arenes, pillar[n]arenes can be designed for organic catalysis and reductant application in solution. Owing to their pillar-shaped structures, the surface modification of pillar[n]arenes is one of the useful applications. The regular polygonal prism shape contributes to the construction of highly assembled structures such as 2D sheets and 3D sphere assemblies. The simple and highly symmetrical shape of pillar[n]arenes should have unlimited potential and serve as ideal building blocks to create various supramolecular assemblies. This year, 2018, marks one decade of research into pillar[n]arenes; thus we believe this research has laid a firm foundation, and we now eagerly anticipate the new discoveries in pillar[n]arene chemistry in the future.

References

1. G. Crini, Chem. Rev. **114**, 10940–10975 (2014)
2. M.V. Rekharsky, Y. Inoue, Chem. Rev. **98**, 1875–1918 (1998)
3. K. Uekama, F. Hirayama, T. Irie, Chem. Rev. **98**, 2045–2076 (1998)
4. G. Wenz, B.H. Han, A. Müller, Chem. Rev. **106**, 782–817 (2006)
5. S.A. Nepogodiev, J.F. Stoddart, Chem. Rev. **98**, 1959–1976 (1998)
6. A. Harada, A. Hashidzume, H. Yamaguchi, Y. Takashima, Chem. Rev. **109**, 5974–6023 (2009)
7. A. Ikeda, S. Shinkai, Chem. Rev. **97**, 1713–1734 (1997)
8. N. Morohashi, F. Narumi, N. Iki, T. Hattori, S. Miyano, Chem. Rev. **106**, 5291–5316 (2006)
9. D.M. Homden, C. Redshaw, Chem. Rev. **108**, 5086–5130 (2008)
10. J. Vicens VB, *Calixarenes: A Versatile Class of Macrocyclic Compounds* (Kluwer Academic, Dordrecht, 1991)
11. C.D. Gutsche, *Calixarenes* (The Royal Society of Chemistry, Cambridge, 1989)
12. D.S. Guo, Y. Liu, Acc. Chem. Res. **47**, 1925–1934 (2014)
13. J.W. Lee, S. Samal, N. Selvapalam, H.J. Kim, K. Kim, Acc. Chem. Res. **36**, 621–630 (2003)
14. J. Lagona, P. Mukhopadhyay, S. Chakrabarti, L. Isaacs, Angew. Chem. Int. Ed. **44**, 4844–4870 (2005)
15. A.E. Kaifer, Acc. Chem. Res. **47**, 2160–2167 (2014)
16. C.J. Pedersen, J. Am. Chem. Soc. **89**, 7017–7036 (1967)
17. M.J. Langton, P.D. Beer, Acc. Chem. Res. **47**, 1935–1949 (2014)
18. T. Ogoshi, S. Kanai, S. Fujinami, T. Yamagishi, Y. Nakamoto, J. Am. Chem. Soc. **130**, 5022–5023 (2008)
19. T. Ogoshi, *Pillararenes* (The Royal Society of Chemistry, Cambridge, 2016)
20. T. Ogoshi, T. Yamagishi, Y. Nakamoto, Chem. Rev. **116**, 7937–8002 (2016)
21. N.L. Strutt, H.C. Zhang, S.T. Schneebeli, J.F. Stoddart, Acc. Chem. Res. **47**, 2631–2642 (2014)
22. M. Xue, Y. Yang, X.D. Chi, Z.B. Zhang, F. Huang, Acc. Chem. Res. **45**, 1294–1300 (2012)
23. C. Li, Chem. Commun. **50**, 12420–12433 (2014)
24. T. Ogoshi, T. Yamagishi, Eur. J. Org. Chem. 2961–2975 (2013)
25. G.C. Yu, B. Hua, C.Y. Han, Org. Lett. **16**, 2486–2489 (2014)
26. B. Jiang, W. Wang, Y. Zhang, Y. Lu, C.W. Zhang, G.Q. Yin, X.L. Zhao, L. Xu, H.W. Tan, X.P. Li, G.X. Jin, H.B. Yang, Angew. Chem. Int. Ed. **56**, 14438–14442 (2017)
27. T. Ogoshi, T. Akutsu, Y. Shimada, T. Yamagishi, Chem. Commun. **52**, 6479–6481 (2016)
28. T. Ogoshi, N. Ueshima, T. Yamagishi, Org. Lett. **15**, 3742–3745 (2013)
29. D.G. Liz, A.M. Manfredi, M. Medeiros, R. Montecinos, B. Gomez-Gonzalez, L. Garcia-Rio, F. Nome, Chem. Commun. **52**, 3167–3170 (2016)
30. T. Ogoshi, Y. Hasegawa, T. Aoki, Y. Ishimori, S. Inagi, T. Yamagishi, Macromolecules **44**, 7639–7644 (2011)
31. T. Ogoshi, S. Takashima, T. Yamagishi, J. Am. Chem. Soc. **137**, 10962–10964 (2015)
32. T. Ogoshi, S. Takashima, T. Yamagishi, J. Am. Chem. Soc. **140**, 1544–1548 (2018)
33. J.H. Bi, X.F. Zeng, D.M. Tian, H.B. Li, Org. Lett. **18**, 1092–1095 (2016)
34. R. Wang, Y. Sun, F. Zhang, M.M. Song, D.M. Tian, H.B. Li, Angew. Chem. Int. Ed. **56**, 5294–5298 (2017)
35. T. Ogoshi, K. Yoshikoshi, R. Sueto, H. Nishihara, T. Yamagishi, Angew. Chem. Int. Ed. **54**, 6466–6469 (2015)
36. C. Tsuneishi, Y. Koizumi, R. Sueto, H. Nishiyama, K. Yasuhara, T. Yamagishi, T. Ogoshi, I. Tomita, S. Inagi, Chem. Commun. **53**, 7454–7456 (2017)
37. T. Ogoshi, R. Sueto, K. Yoshikoshi, K. Yasuhara, T. Yamagishi, J. Am. Chem. Soc. **138**, 8064–8067 (2016)
38. S. Fu, G. An, H.C. Sun, Q. Luo, C.X. Hou, J.Y. Xu, Z.Y. Dong, J.Q. Liu, Chem. Commun. **53**, 9024–9027 (2017)
39. L.L. Tan, H.W. Li, Y.C. Tao, S.X.A. Zhang, B. Wang, Y.W. Yang, Adv. Mater. **26**, 7027–7031 (2014)

40. T. Ogoshi, R. Sueto, K. Yoshikoshi, T. Yamagishi, Chem. Commun. **50**, 15209–15211 (2014)
41. T. Ogoshi, R. Sueto, K. Yoshikoshi, Y. Sakata, S. Akine, T. Yamagishi, Angew. Chem. Int. Ed. Engl. **54**, 9849–9852 (2015)
42. T. Ogoshi, K. Saito, R. Sueto, R. Kojima, Y. Hamada, S. Akine, A.M.P. Moeljadi, H. Hirao, T. Kakuta, T. Yamagishi, Angew. Chem. Int. Ed. **57**, 1592–2595 (2018)
43. T. Ogoshi, Y. Shimada, Y. Sakata, S. Akine, T. Yamagishi, J. Am. Chem. Soc. **139**, 5664–5667 (2017)

Chapter 3
Inherently Chiral Calix[4]arenes as Supramolecular Catalysts

Seiji Shirakawa and Shoichi Shimizu

Abstract The chemistry of chiral calixarenes is recognized as an important research field to the development of new chiral receptors for asymmetric recognition. This provides a potent tool for understanding the stereochemistry of biochemical systems. Hence, many chiral calixarenes containing chiral residue at the calixarene skeleton have been prepared for asymmetric molecular recognition and asymmetric catalysis. A more challenging and attractive approach to the design of chiral calixarenes is to make the calixarene "inherently chiral" by creating an asymmetric array of achiral substituents on the calixarene platform. In the present chapter, we summarized our own approach for the design, synthesis, and optical resolution of wide-rim functionalized calix[4]arenes and these chiral recognition abilities. The potential utility of these newly designed inherently chiral calix[4]arenes as chiral supramolecular organocatalysts has been demonstrated.

Keywords Calixarenes · Organocatalysts · Molecular recognition · Supramolecular catalysts · Asymmetric synthesis

3.1 Introduction

Calixarenes are one of the most representative host molecules in supramolecular chemistry [1–6]. A wide variety of artificial molecular receptors have been designed and synthesized by the use of calixarene platforms. Among calixarene chemistry, design of chiral calixarenes has received much attention in recent years [7–9]. The chemistry of chiral calixarenes is recognized as an important research field to the

S. Shirakawa (✉)
Department of Environmental Science, Graduate School of Fisheries and Environmental Sciences, Nagasaki University, Nagasaki, Japan
e-mail: seijishirakawa@nagasaki-u.ac.jp

S. Shimizu
Department of Applied Molecular Chemistry, College of Industrial Technology, Nihon University, Narashino, Chiba, Japan

S. Shirakawa (ed.), *Designed Molecular Space in Material Science and Catalysis*,
https://doi.org/10.1007/978-981-13-1256-4_3

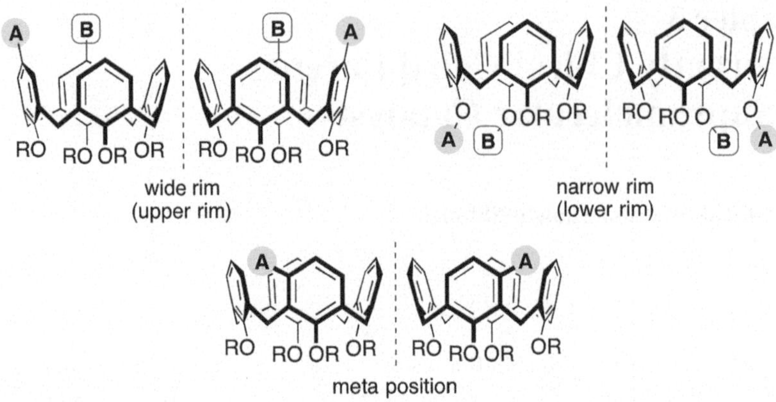

Fig. 3.1 Inherently chiral calix[4]arenes

development of new chiral receptors for asymmetric recognition. This provides a potent tool for understanding the stereochemistry of biochemical systems. Hence, many chiral calixarenes containing chiral residue on the calixarene skeleton have been prepared for asymmetric molecular recognition and asymmetric catalysis. A more challenging and attractive approach to the design of chiral calixarenes is to make the calixarene "inherently chiral" by creating an asymmetric array of achiral substituents on the calixarene platform (Fig. 3.1) [10–15]. Over more than the past two decades, many inherently chiral calixarenes have been synthesized, and some of them have been resolved into individual enantiomers [16–22]. In spite of these efforts, limited examples of enantiomeric recognition and asymmetric catalysis with optically active inherently chiral calixarenes have been reported. These limited results might arise from difficulties associated with both the design of a synthetic route to functionalized inherently chiral calixarenes and separation of enantiomers of the synthesized racemic calixarenes into optically pure forms. In the present chapter, we summarize our own approach for the design, synthesis, and optical resolution of functionalized calix[4]arenes and these chiral recognition abilities. The potential utility of these newly designed inherently chiral calix[4]arenes as chiral organocatalysts has been demonstrated [23–28].

3.2 Inherently Chiral Calix[4]arenes with ABCC Substitution Pattern

3.2.1 Aminophenol-Type Inherently Chiral Calix[4]arenes

Our initial design of the inherently chiral calix[4]arene is shown in Fig. 3.2 [23, 24]. The chiral calix[4]arene **1** possesses amino and hydroxyl groups, which are involved in molecular recognition, at proximal positions on the wide rim. Such

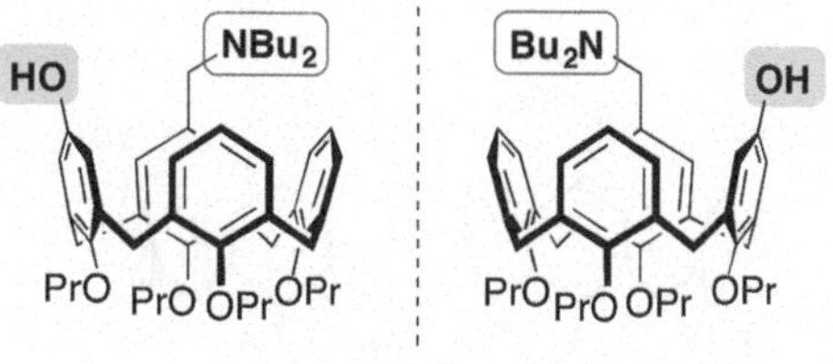

Fig. 3.2 Aminophenol-type inherently chiral calix[4] arene **1**

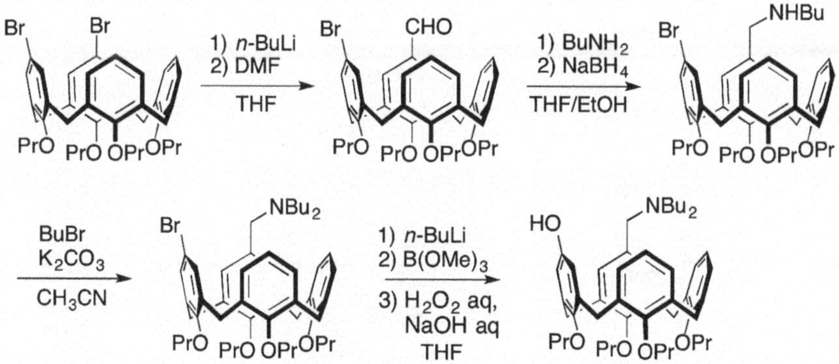

Scheme 3.1 Synthesis of inherently chiral calix[4]arene **1**

aminophenol (-alcohol) structures are often present in useful chiral building blocks and natural products, such as ephedrine, cinchonidine, and prolinol. The conformation of chiral calix[4]arene **1** is fixed in the cone conformer by attaching propyl groups at narrow rim, which provides a rigid and stable chiral environment.

The newly designed inherently chiral calix[4]arene **1** can be prepared from the known proximally *p*-dibrominated calix[4]arene in a four-step sequence (Scheme 3.1) [29]. The *p*-dibromocalix[4]arene was treated with 1.0 equivalent of *n*-BuLi, and subsequent addition of *N*,*N*-dimethylformamide afforded mono-formylated inherently chiral calix[4]arene. The reductive amination of the formyl group in the resulting calix[4]arene with *n*-butylamine gave the secondary amine compound. The inherently chiral calix[4]arene containing a secondary amine was transformed with *n*-butyl bromide to the tertiary amine compound. Lithiation of the bromine substituent on the resulting calix[4]arene and the trapping with $B(OMe)_3$ gave the corresponding boronate. The boronate was oxidized by H_2O_2 in one-pot, giving the target calix[4]arene **1** as a racemate.

Efficient optical resolution of the racemic calix[4]arene **1** was achieved by recrystallization after complexation with chiral mandelic acid (Scheme 3.2). The optical purity of the separated chiral calix[4]arene **1** was determined by means of chiral HPLC analysis with a SUMICHIRAL OA-4800 column. After repeating the recrystallization of calix[4]arene **1**-chiral mandelic acid complex, 2–3 times, enantiomerically pure inherently chiral calix[4]arene **1** could be obtained. This is a

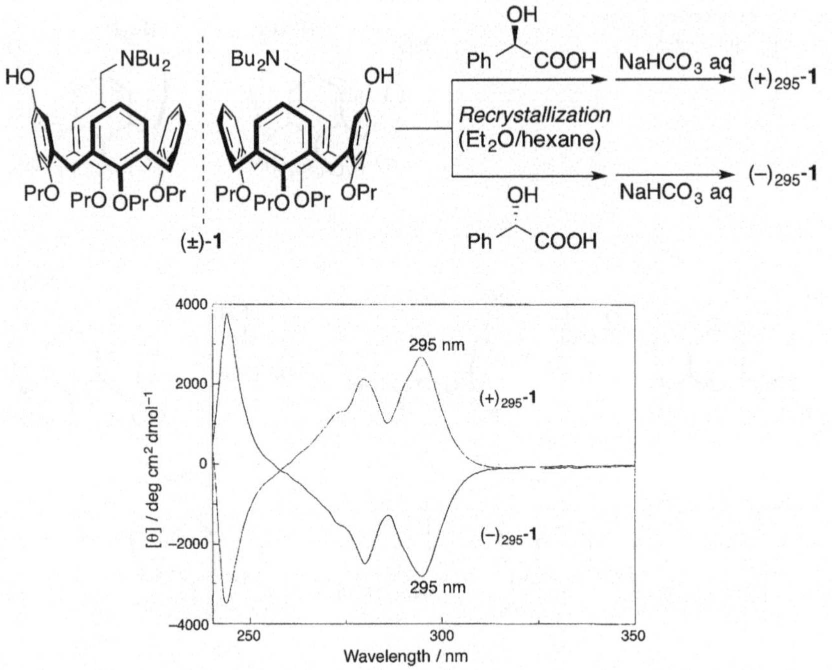

Scheme 3.2 Optical resolution and CD spectra of inherently chiral calix[4]arene **1**

valuable example of the optical resolution of an inherently chiral calix[4]arene by diastereomeric complexation without covalent bonding between chiral molecules. The circular dichroism (CD) spectra of the separated enantiomers of **1** showed perfect mirror images.

The chiral recognition ability of optically pure chiral calix[4]arene **1** was investigated in NMR experiments (Fig. 3.3). The ^{1}H NMR studies of the chiral calix[4] arene $(+)_{295}$-**1** were performed with racemic mandelic acid in $CDCl_3$ [30–35]. As a result of diastereomeric complexation of chiral calix[4]arene **1** and equimolar amounts of racemic mandelic acid, clear signal splitting (4.93 and 4.96 ppm; Fig. 3.3b) with an upfield shift of the benzylic proton of racemic mandelic acid (5.26 ppm; Fig. 3.3a) was observed. These results suggested that inherently chiral calix[4]arene **1** could be used as a chiral NMR solvating agent for determination of the enantiopurity of mandelic acid. The association constants of $(+)_{295}$-**1** with (*R*)- or (*S*)-mandelic acid were also measured by a titration experiment using UV-Vis spectral measurements. The association constant, K_a, of $(+)_{295}$-**1** with (*S*)-mandelic acid was 2.2 times larger than that with (*R*)-mandelic acid.

The application of inherently chiral calixarenes as chiral catalysts is a worthy challenge in organic syntheses. As a valuable example of an inherently chiral calixarene catalyst, the functionalized calixarene **1** was applied to the asymmetric conjugate addition of thiophenol to cyclohexenone (Scheme 3.3) [36–43]. Both enantiomers, $(+)_{295}$-**1** and $(-)_{295}$-**1**, promoted the reaction efficiently and gave a

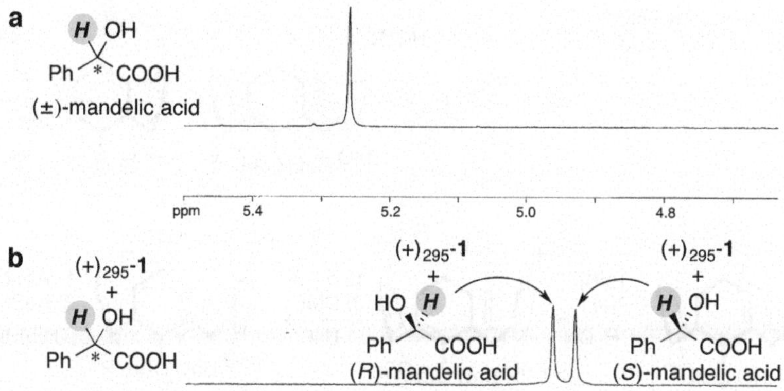

Fig. 3.3 ^{1}H NMR spectra of racemic mandelic acid in the absence and presence of $(+)_{295}$-**1** in $CDCl_3$

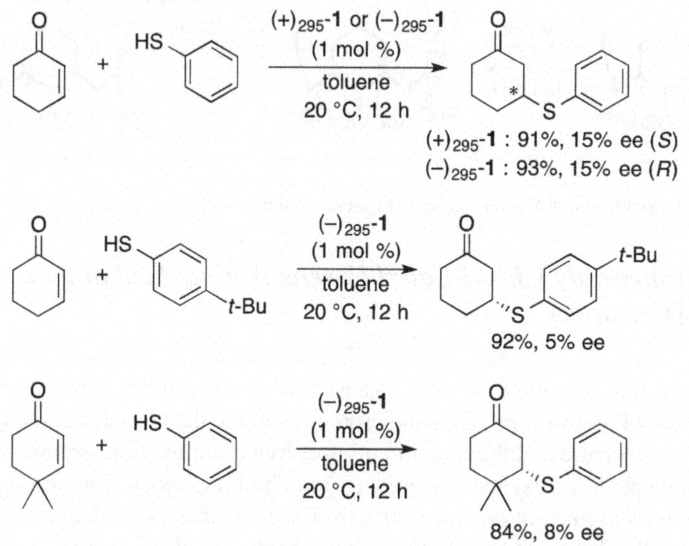

Scheme 3.3 Asymmetric conjugate additions catalyzed by inherently chiral calix[4]arene **1**

conjugate addition product in excellent yields. The chiral induction of the product was observed as 15% ee, and the configuration of the major enantiomer was *S* with $(+)_{295}$-**1**, *R* with $(-)_{295}$-**1**, respectively. Although observed enantioselectivities of the product were low, certain chiral inductions were observed. It is noteworthy that this is the important example of an inherently chiral calixarene with no chiral residue being applied to asymmetric catalysis. Other substrates for asymmetric conjugate additions catalyzed by chiral calixarene **1** were also examined, and selected examples are shown in Scheme 3.3.

Fig. 3.4 Inherently chiral calix[4]arene amino acid

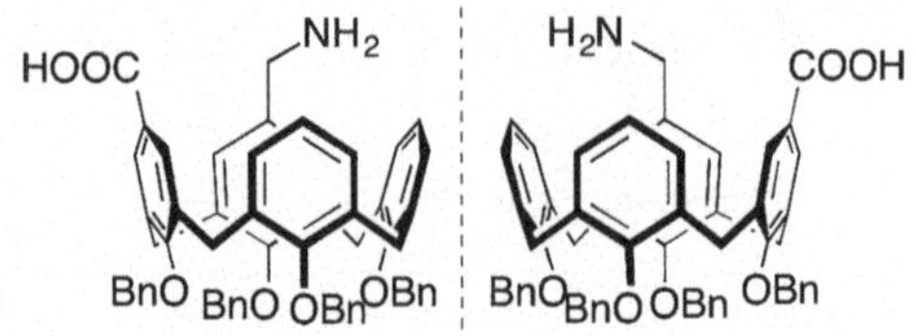

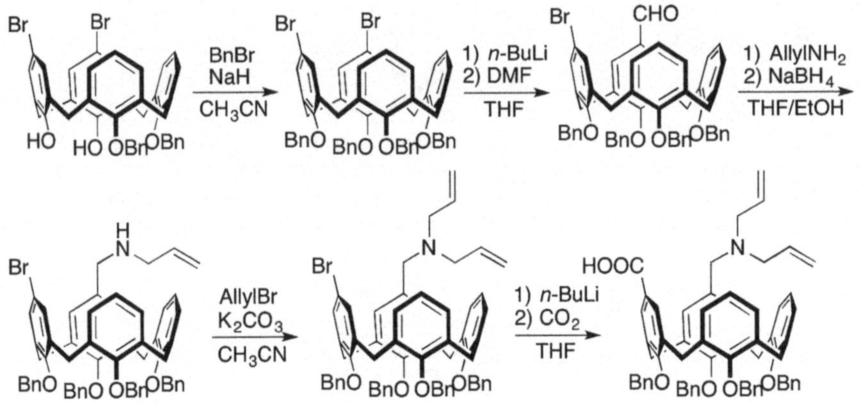

(±)-**2**

Scheme 3.4 Synthesis of *N*-protected calix[4]arene amino acid **2**

3.2.2 Inherently Chiral Calix[4]arene Amino Acid and its Derivatives

Amino acids are a biologically significant class of compounds, since they are the basic building blocks for peptides and proteins, and are the biopolymers responsible for both the structure and the function of most living things. Furthermore, optically active amino acids are extensively used as chiral building blocks for the synthesis of chiral catalysts in modern organic synthesis. Based on these knowledge, we became interested in the design and synthesis of an inherently chiral calix[4]arene amino acid as a chiral building block for subsequent transformation to various types of inherently chiral calix[4]arenes (Fig. 3.4) [25–27].

The *N*-protected calix[4]arene amino acid **2** can be synthesized from the reported proximally *p*-dibrominated calix[4]arene dibenzyl ether (Scheme 3.4) [29]. *O*-Alkylation of the *p*-dibrominated calix[4]arene dibenzyl ether with benzyl bromide in the presence of NaH gave the proximally *p*-dibrominated calix[4]arene tetrabenzyl ether in the cone conformation as a key intermediate. The *p*-dibromocalix[4]arene in cone conformation was transformed to the mono-formylated compound by treatment with a 1.1 equivalent of *n*-BuLi and the subsequent addition of *N*,*N*-dimethylformamide. The reductive amination of the formyl group on the calixarene with allylamine gave the secondary amine. The inherently chiral calix[4]arene containing a secondary amine was transformed with allyl bromide to the tertiary amine. Lithiation of the

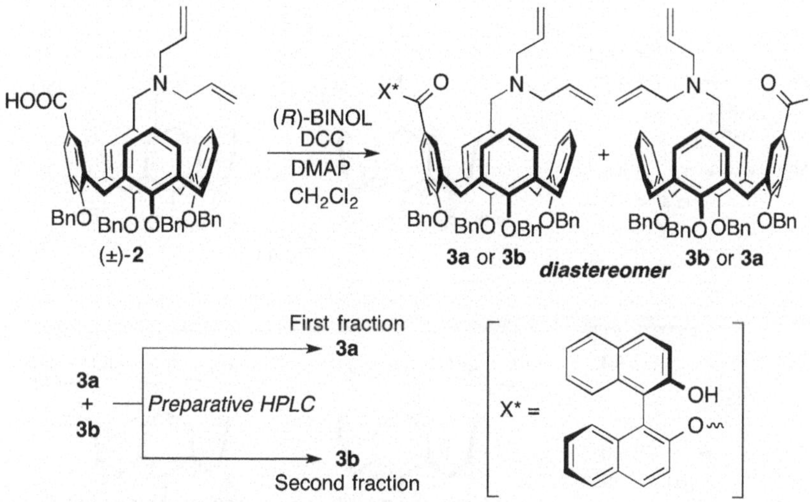

Scheme 3.5 Resolution of a calix[4]arene amino acid derivative

remaining bromine substituent on wide rim of the calix[4]arene, and the trapping with CO_2, gave the target *N*-protected calix[4]arene amino acid **2** as a racemate.

The efficient resolution of an inherently chiral calix[4]arene amino acid derivative was achieved by preparative high-performance liquid chromatography (HPLC) after conversion into diastereomeric (*R*)-BINOL esters **3a** and **3b** (Scheme 3.5) [44–46]. Thus, treatment of racemic *N*-protected calix[4]arene amino acid **2** with (*R*)-BINOL in the presence of *N*,*N*′-dicyclohexylcarbodiimide (DCC) and 4-(*N*,*N*-dimethylamino)pyridine (DMAP) gave the ~1:1 diastereomeric mixture of **3a** and **3b**. The mixture of chiral calix[4]arene **3a** and **3b** was loaded onto the preparative HPLC column, and diastereomerically pure **3a** and **3b** were obtained efficiently.

In an efficient synthetic scheme for the amino acid-type inherently chiral calix[4]arene in hand, the derivatization of separated diastereomers **3a** and **3b** to different types of inherently chiral calix[4]arenes was examined (Scheme 3.6). The chiral calix[4]arene **4a** containing an amino alcohol structure was synthesized by the reduction of **3a** or **3b** with $LiAlH_4$, and optically pure chiral calix[4]arenes (−)-**4a** and (+)-**4a** were obtained. Additionally, amino alcohol-type inherently chiral calix[4]arene **4b** possessing a diarylmethanol structure, which is often seen in efficient chiral catalysts, such as α,α-diarylprolinols and TADDOLs [47–49], was also prepared by the treatment of **3a** or **3b** with the corresponding aryllithium. The prepared chiral calix[4]arenes **4a** and **4b** were applied to the asymmetric conjugate addition with thiophenol as a model reaction, which is known to be catalyzed by chiral amino alcohols [36–43]. Both enantiomers (+)-**4a** and (−)-**4a** promoted the reaction efficiently and gave a conjugate addition product in good yields. The chiral induction of the product was observed as 15% ee and the configuration of the major enantiomer was *S* with (+)-**4a** and *R* with (−)-**4a**. The observed enantioselectivities were comparable with previous results using an inherently chiral calix[4]arene

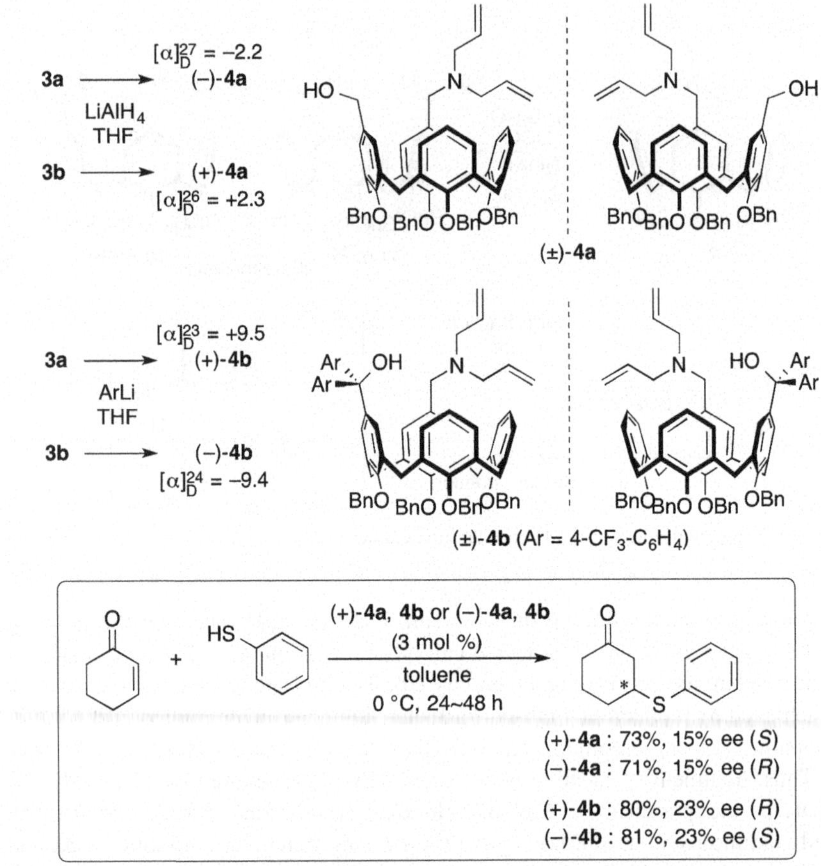

Scheme 3.6 Synthesis and catalysis of amino alcohol-type inherently chiral calix[4]arenes **4a** and **4b**

1 with an aminophenol structure. We next employed the chiral calix[4]arene **4b** containing a diarylmethanol structure for the conjugate addition of thiophenol. Pleasingly, a positive effect of the additional diaryl group was observed, and the conjugate addition product was obtained in 23% ee with good yields.

Other substrates for conjugate additions catalyzed by amino alcohol-type inherently chiral calix[4]arene **4b** were also examined, and selected examples are shown in Scheme 3.7. Substituted thiophenols could be applied to the reaction system to give the products in good yields with moderate enantioselectivities (17~23% ee). Another cyclic enone with a different ring size and an acyclic enone were also applied to the reaction, to give the products with moderate enantioselectivities (13~20% ee).

Inherently chiral calix[4]arenes containing a quaternary ammonium moiety **5a** and **5b** were prepared in optically pure form by the *N*-allylation of chiral calix[4] arene **4a** and **4b** (Scheme 3.8). The inherently chiral calix[4]arenes **5**, containing a

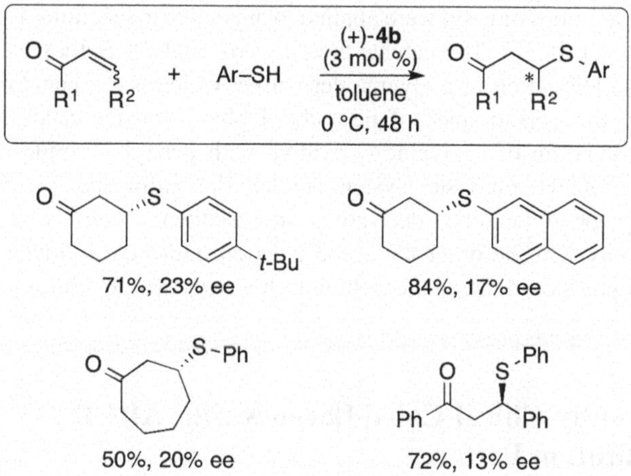

Scheme 3.7 Asymmetric conjugate additions catalyzed by inherently chiral calix[4]arene **4b**

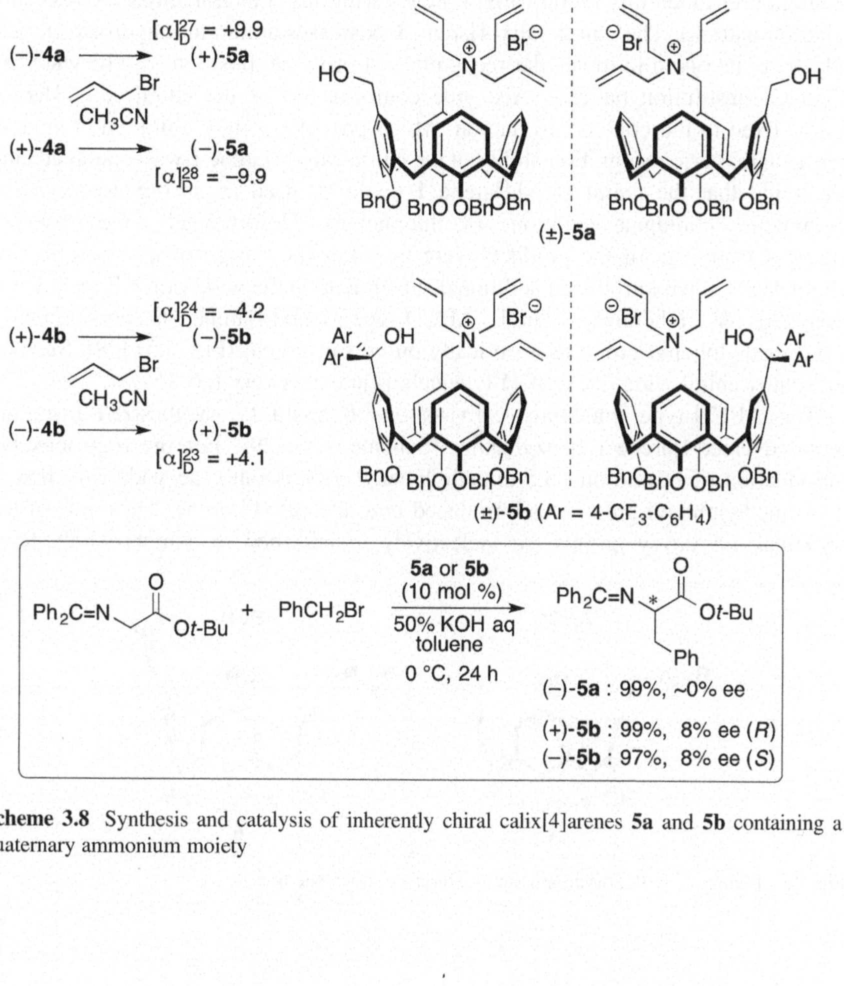

Scheme 3.8 Synthesis and catalysis of inherently chiral calix[4]arenes **5a** and **5b** containing a quaternary ammonium moiety

quaternary ammonium moiety, were applied to asymmetric reactions as chiral phase-transfer catalysts [50–53]. To evaluate the effect of catalysts **5**, they were applied to the asymmetric alkylation of a glycine derivative, which is the benchmark reaction for examining the performance of new chiral phase-transfer catalysts. Thus, the asymmetric alkylation of a glycine derivative with benzyl bromide in a toluene–50% aqueous KOH biphasic system under the influence of **5a** gave the corresponding phenylalanine derivative in excellent yield with almost no enantioselectivity. On the other hand, the reaction under the influence of **5b** gave the alkylation product in excellent yield with low, but certain, chiral induction.

3.3 Inherently Chiral Calix[4]arenes with ABCD Substitution Pattern

In the previous section, we showed efficient methods for the synthesis and optical resolution of inherently chiral calix[4]arenes with dual functionalities (ABCC substitution pattern). The chiral calix[4]arene **1** possesses amino and hydroxy groups, which are involved in molecular recognition at proximal positions on the wide rim (ABCC substitution pattern). Also, the conformation of the chiral calix[4]arene **1** was fixed in the cone conformation, which provides a rigid chiral environment. The enantiomeric recognition ability of the chiral calix[4]arene **1** was examined, and we found that the chiral calix[4]arene **1** could be used as an organocatalyst in asymmetric conjugate additions of thiophenols. Unfortunately, the observed enantioselectivities of the products were low. For the design of a more effective chiral catalyst, we introduced additional substituent on the wide rim. Ultimately, we designed an inherently chiral ABCD-type calix[4]arene **6** containing a 3,5-dimethylphenyl group as a sterically bulky substituent (Fig. 3.5) [28]. Such an inherently chiral wide rim ABCD-type calix[4]arene is very rare [54].

The ABCD-type inherently calix[4]arene **6** could be synthesized from the reported calix[4]arene dibenzyl ether (Scheme 3.9) [29]. For the regioselective introduction of bromo and 3,5-dimethylphenyl groups onto the wide rim, first, a 3,5-dimethylphenyl group was introduced onto the calix[4]arene. Thus, one of the proximal dihydroxy groups was selectively transformed to a mono-*O*-alkylated

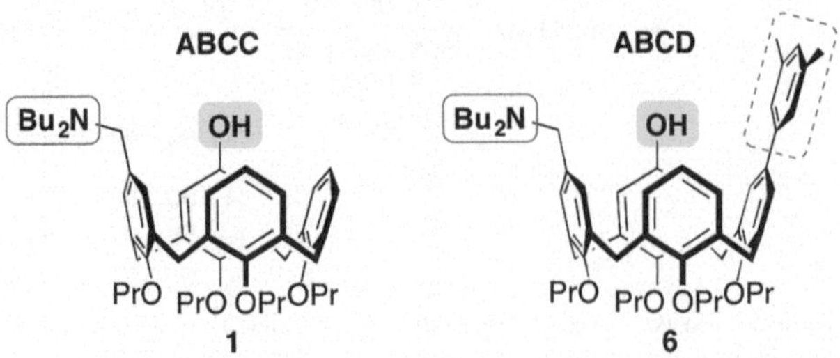

Fig. 3.5 Design of ABCD-type inherently chiral calix[4]arene **6**

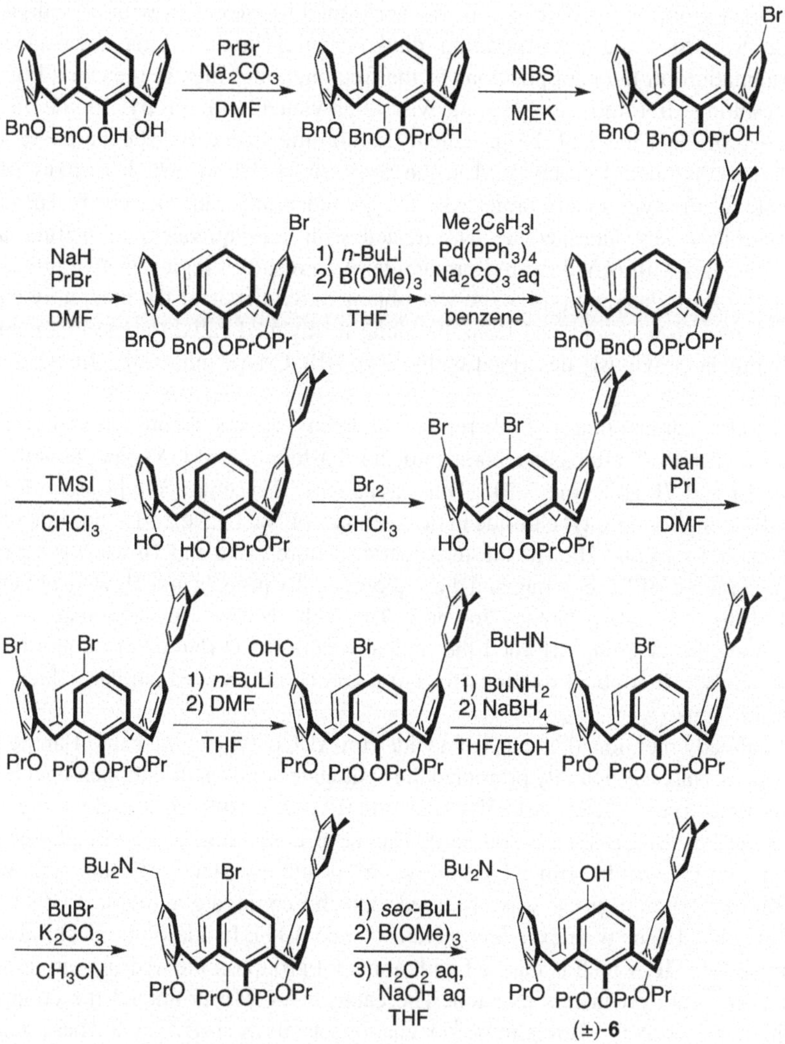

Scheme 3.9 Synthesis of inherently chiral calix[4]arene **6**

product by treatment with propyl bromide in the presence of sodium carbonate as a base. Use of sodium carbonate as the base was a key factor in the selective mono-*O*-alkylation. The para position of the free phenolic ring of tri-*O*-alkylated calix[4] arene was selectively brominated. After *O*-propylation of the resulting calix[4]arene, the tetraalkoxycalix[4]arene was treated with *n*-BuLi and the resulting lithiation compound was transformed to the boronate by treatment with $B(OMe)_3$. Then, the 3,5-dimethylphenyl group was installed using a Suzuki-Miyaura coupling reaction between the boronate of calix[4]arene and 1-iodo-3,5-dimethylbenzene to afford the calix[4]arene possessing a 3,5-dimethylphenyl group. The selective dealkylation of the benzyl group at the narrow rim was performed by treatment with trimethylsilyl iodide to afford the corresponding dialkoxycalix[4]arene. After regioselective dibromination on the para positions of the free phenolic rings in the calix[4]arene, the resulting dibromo-calix[4]arene was *O*-alkylated with propyl iodide in the presence of NaH to yield the proximally *p*-dibrominated calix[4]arene possessing a 3,5-dimethylphenyl group fixed in the cone conformation, which is a key intermediate for the synthesis of target ABCD-type inherently calix[4]arene **6**. The calix [4]arene as a key intermediate was treated with 1.1 equivalent of *n*-BuLi and, subsequently, with *N*,*N*-dimethylformamide. As a result of regioselective lithiation, the mono-formylated calix[4]arene was obtained. Synthesis of the inherently chiral calix[4]arene **6** was achieved from the mono-formylated calix[4]arene in a manner similar to the previously described synthesis of ABCC-type inherently chiral calix[4] arene **1**.

The efficient resolution of inherently chiral calix[4]arene **6** could be achieved by preparative HPLC after conversion into the diastereomeric (1*S*)-camphorsulfonyl esters **7a** and **7b** (Scheme 3.10). Thus, treatment of racemic calix[4]arene **6** with (1*S*)-10-camphorsulfonyl chloride in the presence of NaH gave a ~1:1 mixture of the diastereomers **7a** and **7b**. The diastereomeric mixture of **7a** and **7b** was loaded onto the preparative HPLC column, and diastereomerically pure **7a** and **7b** were obtained. Hydrolysis of calix[4]arene **7a** and **7b** with NaOH for removal of the camphorsulfonyl group afforded the optically pure calix[4]arene (+)-**6** and (−)-**6**, respectively. The optical rotations for (+)-**6** and (−)-**6** showed similar values with opposite signs.

As already mentioned, the use of a wide-rim ABCC-type chiral calix[4]arene **1** as an organocatalyst efficiently promoted the conjugate addition of thiophenol with low enantioselectivity (15% ee). For the improvement of enantioselectivity, we employed an ABCD-type chiral calix[4]arene **6** containing a 3,5-dimethylphenyl group at the wide rim. Pleasingly, a positive effect of the additional 3,5-dimethylphenyl group was observed, and the conjugate addition product was obtained in 31% ee with excellent yield (Scheme 3.11). In addition, the chiral calix [4]arene **6-OMe**, which is masked with a methyl group on the hydroxy group, was tested as a catalyst in this reaction. The catalyst **6-OMe** promoted the conjugate addition moderately with almost no enantioselectivity (~0% ee). These results clearly indicate that both the amino and hydroxy groups of catalyst **6** are important for both reactivity and selectivity in the conjugate additions of thiophenols. Although the details of the mechanism of the asymmetric conjugate addition are not clear, it is expected that the amino group of catalyst **6** activates thiophenol and forms the ammonium thiolate complex and that the hydroxy group (and/or protonated amine moiety) of catalyst **6** activates cyclohexenone via a hydrogen bond between the carbonyl group of the substrate and the hydroxy group (and/or protonated amine moiety) of the catalyst [42]. Moreover, the 3,5-dimethylphenyl group of catalyst **6** may selectively block one of the enantio-faces of cyclohexenone, and the thiolate anion attacks to the cyclohexenone from the opposite face. So, this hypothesis indicates the direction that design of more efficient inherently chiral calixarene catalysts should take.

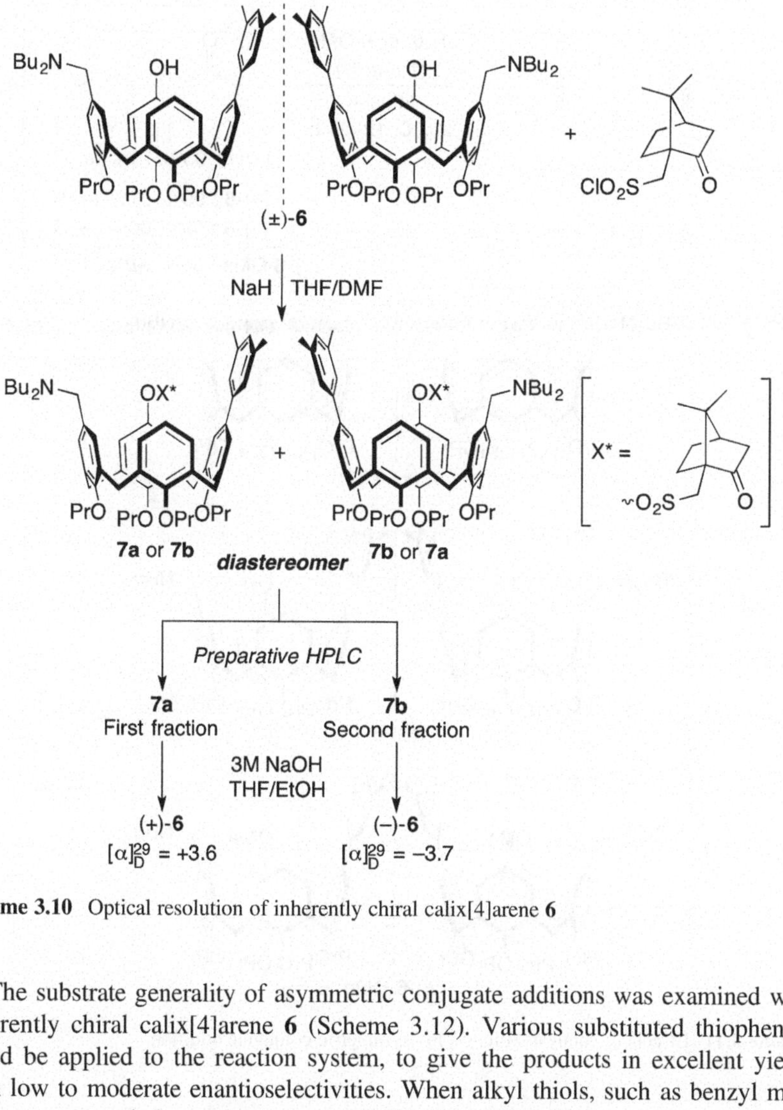

Scheme 3.10 Optical resolution of inherently chiral calix[4]arene **6**

The substrate generality of asymmetric conjugate additions was examined with inherently chiral calix[4]arene **6** (Scheme 3.12). Various substituted thiophenols could be applied to the reaction system, to give the products in excellent yields with low to moderate enantioselectivities. When alkyl thiols, such as benzyl mercaptan, were used, the product was obtained in low yield with low enantioselectivity. Furthermore, this reaction system could be applied to other cyclic and acyclic enones. Substituted cyclohexenone gave the product with a degree of enantioselectivity comparable to that observed for the reaction of 2-cyclohexen-1-one. Other cyclic enones with different ring sizes gave the products in lower enantioselectivities. The reaction of chalcone, which is an acyclic enone, afforded the product in excellent yield with moderate enantioselectivity.

Based on the design of ABCD-type inherently chiral calix[4]arene **6** possessing aminophenol structure, we also examined the synthesis of phosphine-containing calix[4]arene (Scheme 3.13). A proximally *p*-dibrominated calix[4]arene possessing

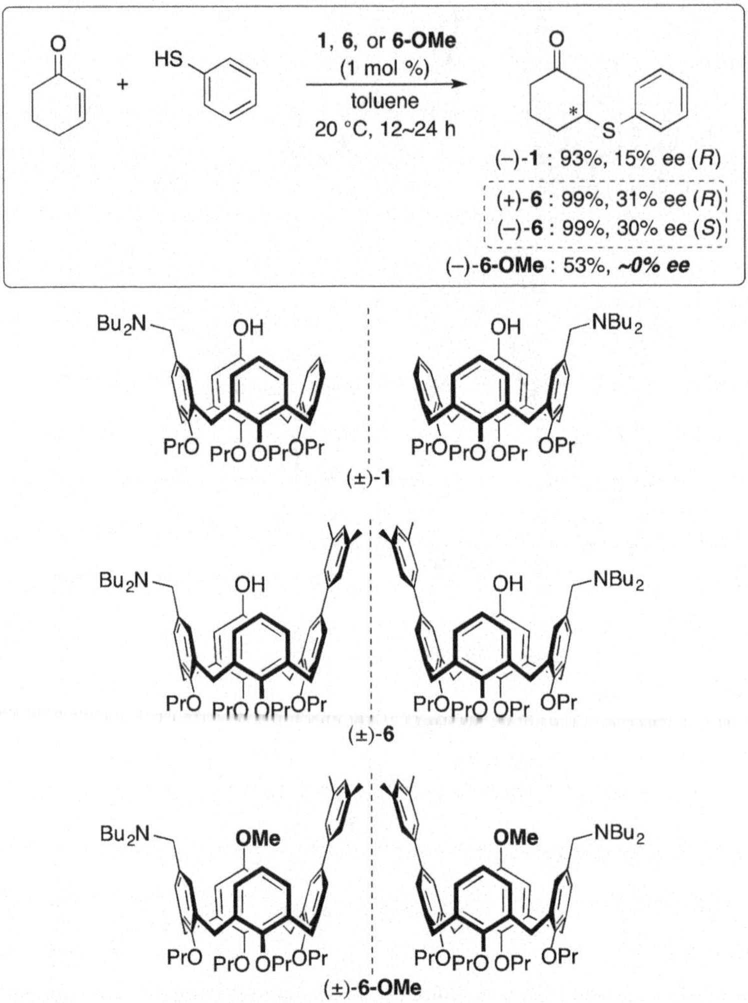

Scheme 3.11 Effects of catalyst structure in asymmetric conjugate addition

a 3,5-dimethylphenyl group as a key intermediate was treated with 1.1 equivalent of *n*-BuLi and, subsequently, with chlorodiphenylphosphine to give the corresponding phosphine derivative. The remained bromo group of the phosphino calix[4]arene was lithiated by the use of *sec*-BuLi. The resulting lithiated calix[4]arene was treated with $B(OMe)_3$ and gave the corresponding boronate. The boronate was oxidized by H_2O_2 in one-pot, giving the calix[4]arene **8** containing a phosphine oxide moiety as a racemate.

The resolution of inherently chiral calix[4]arene **8** could be achieved by column chromatography on silica gel after conversion into the diastereomeric (*S*)-naproxen esters **9a** and **9b** (Scheme 3.14). Thus, treatment of racemic calix[4]arene **8** with (*S*)-

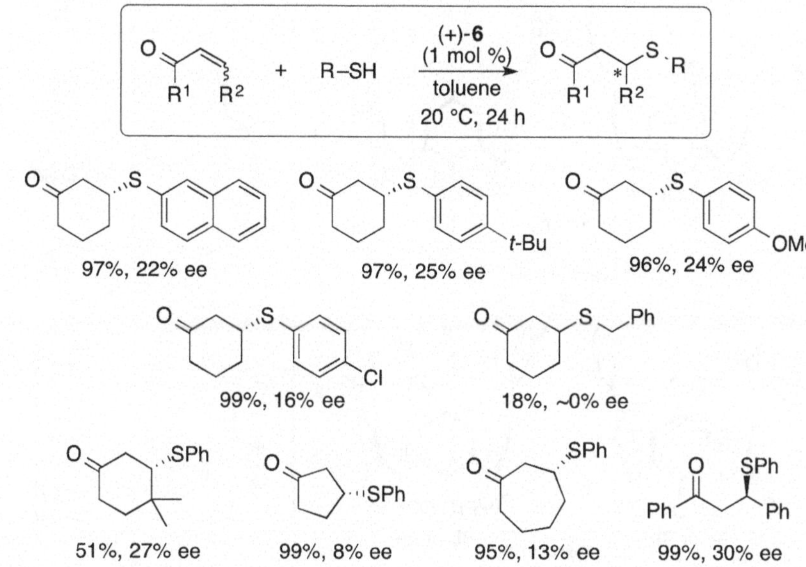

Scheme 3.12 Asymmetric conjugate additions catalyzed by inherently chiral calix[4]arene **6**

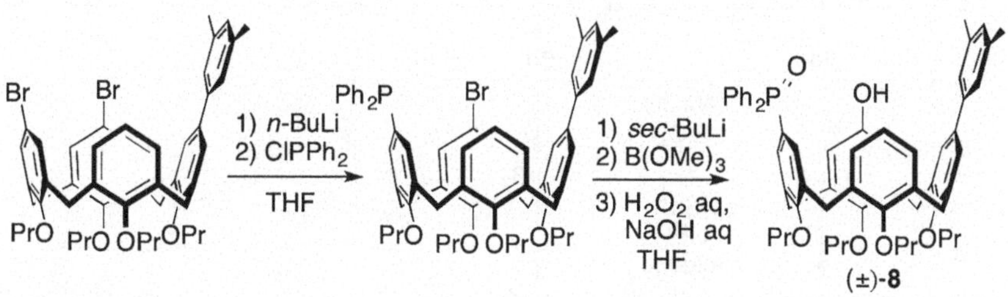

Scheme 3.13 Synthesis of inherently chiral calix[4]arene containing a phosphorous moiety

naproxen chloride in the presence of NaH gave a ~1:1 mixture of the diastereomers **9a** and **9b**. The diastereomeric mixture of **9a** and **9b** was separated by column chromatography on silica gel, and diastereomerically pure **9a** and **9b** were obtained. Hydrolysis of calix[4]arene **9a** and **9b** with NaOH for removal of the naproxen moiety afforded the optically pure calix[4]arene (+)-**8** and (−)-**8**, respectively. The reduction of phosphine oxides (+)-**8** and (−)-**8** with trichlorosilane gave inherently chiral calix[4]arenes (+)-**10** and (−)-**10** containing diphenylphosphino and hydroxy groups.

The optically pure, chiral calix[4]arene **10** was applied to asymmetric aza-Morita-Baylis-Hillman reaction as an organocatalyst (Scheme 3.15) [55, 56]. This is the first example of an inherently chiral calixarene being applied to asymmetric catalysis for the aza-Morita-Baylis-Hillman reaction. Although observed enantioselectivity of the aza-Morita-Baylis-Hillman product was low, certain chiral induction was observed.

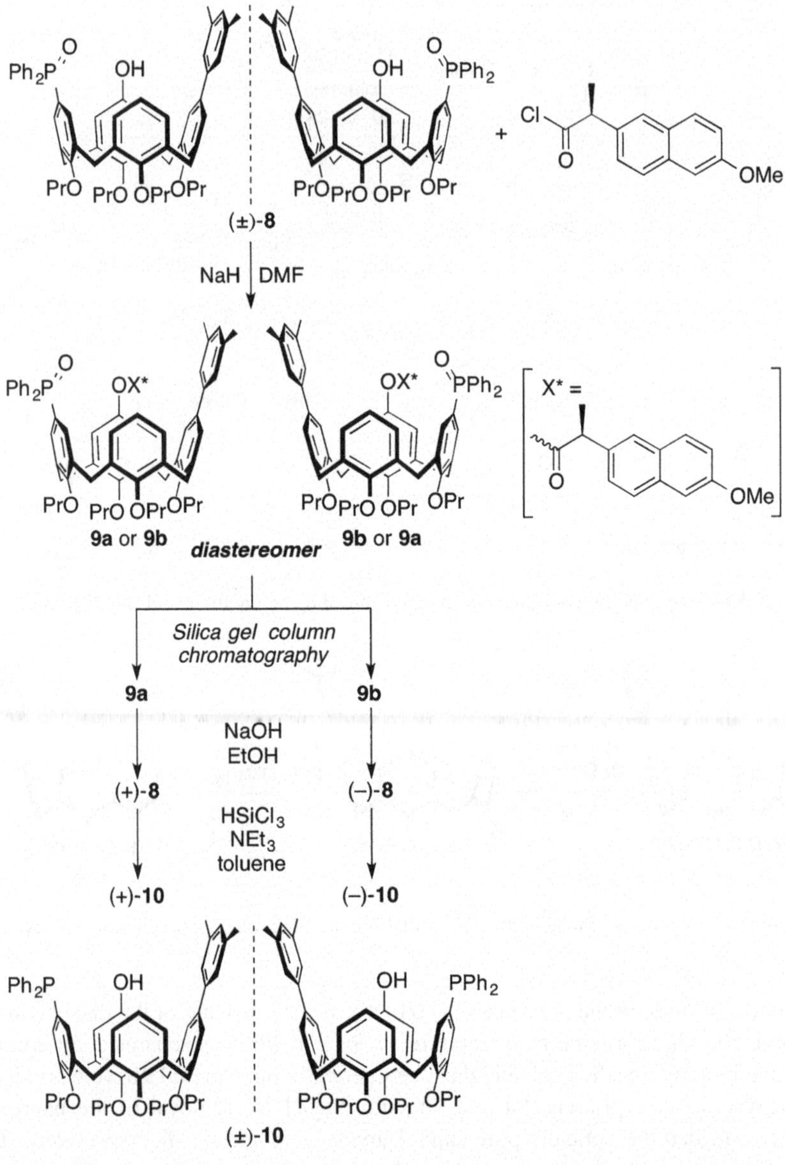

Scheme 3.14 Optical resolution of inherently chiral calix[4]arene containing a phosphorus moiety

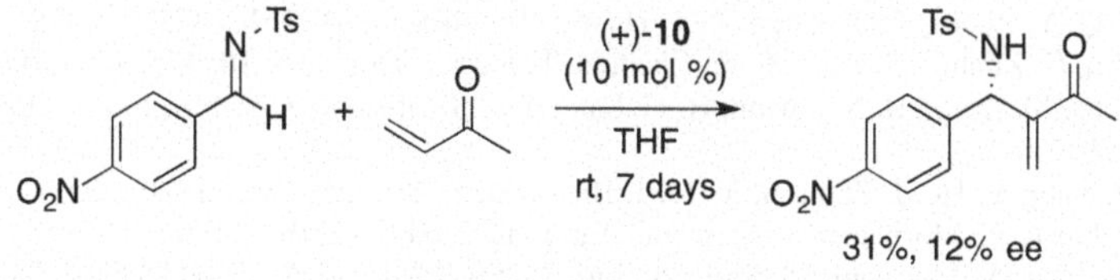

Scheme 3.15 Asymmetric aza-Morita-Baylis-Hillman reaction catalyzed by inherently chiral calix [4]arene **10**

3.4 Conclusions

In this chapter, we have showcased efficient synthetic routes to functionalized inherently chiral calix[4]arenes at the wide rim in the cone conformation. The synthetic methods show potential for the synthesis of a wide variety of functionalized inherently chiral calix[4]arenes. For instance, the inherently chiral calix[4]arenes containing an aminophenol (-alcohol) structure were synthesized and resolved into optically pure enantiomers. The effects of the inherently chiral calix[4] arenes as chiral organocatalysts in the asymmetric conjugate additions were examined, and the reactions were efficiently promoted by the calix[4]arene catalysts. The asymmetric inductions observed for the conjugate additions were still moderate at the best. However, we believe that these results revealed the potential utility of inherently chiral calixarenes as supramolecular catalysts.

References

1. C.D. Gutsche, Calixarenes, in *Monographs in Supramolecular Chemistry*, ed. by J.F. Stoddart (The Royal Society of Chemistry, Cambridge, 1989)
2. S. Shinkai, Tetrahedron **49**, 8933–8968 (1993)
3. V. Böhmer, Angew. Chem. Int. Ed. Engl. **34**, 713–745 (1995)
4. A. Ikeda, S. Shinkai, Chem. Rev. **97**, 1713–1734 (1997)
5. C.D. Gutsche, Calixarene revisited, in *Monographs in Supramolecular Chemistry*, ed. by J.F. Stoddart (The Royal Society of Chemistry, Cambridge, 1998)
6. N. Morohashi, F. Narumi, N. Iki, T. Hattori, S. Miyano, Chem. Rev. **106**, 5291–5316 (2006)
7. M. Vysotsky, C. Schmidt, V. Böhmer, Adv. Supramol. Chem. **7**, 139–233 (2000)
8. A. Sirit, M. Yilmaz, Turk. J. Chem. **33**, 159–200 (2009)
9. J. Luo, Y.-S. Zheng, Curr. Org. Chem. **16**, 483–506 (2012)
10. V. Böhmer, D. Kraft, M. Tabatabai, J. Incl. Phenom. Mol. Recognit. Chem. **19**, 17–39 (1994)
11. H. Otsuka, S. Shinkai, Supramol. Sci. **3**, 189–205 (1996)
12. A. Dalla Cort, L. Mandolini, C. Pasquini, L. Schiaffino, New J. Chem. **28**, 1198–1199 (2004)
13. S.-Y. Li, Y.-W. Xu, J.-M. Liu, C.-Y. Su, Int. J. Mol. Sci. **12**, 429–455 (2011)
14. Y.-S. Zheng, J. Luo, J. Incl. Phenom. Macrocycl. Chem. **71**, 35–56 (2011)
15. G.E. Arnott, Chem. Eur. J. **24**, 1744–1754 (2018)
16. K. Araki, K. Inada, S. Shinkai, Angew. Chem. Int. Ed. Engl. **35**, 72–74 (1996)
17. C. Dieleman, S. Steyer, C. Jeunesse, D. Matt, J. Chem. Soc. Dalton Trans., 2508–2517 (2001)
18. F. Narumi, T. Hattori, N. Matsumura, T. Onodera, H. Katagiri, C. Kabuto, H. Kameyama, S. Miyano, Tetrahedron **60**, 7827–7833 (2004)

19. J. Luo, Q.-Y. Zheng, C.-F. Chen, Z.-T. Huang, Tetrahedron **61**, 8517–8528 (2005)
20. Z.-X. Xu, C. Zhang, Y. Yang, C.-F. Chen, Z.-T. Huang, Org. Lett. **10**, 477–479 (2008)
21. M. Paletta, M. Klaes, B. Neumann, H.-G. Stammler, S. Grimme, J. Mattay, Eur. J. Org. Chem., 555–562 (2008)
22. M.-L. Chang, Y. He, J. Zhou, S.-Y. Li, J. Braz. Chem. Soc. **28**, 1363–1370 (2017)
23. S. Shirakawa, A. Moriyama, S. Shimizu, Org. Lett. **9**, 3117–3119 (2007)
24. S. Shirakawa, A. Moriyama, S. Shimizu, Eur. J. Org. Chem., 5957–5964 (2008)
25. S. Shirakawa, Y. Tanaka, T. Kobari, S. Shimizu, New J. Chem. **32**, 1835–1838 (2008)
26. S. Shirakawa, S. Shimizu, Eur. J. Org. Chem., 1916–1924 (2009)
27. S. Shirakawa, S. Shimizu, New J. Chem. **34**, 1217–1222 (2010)
28. S. Shirakawa, T. Kimura, S.-i. Murata, S. Shimizu, J. Org. Chem. **74**, 1288–1296 (2009)
29. S. Shimizu, A. Moriyama, K. Kito, Y. Sasaki, J. Org. Chem. **68**, 2187–2194 (2003)
30. I. Takahashi, K. Odashima, K. Koga, Tetrahedron Lett. **25**, 973–976 (1984)
31. M.-H. Xu, J. Lin, Q.-S. Hu, L. Pu, J. Am. Chem. Soc. **124**, 14239–14246 (2002)
32. Y. He, Y. Xiao, L. Meng, Z. Zeng, X. Wu, C.-T. Wu, Tetrahedron Lett. **43**, 6249–6253 (2002)
33. X.-X. Liu, Y.-S. Zheng, Tetrahedron Lett. **47**, 6357–6360 (2006)
34. A. González-Álvarez, I. Alfonso, V. Gotor, Tetrahedron Lett. **47**, 6397–6400 (2006)
35. T. Ema, D. Tanida, T. Sakai, J. Am. Chem. Soc. **129**, 10591–10596 (2007)
36. H. Hiemstra, H. Wynberg, J. Am. Chem. Soc. **103**, 417–430 (1981)
37. K. Suzuki, A. Ikegami, T. Mukaiyama, Bull. Chem. Soc. Jpn. **55**, 3277–3282 (1982)
38. F. Wang, M. Tada, Agric. Biol. Chem. **54**, 2989–2992 (1990)
39. M. Ostendorf, S. van der Neut, F.P.J.T. Rutjes, H. Hiemstra, Eur. J. Org. Chem., 105–113 (2000)
40. M. Ostendorf, J. Dijkink, F.P.J.T. Rutjes, H. Hiemstra, Eur. J. Org. Chem., 115–124 (2000)
41. P. McDaid, Y. Chen, L. Deng, Angew. Chem. Int. Ed. **41**, 338–340 (2002)
42. M.N. Grayson, K.N. Houk, J. Am. Chem. Soc. **138**, 1170–1173 (2016)
43. M.N. Grayson, K.N. Houk, J. Am. Chem. Soc. **138**, 9041–9044 (2016)
44. Y.D. Cao, J. Luo, Q.-Y. Zheng, C.-F. Chen, M.-X. Wang, Z.-T. Huang, J. Org. Chem. **69**, 206–208 (2004)
45. J. Luo, Q.-Y. Zheng, C.-F. Chen, Z.-T. Huang, Chem. Eur. J. **11**, 5917–5928 (2005)
46. S.-Y. Li, Q.-Y. Zheng, C.-F. Chen, Z.-T. Huang, Tetrahedron Asymmetry **16**, 641–645 (2005)
47. D. Seebach, A.K. Beck, A. Heckel, Angew. Chem. Int. Ed. **40**, 92–138 (2001)
48. H. Pellissier, Tetrahedron **64**, 10279–10317 (2008)
49. A. Lattanzi, Chem. Commun., 1452–1463 (2009)
50. M.J. O'Donnell, Aldrichimica Acta **34**, 3–15 (2001)
51. T. Ooi, K. Maruoka, Angew. Chem. Int. Ed. **46**, 4222–4266 (2007)
52. S. Shirakawa, K. Maruoka, Angew. Chem. Int. Ed. **52**, 4312–4348 (2013)
53. S. Kaneko, Y. Kumatabara, S. Shirakawa, Org. Biomol. Chem. **14**, 5367–5376 (2016)
54. M. Ciaccia, I. Tosi, R. Cacciapaglia, A. Casnati, L. Baldini, S. Di Stefano, Org. Biomol. Chem. **11**, 3642–3648 (2013)
55. Y. Wei, M. Shi, Chem. Rev. **113**, 6659–6690 (2013)
56. H. Pellissier, Tetrahedron **73**, 2831–2861 (2017)

Chapter 4
Synthesis of the Longest Carbohelicene by Multiple Oxidative Photocyclizations of Arylene–Vinylene Oligomers

Takashi Murase

Abstract Oxidative photocyclization of stilbene-type precursors is still a powerful method for the synthesis of longer helicenes. Only the simplest subunits [1] and [2] (phenylene and naphthylene units) are arranged in the precursor to avoid unfavorable side reactions in photocyclization. Based on the synthetic guideline for such arylene–vinylene oligomers, [9]helicene was efficiently prepared by triple photocyclization of the precursor with a [2]+[1]+[1]+[2] sequence (where "+" denotes vinylene linkers). Moreover, the longest [16]helicene was prepared by sextuple photocyclization of the precursor with a [2]+[1]+[1]+[2]+[1]+[1]+[2] sequence. Other precursors with different sequences lowered the product yield in spite of the decreased number of cyclization. X-ray crystallographic analysis clearly showed the triple-layer structure of [16]helicene. The judiciously designed precursor has an intrinsic propensity to exhibit spontaneous helical folding under photoirradiation, which helps the formation of helicene skeletons.

Keyword Helicene · Photocyclization · Polycyclic aromatic compounds · Stilbene

4.1 Introduction

Discrete aromatic stacks are fascinating synthetic targets for understanding the physical properties and developing unique functions derived from their π-layered structures. [*n*]Helicenes are helical polyaromatic compounds formed by *n* number of *ortho*-fused aromatic rings [1–13]. In addition to the frequently discussed intrinsic chirality resulting from left- or right-handed helicity, transannular interactions between the overlapped aromatic rings become prominent in [*n*]helicenes with increasing *n* number. Accordingly, longer helicenes can be regarded as discrete aromatic stacks whose rigid covalent frameworks possess unique π systems that

T. Murase (✉)
Department of Material and Biological Chemistry, Faculty of Science, Yamagata University, Yamagata-shi, Yamagata, Japan
e-mail: tmurase@sci.kj.yamagata-u.ac.jp

S. Shirakawa (ed.), *Designed Molecular Space in Material Science and Catalysis*,
https://doi.org/10.1007/978-981-13-1256-4_4

spread laterally (π-conjugation) and vertically (π-stacking). In carbo[*n*]helicenes with an all-carbon skeletons (hereafter simply referred to as [*n*]helicenes), a double-layer structure appears when $n \geq 7$, and a triple-layer structure appears when $n \geq 13$. Here is a fundamental question and challenge: "How long can we synthetically elongate helicenes?" It is easily anticipated that the synthesis of longer helicenes becomes increasingly difficult and tough on account of steric congestion of multiple benzene rings. The current longest helicene is [16]helicene, which was prepared in 2015, and the record for the longest [*n*]helicene was updated for the first time in 40 years [14]. This review details how [16]helicene was designed and synthesized, showing the brief history of longer carbohelicenes. A myriad of helicenes and their homologues have been synthesized to date, but only non-substituted helicenes are illustrated here, due to limitations of space.

4.2 History of Longer Carbohelicenes

The history of longer carbohelicenes dates back to the first synthesis of [4]helicene by Weitzenböck and Lieb in 1913 (Fig. 4.1) [15]. However, in the strict sense, this is not the first report on helicenes; in 1903 Meisenheimer and Witte reported aza[5] helicenes possessing nitrogen atoms on the backbones [16]. In 1918, Weitzenböck and Klinger firstly prepared carbo[5]helicene [17]. A milestone in helicene chemistry was the first synthesis and the chiral resolution of [6]helicene by Newman and co-workers in 1955 and 1966 [18, 19]. This synthesis started from the Knoevenagel condensation of 1-naphthaldehyde with ethyl malonate and required ten steps, in which two central rings of [6]helicene were sequentially cyclized by Friedel–Crafts reactions.

In 1964, Mallory and co-workers discovered the photochemical conversion of stilbene to phenanthrene in the presence of iodine as an oxidant [20]. (Z)-Stilbene (*cis*-stilbene) undergoes 6π-electrocyclization under UV light irradiation to form 4*a*,4*b*-dihydrophenanthrene, which is unstable and further oxidized to phenanthrene. Martin and co-workers applied the Mallory's method to helicenes and prepared the first [7]helicene with a double-layer structure in 1967 [21]. Since then, a series of [*n*] helicenes ($n = 7$–14) has been prepared by the oxidative photocyclization of stilbene derivatives (Table 4.1) [22–29]. In 1991, Katz and co-workers improved this method by adding propylene oxide as an acid scavenger [30]. Nowadays, the Katz's method is a standard method for the photochemical synthesis of helicenes.

The merits of oxidative photocyclization are (1) easy access to the precursor stilbenes either by Wittig reaction or by Mizoroki–Heck reaction and (2) facile *E*/*Z*-photoisomerization that allows the use of *E*/*Z* mixtures as precursors. However, oxidative photocyclization must be conducted under highly dilute conditions to prevent [2π + 2π] photodimerization, and therefore a large-scale synthesis of helicenes is usually troublesome in the photochemical procedures. To overcome the major drawback, a variety of non-photochemical procedures such as Diels–Alder reactions [31, 32], coupling reactions using a transition-metal catalyst [33], ring-closing metathesis [34], and [2π + 2π + 2π] cycloisomerization of triynes [35] have

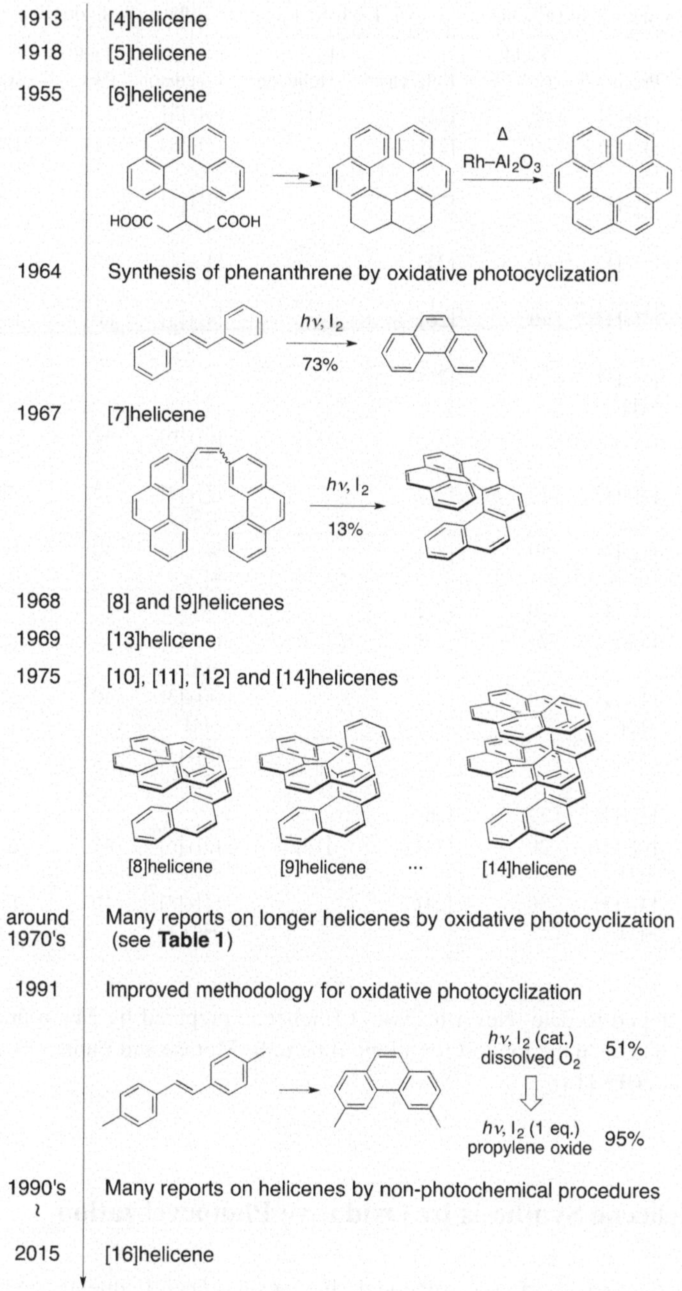

Fig. 4.1 The history of longer [*n*]helicenes

Table 4.1 Synthesis of [*n*]helicenes ($n = 8–14$) from precursor stilbenes with different sequences

[*n*] Helicene	Precursor	Yield (%)	References	[*n*] Helicene	Precursor	Yield (%)	References
[8]	[4]+[3]	62	[22]	[11]	[6]+[4]	45	[26]
	[4]+[3]	80	[23]		[3]+[3] +[3]	54	[28]
	[4]+[3]	85	[24]		[4]+[1] +[4]	84	[28]
	[6]+[1]	40	[25]		[4]+[1] +[4]	60	[24]
	[6]+[1]	40	[26]		[4]+[3] +[2]	80	[24]
	[4]+[1] +[1]	30	[24]				
				[12]	[2]+[6] +[2]	42	[28]
[9]	[4]+[4]	48	[22]		[3]+[4] +[3]	32	[28]
	[4]+[4]	50	[23]		[4]+[3] +[3]	30	[24]
	[4]+[4]	70	[27]				
	[6]+[2]	74	[26]	[13]	[4]+[3] +[4]	52	[29]
	[4]+[1] +[2]	50	[24]		[4]+[3] +[4]	40	[24]
					[6]+[3] +[2]	2	[26]
[10]	[6]+[3]	79	[26]				
	[4]+[1] +[3]	20	[24]	[14]	[3]+[6] +[3]	45	[28]
	[4]+[3] +[1]	30	[24]		[4]+[4] +[4]	10	[28]

been developed to date. Nevertheless, [14]helicene prepared by Matin and Bae in 1975 [28] had been the longest for a long time until Murase and Fujita reported [16] helicene in 2015 [14].

4.3 Helicene Synthesis by Oxidative Photocyclization

Oxidative photocyclization of stilbene derivatives has been frequently employed in the construction of helicene frameworks. In addition to the aforementioned merits of this procedure, predictable arene positions connected by the cyclization are useful for designing the precursor stilbenes.

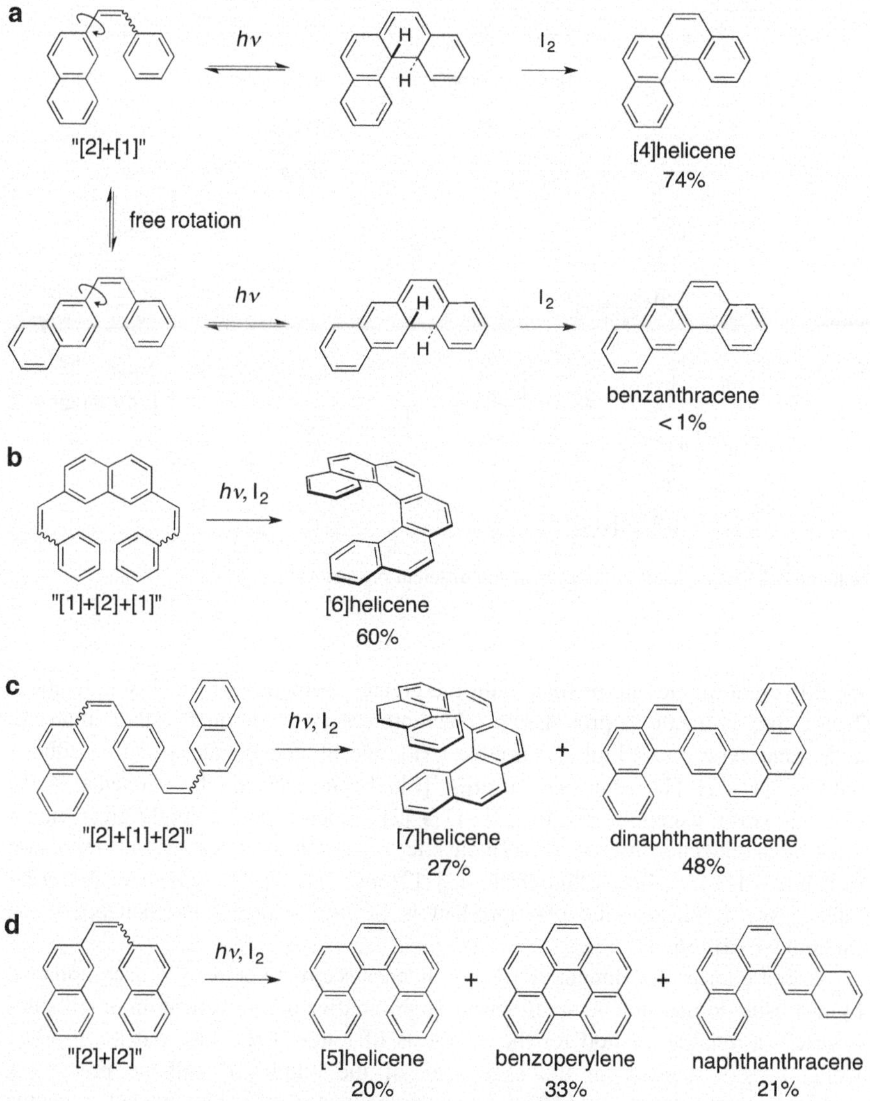

Scheme 4.1 Oxidative photocyclization of stilbenes with different sequences: (**a**) [2]+[1], (**b**) [1]+[2]+[1], (**c**) [2]+[1]+[2] and (**d**) [2]+[2]

For example, [4]helicene is synthesized from the stilbene with a [2]+[1] sequence, where "[*n*]" and "+" denote *ortho*-fused [*n*]helicene subunits and vinylene (–CH=CH–) linkers, respectively. In theory, by free rotation around the naphthyl–CH bond, both [4]helicene and benzanthracene can be generated by the oxidative photocyclization. In fact, however, [4]helicene is exclusively generated in a high yield (Scheme 4.1a) [36]. This result indicates the preferred route corresponding to

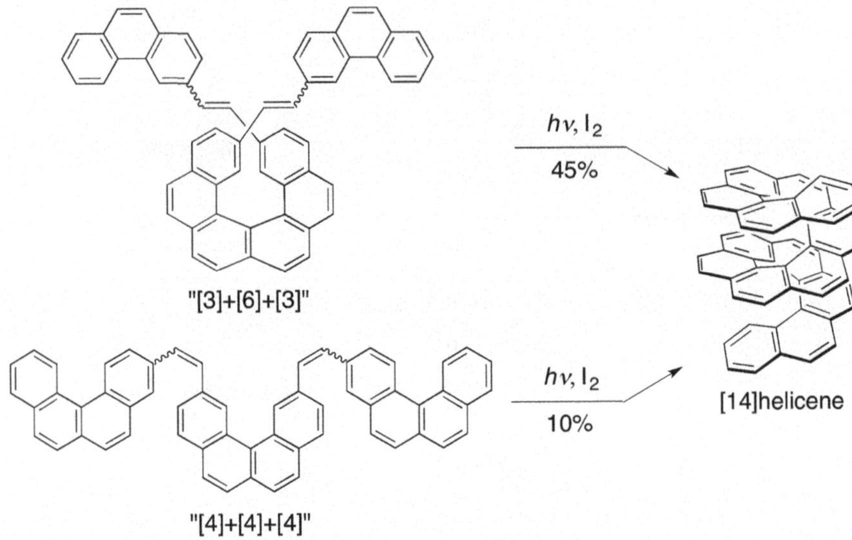

Scheme 4.2 Synthesis of [14]helicene from different precursors

the dihydroaromatic intermediate with the greater aromatic resonance stabilization. Thus, the cyclization provides a phenanthrene-type product rather than an anthracene-type one. Similarly, double oxidative photocyclization of the stilbene with a [1]+[2]+[1] sequence affords [6]helicene selectively (Scheme 4.1b) [37]. However, exceptionally, the [2]+ [1]+ [2] sequence preferentially gives planar dinaphthanthracene instead of [7]helicene to circumvent the steric repulsion (Scheme 4.1c) [38–40]. Although the [2]+[2] and [1]+[1]+[1] sequences preferentially give [5]helicene, the photoproduct is further oxidized into benzoperylene (Scheme 4.1d) [36, 41, 42].

At first glance, the elongation of [*n*]helicenes seems to be readily accomplished by just using longer precursor stilbenes. The oxidative photocyclization of stilbenes is surely a reliable method for the synthesis of longer helicenes, but the product yields largely depend on the sequences of the employed stilbene precursors (Table 4.1). For example, [14]helicene is prepared from the [3]+[6]+[3] sequence more efficiently than the [4]+[4]+[4] sequence (Scheme 4.2) [28]. Moreover, the [6]+[6] sequence failed to furnish [13]helicene under any conditions, despite favorable theoretical predictions, presumably due to the steric congestion around the C=C bond (Scheme 4.3) [43]. These results indicate that the connection of shorter helicene subunits followed by the cyclization is not always a suitable choice for the preparation of longer helicenes. In addition to the synthetic difficulty, the lack of a guideline for designing precursor stilbenes has hampered the elongation of [*n*] helicenes with $n \geq 15$ for a long time.

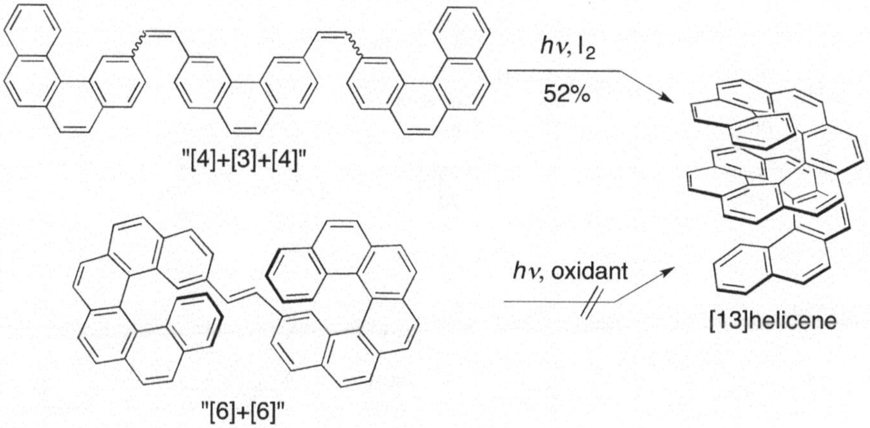

Scheme 4.3 Synthesis of [13]helicene from different precursors

4.4 Synthetic Guideline for Precursor Stilbenes Toward Longer Helicenes

Conventional precursors that have been employed for oxidative photocyclization have shorter helicene subunits with three and more *ortho*-fused benzene rings (i.e., [3], [4] and [6] units; Table 4.1). Since those helicene subunits are also prepared by oxidative photocyclization, the same photochemical procedure should be repeated to obtain the final longer helicenes. This synthetic disadvantage is eliminated by using only the simplest subunits [1] and [2] (namely, phenylene and naphthylene units) and connecting them through multiple C=C bonds in the precursor stilbenes. Such precursors can find a suitable pathway to fold into longer helicenes because the photocyclization before oxidation is a reversible process. However, as mentioned above, the [1]+[1]+[1], [2]+[2] and [2]+[1]+[2] sequences should not be included in the precursors because these sequences reduce the yields of helicenes. Based on these considerations, the following guideline for the design of precursor stilbenes is deduced: "[2] units must be separated by two [1] units". Namely, only ···+[2]+[1]+[1]+[2]+[1]+[1]+[2]+··· sequences can undergo oxidative photocyclizations without unfavorable benzoperylene or anthracene formation.

4.5 [9]Helicene

The design guideline was confirmed by triple photocyclization of the precursor stilbene with a [2]+[1]+[1]+[2] sequence in the synthesis of [9]helicene. The precursor was easily prepared in five steps and in large amounts (Scheme 4.4). The

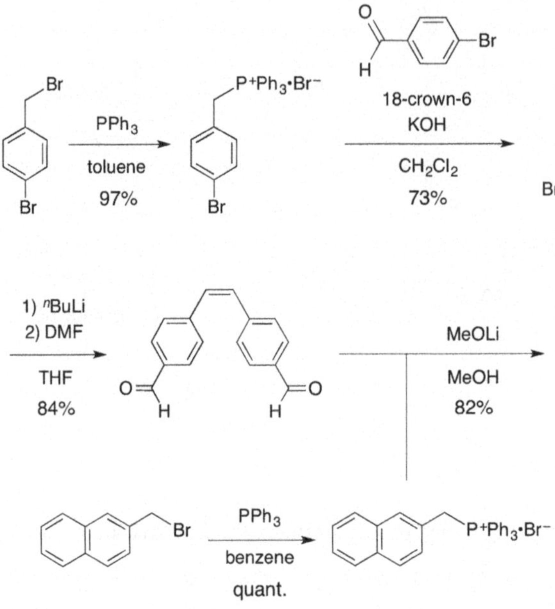

Scheme 4.4 Synthesis of the [2]+[1]+[1]+[2] precursor

solubility of the [2]+[1]+[1]+[2] precursor in chloroform largely depends on the number and position of (*E*)-olefins: (*Z*,*Z*,*Z*) > (*Z*,*Z*,*E*) > > (*E*,*Z*,*E*). Considering the solubility and easy handling, it is desirable to prepare the precursor by (*Z*)-selective olefinations, even though *E*/*Z* mixtures in the precursor are no problem in the last photocyclization.

When a solution of the [2]+[1]+[1]+[2] precursor in toluene (1.0 mM, 4 mL) was irradiated with a 450-W high-pressure Hg lamp at 90 °C in the presence of I_2 (oxidant, 3 equiv) and propylene oxide (acid scavenger, 50 equiv), a yellow solid was immediately precipitated, and [9]helicene was formed in 3% NMR yield after 1-h photoirradiation (Table 4.2, Entry 1). Rapid *E*/*Z* photoisomerization brings about precipitation of the [2]+[1]+[1]+[2] precursor as a planar (*E*,*E*,*E*) form even at high temperature. This phenomenon is characteristic of the oxidative photocyclization of precursor stilbenes bearing multiple C=C bonds. However, photoirradiation for a longer time suddenly increased the product yield, due to the participation of the yellow precipitate in the helicene formation. The optimized irradiation time and concentration were 6 h and 0.20 mM, respectively, and [9]helicene was obtained in 74% NMR yield (67% isolated yield) (Entry 4). Further prolonged irradiation gradually decreased the product yield probably due to the photodecomposition (Entry 5,6). The product yield in the triple photocyclization is comparable to those in the previous single or double photocyclization (Scheme 4.5) [24, 26, 27]. This is the first demonstration of triple photocyclization in the helicene synthesis and verifies the synthetic guideline for precursor stilbenes.

Table 4.2 Optimization of photoirradiation conditions for the [2]+[1]+[1]+[2] precursor

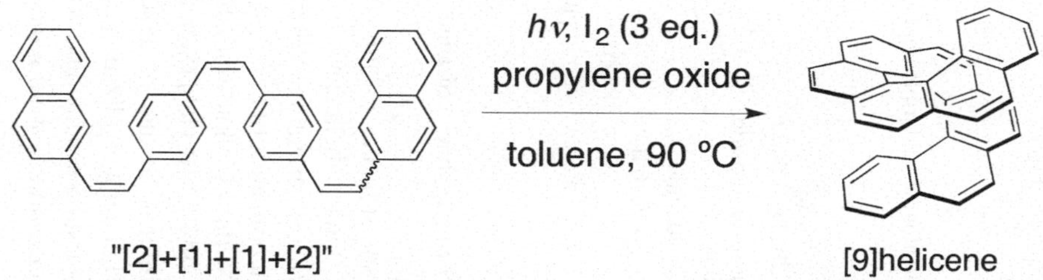

Entry	Scale (mL)	Concentration (mM)	Time (h)	NMR yield (%)
1	4	1.0	1	3
2	4	1.0	12	54
3	20	0.20	4	8
4	20	0.20	6	74
5	20	0.20	8	65
6	20	0.20	12	58

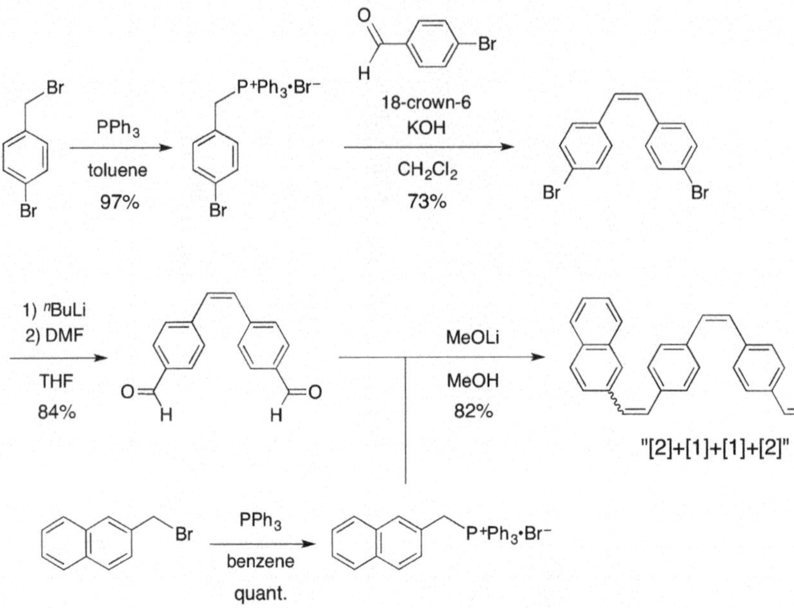

Scheme 4.5 Synthesis of [9]helicene by single, double, and triple photocyclizations

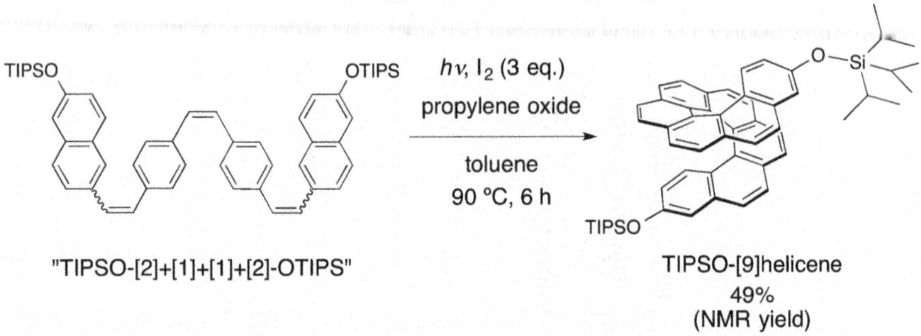

Scheme 4.6 Synthesis of TIPSO-[9]helicene by triple photocyclization

Decreasing the solubility of precursor stilbenes during photoirradiation is the major concern in the synthesis of longer helicenes by multiple oxidative photocyclizations through the rapid *E*/*Z* isomerization. When the triisopropylsilyl ether (TIPSO) groups were introduced at both ends of the [2]+[1]+[1]+[2] precursor, the solubility was significantly increased during photoirradiation. After 6-h irradiation, TIPSO-[9]helicene was formed in 49% NMR yield (Scheme 4.6). Although the NMR yield of TIPSO-[9]helicene is inferior to that of unsubstituted [9]helicene (74% NMR yield, 6-h irradiation), the triple photocyclization did proceed even in the presence of bulky substituents. Precursor stilbenes handled with ease pave the way for one-pot multiple oxidative photocyclizations toward longer helicenes.

4.6 [16]Helicene

4.6.1 *TIPSO-[2]+[1]+[1]+[2]+[1]+[1]+[2]-OTIPS Precursor*

According to the synthetic guideline, the [2]+[1]+[1]+[2]+[1]+[1]+[2] precursor was designed to perform sextuple photocyclization toward [16]helicene. The solubility of the pristine precursor is expected to be considerably poor. Therefore, bulky TIPSO groups and non-planar, *Z*-configured olefin parts were introduced in advance. The TIPSO-[2]+[1]+[1]+[2]+[1]+[1]+[2]-OTIPS sequence was retrosynthetically divided into three components as Wittig reaction partners: terminal TIPSO-[2], [1]+[1] and central [2] units (**1**, **2**, and **3**, respectively; Scheme 4.7). The terminal TIPSO-[2] unit **1** was prepared in high yield in four steps from 6-hydroxy-2-naphthaldehyde (**4**) (TIPSCl/imidazole, $NaBH_4$, CBr_4/PPh_3, and then PPh_3). The [1]+[1] unit **2** was prepared by the mono-formylation of (*Z*)-4,4′-dibromostilbene (**8**) (*n*-BuLi/DMF). The central [2] unit **3** as a bisphosphonium salt was prepared quantitatively from 2,7-bis(bromomethyl)naphthalene (**9**).

The connection order of three components **1–3** is important; when the central [2] unit **3** was firstly coupled with the [1]+[1] unit **2** by double Wittig reaction, the resulting [1]+[1]+[2]+[1]+[1] sequence **13** is an insoluble yellow solid, and the further treatment was impossible (Scheme 4.8a). Alternatively, the terminal TIPSO-[2] unit **1** was coupled with the [1]+[1] unit **2** in advance to prepare the TIPSO-[2]+[1]+[1] sequence **14** as a pale yellow oil (Scheme 4.8b). Then, the bromide **14** was again formylated to TIPSO-[2]+[1]+[1] sequence **15**. Finally, the central [2] unit **3** was coupled with TIPSO-[2]+[1]+[1] sequence **15** by double Wittig reactions to furnish the TIPSO-[2]+[1]+[1]+[2]+[1]+[1]+[2]-OTIPS sequence **16** as a yellow sticky solid in 85% yield. The precursor was soluble in chloroform and toluene. The *E*/*Z* ratio was estimated from the integral value of the ^1H NMR signals of *Z*-olefins that appeared in a relatively upfield region ($\delta = 6.8$–6.4 ppm) and are distinguished from the signals of *E*-olefins and aromatic protons ($\delta = 7.9$–7.0 ppm).

4.6.2 *Sextuple Photocyclization*

A toluene solution of TIPSO-[2]+[1]+[1]+[2]+[1]+[1]+[2]-OTIPS precursor **16** was irradiated with a 450-W high-pressure Hg lamp at 90 °C in the presence of I_2 (oxidant, 6 equiv.) and propylene oxide (acid scavenger, 100 equiv.) (Scheme 4.9). The experimental setup is shown in Fig. 4.2; up to 11 sample tubes (20 mL each) were set around the UV light source and immersed in an oil bath prior to the irradiation. To grasp the behavior of the precursor and photoreaction intermediates, the time course of the sextuple photocyclization was monitored by HPLC. After 24-h irradiation, the precursor was completely consumed and numerous uncharacterized intermediates appeared. However, further irradiation (48-h irradiation in total) gave rise to one prominent peak. The corresponding fraction was isolated by preparative

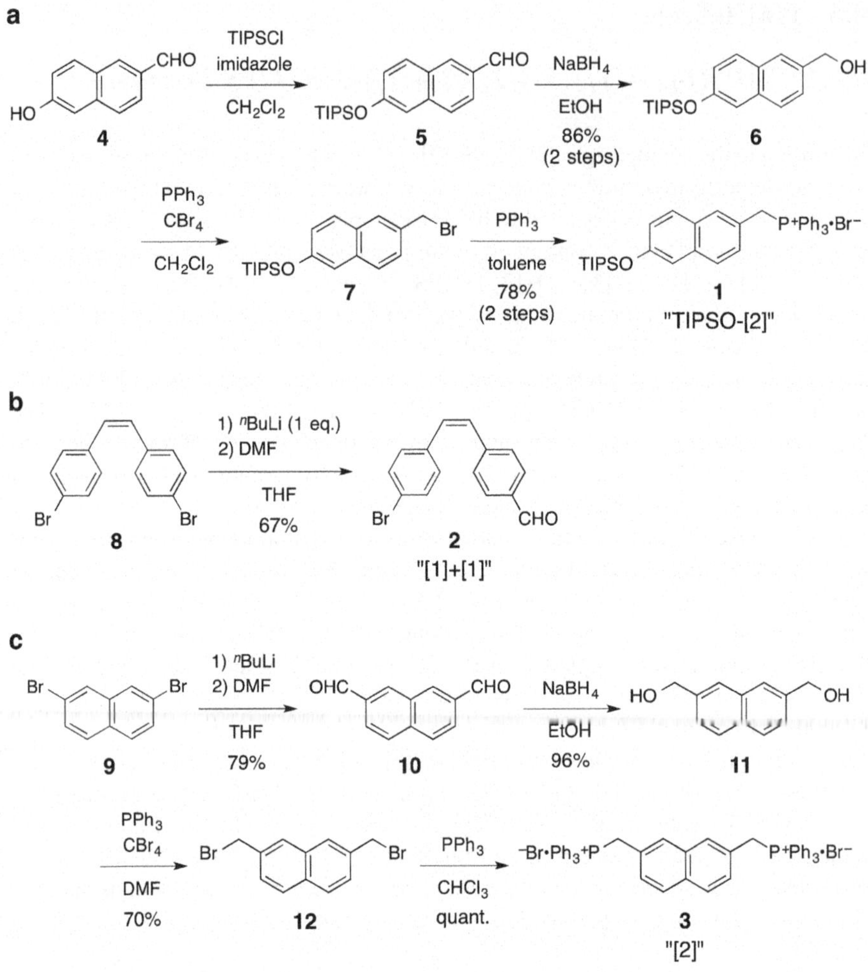

Scheme 4.7 Synthesis of the three components: (**a**) the terminal TIPSO-[2] unit **1**, (**b**) the [1] +[1] unit **2**, and (**c**) the central [2] unit **3**

HPLC and a yellow solid was obtained in 7% yield. In size-exclusion chromatography, the increased retention time, compared to the original one, indicated a folded compact structure of the product. The molecular mass of the product and the isotope pattern were consistent with the calculated value for the target TIPSO-[16]helicene, which means that the sextuple photocyclization did proceed to give the product in which 16 benzene rings are fused.

The ^{1}H NMR spectrum of TIPSO-[16]helicene is shown in Fig. 4.3. The terminal aromatic protons (H_p and H_q) appeared at the quite upfield region ($\delta = 5.51$ and 5.78 ppm, respectively), indicating that these aromatic protons are located above the other aromatic rings. Especially, the doublet of doublets signal H_p is characteristics and assigned to the inner *ortho*-protons in the terminal benzenes. The two methyl

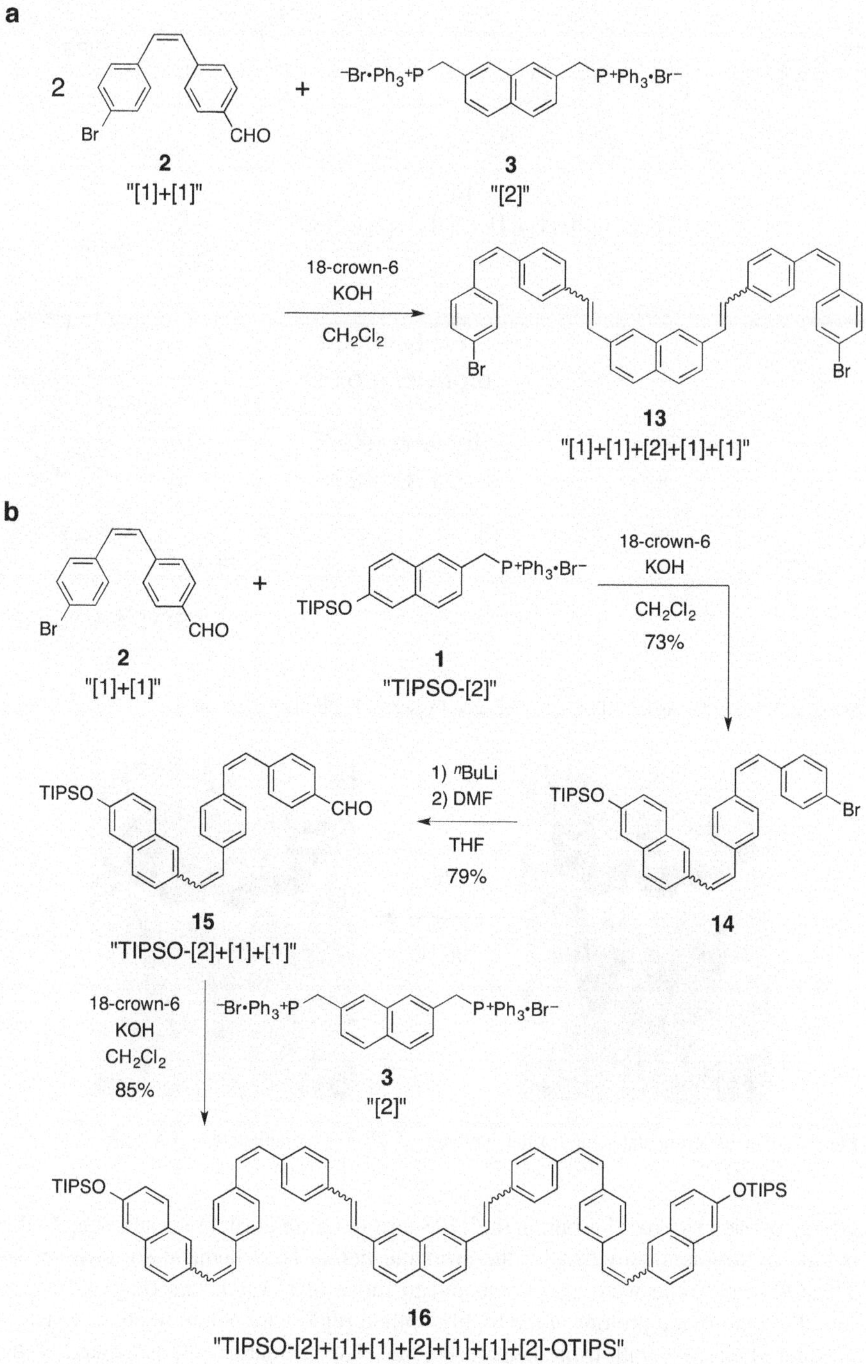

Scheme 4.8 Synthesis of the TIPSO-[2]+[1]+[1]+[2]+[1]+[1]+[2]-OTIPS precursor **16**: (**a**) unsuccessful route due to the insoluble [1]+[1]+[2]+[1]+[1] sequence **13** and (**b**) the successful route

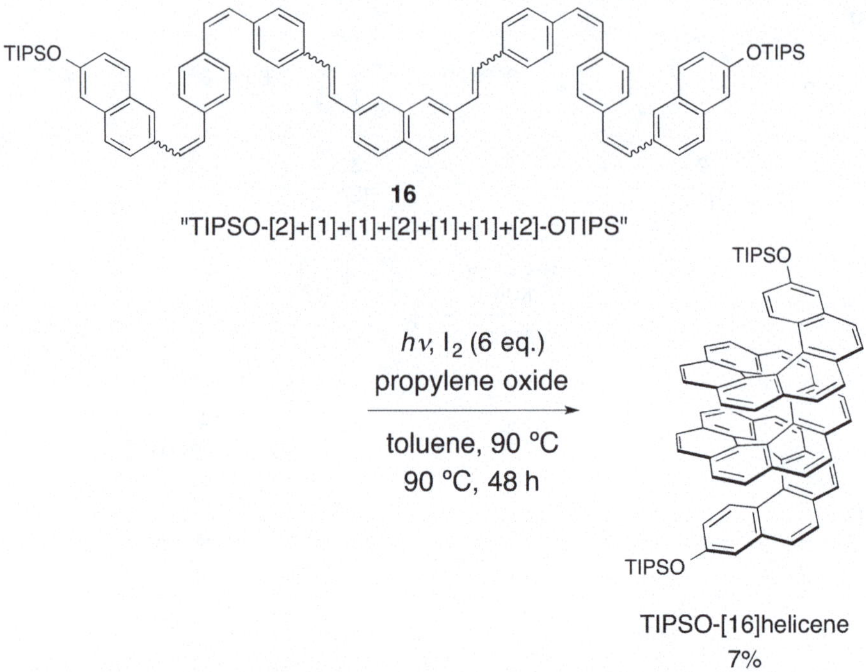

Scheme 4.9 Synthesis of TIPSO-[16]helicene by sextuple photocyclization

Fig. 4.2 The experimental setup for the synthesis of TIPSO-[16]helicene by UV light

groups of each isopropyl group in the TIPS terminals are diastereotopic owing to the chiral environment provided by the aromatic helix. The chemical shift values of TIPSO-[16]helicene were also compared to those of [14]helicene (Fig. 4.4). It is found that aromatic protons at the highly upfield region are common characteristics of longer helicenes with triple-layer structures. The aromatic protons located at the middle layer are also shifted upfield due to the shielding effects from the upper and lower layers.

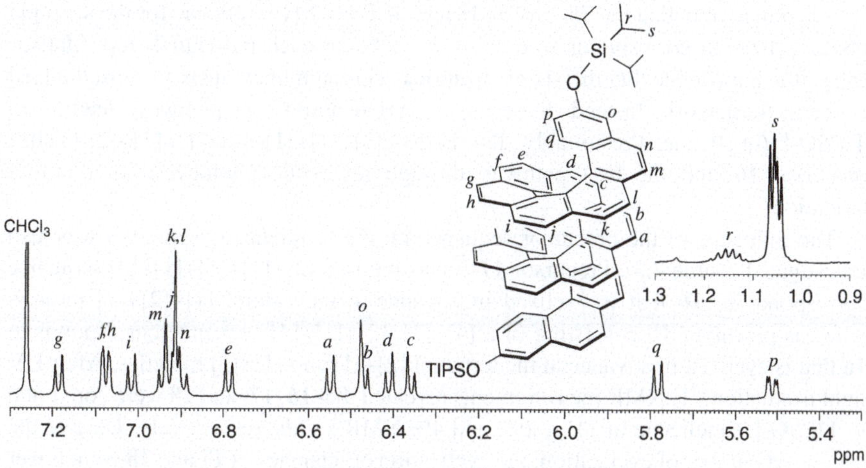

Fig. 4.3 ^{1}H NMR spectrum (500 MHz, CDCl$_3$, 300 K) of TIPSO-[16]helicene

a

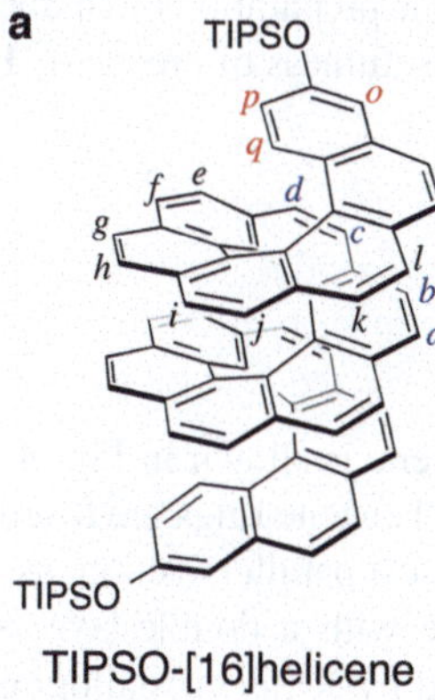

δ / ppm

"sandwiched"	"others"
a 6.55	*e* 6.79
b 6.47	*f* 7.08
c 6.36	*g* 7.18
d 6.41	*h* 7.07
	i 7.01
"terminal"	*j* 6.93
o 6.48	*k,l* 6.91
p 5.51	*m* 6.94
q 5.78	*n* 6.89

b

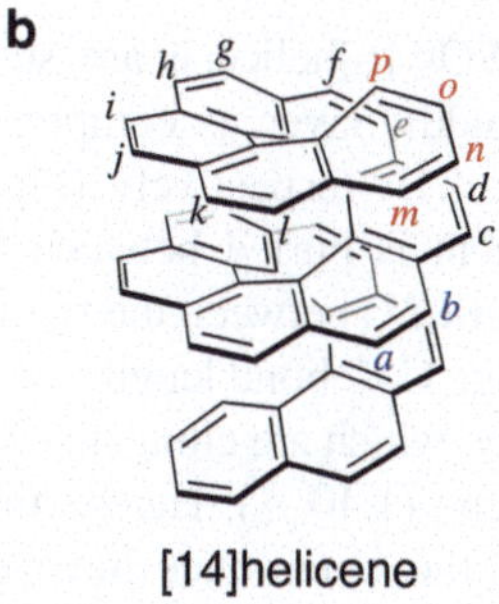

δ / ppm

"sandwiched"	"others"
a 6.27	*c* 6.75
b 6.34	*d* 7.12
	k 6.98
"terminal"	*l* 7.00
m 7.06	*e–j* 7.30–7.03
n 6.69	
o 5.95	
p 6.14	

Fig. 4.4 Chemical shift values of the protons located on the helicene skeleton in the ^{1}H NMR spectrum: (**a**) TIPSO-[16]helicene and (**b**) [14]helicene

It is remarkable that the 7% isolated yield of TIPSO-[16]helicene for the sextuple photocyclization corresponds to a ca. 64% yield for each individual step. Statistically, it is impossible that the six cyclizations occur simultaneously to form the [16] helicene framework. Instead, there are a myriad number of pathways leading to TIPSO-[16]helicene. Presumably, the TIPSO-[2]+[1]+[1]+[2]+[1]+[1]+[2]-OTIPS precursor **16** find the best pathway through reversible photocyclization before oxidation.

The influence of the precursor sequence on the [16]helicene synthesis was also investigated (Table 4.3). Precursor **17** consisting of a [2]+[1]+[6]+[1]+[2] sequence corresponds to **16** that is cyclized in advance at the central [1]+[2]+[1] moiety, whereas precursor **18** consisting of a [4]+[1]+[2]+[1]+[4] sequence corresponds to **16** that is cyclized in advance at the terminal [2]+[1] and [1]+[2] moieties. After UV light irradiation, ^{1}H NMR measurements revealed that **16**, **17**, and **18** were converted to TIPSO-[16]helicene in 11%, 2%, and 4% NMR yields, respectively. Despite the decreased number of cyclization, the cyclization efficiencies of **17** and **18** were lower than that of **16**, and [4]helicene subunits, which give diagnostic ^{1}H NMR signals at around 9 ppm, still remained to be unreacted. In addition to the higher photochemical conversion of **16** into TIPSO-[16]helicene, the preparation of **16** does not require photocyclization. These results strengthen the usefulness of precursor **16** in the [16] helicene synthesis.

4.6.3 X-Ray Crystal Structure

The X-ray crystal structure of TIPSO-[16]helicene is shown in Fig. 4.5. Similar to other reported crystals of carbo[*n*]helicenes, six benzene rings are fused for one turn of the helix. Three π layers are stacked in an almost parallel fashion, as shown in the space-filling representation. Since [11]helicene with a double-layer structure has been the longest helicene that was characterized by X-ray crystallographic analysis [44], the present crystal structure is the first observation of a triple-layer structure of [*n*]helicenes.

The detailed structural parameters of TIPSO-[16]helicene are summarized in Fig. 4.6. Benzene ring **H**, embedded in the middle layer, is compressed by rings **B** and **N** with interplanar angles of 6.2° and 4.1°, respectively (Fig. 4.6a). One methyl group of the terminal TIPSO substituent is buried between the π layers, leading to the increase of the interplanar angle (14.2°) between the terminal benzene ring **A** and the above benzene ring **G**. The average C–C bond lengths of the inner and outer helices are 1.44 Å and 1.36 Å, respectively, which are elongated and shortened compared with the C–C bond length of benzene (1.40 Å). Hence, the inner C–C bonds have single bond characters, whereas the outer ones have double bond characters. The average pitches of the inner and outer helices are 3.29 and 3.66 Å, respectively (Fig. 4.6b), which are the same as and shorter than the corresponding pitches of [9]helicene [14]. These findings demonstrate that the triple-layer structure is more tightly compressed than the double one in [*n*]helicenes.

Table 4.3 Synthesis of TIPSO-[16]helicene from different precursors

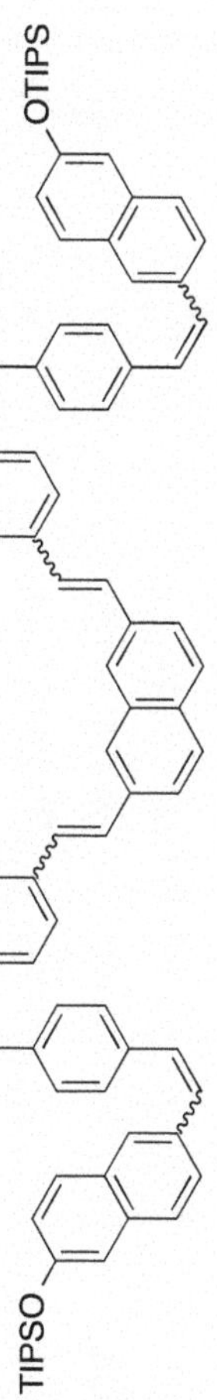

16

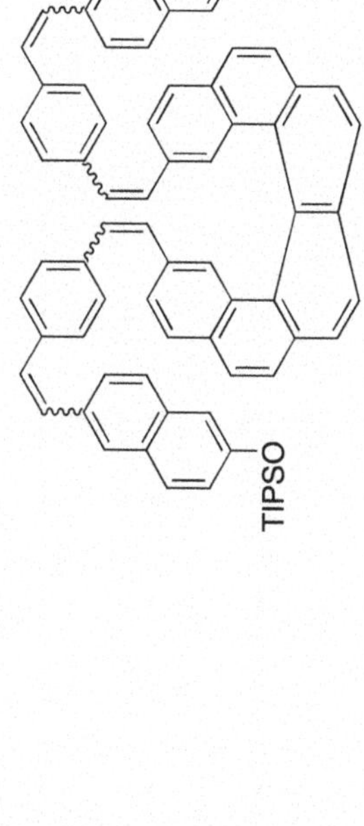

17

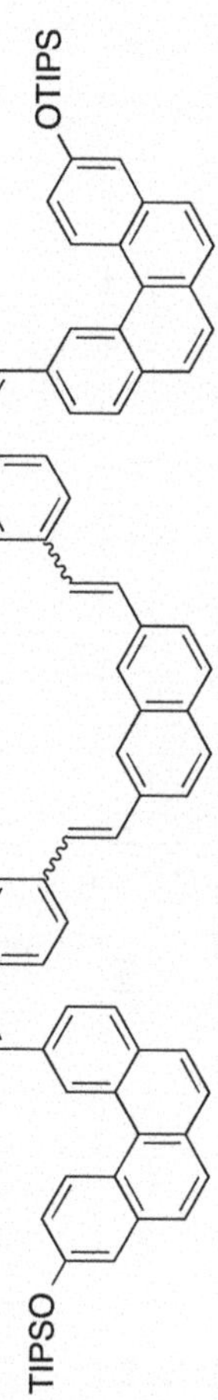

18

(continued)

Table 4.3 (continued)

Precursors	Sequences	Number of cyclization	NMR yields of TIPSO-[16]helicene (%)
16	[2]+[1]+[1]+[2]+[1]+[1]+[2]	6	11
17	[2]+[1]+[6]+[1]+[2]	4	2
18	[4]+[1]+[2]+[1]+[4]	4	4

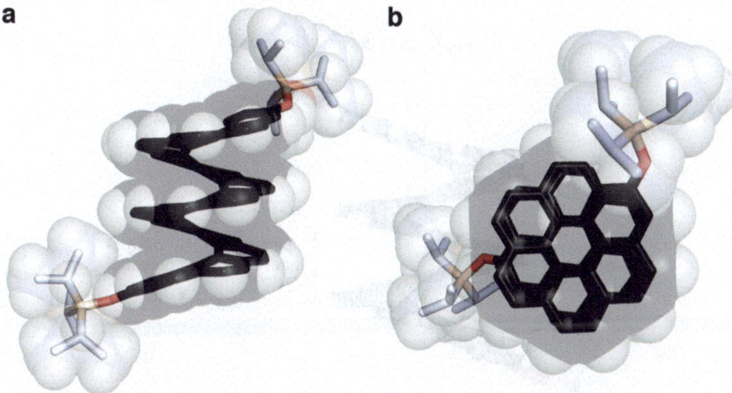

Fig. 4.5 The X-ray crystal structure of TIPSO-[16]helicene: (**a**) side and (**b**) top views

4.6.4 Conversion to Unsubstituted [16]Helicene

The hitherto longest carbohelicene has been prepared as an unsubstituted form. Therefore, TIPSO-[16]helicene was further converted to unsubstituted [16]helicene in there steps without purification of the intermediates (Scheme 4.10). [16]Helicene is a yellow solid and hardly soluble in all common solvents. The ^{1}H NMR spectrum obtained by long accumulation supported the removal of TIPSO groups from the terminal benzene rings of TIPSO-[16]helicene. The molecular mass of [16]helicene was clearly revealed by MALDI-TOF mass spectroscopy.

4.7 Conclusion

Simple arylene–vinylene oligomers containing only [1] and [2] units can be good precursors for the synthesis of longer helicenes when the [1] and [2] units are appropriately arranged in the precursor sequences. Based on the synthetic guideline for precursor stilbenes, the longest [16]helicene was successfully synthesized, and the triple-layer structure was clearly observed by X-ray crystallographic analysis. It is intriguing to investigate whether the synthetic guideline can be applied to the synthesis of much longer carbohelicenes or heterohelicenes. Since oxidative photocyclization should be conducted at highly dilute concentrations, a tiny amount of TIPSO-[16]helicene was obtained in a single experiment. A large-scale photochemical synthesis of longer helicenes is a next challenge, and a continuous flow technique will overcome the drawbacks of multiple photocyclizations in a batch reactor [45–47].

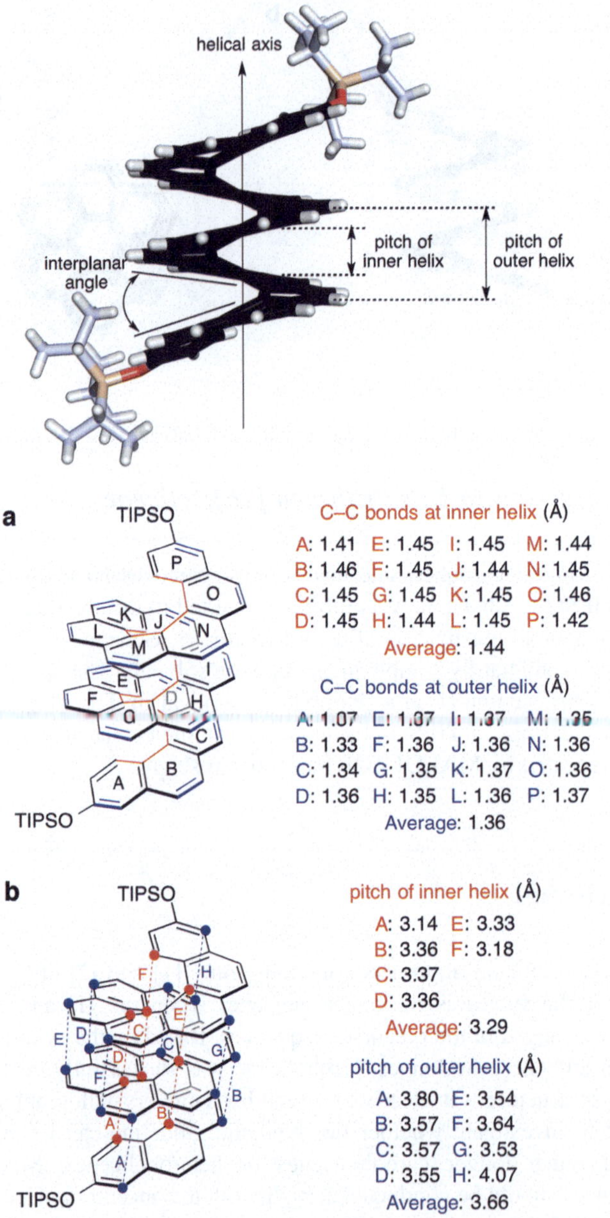

Fig. 4.6 Structural parameters of TIPSO-[16]helicene: (**a**) C–C bond length and (**b**) pitch distance

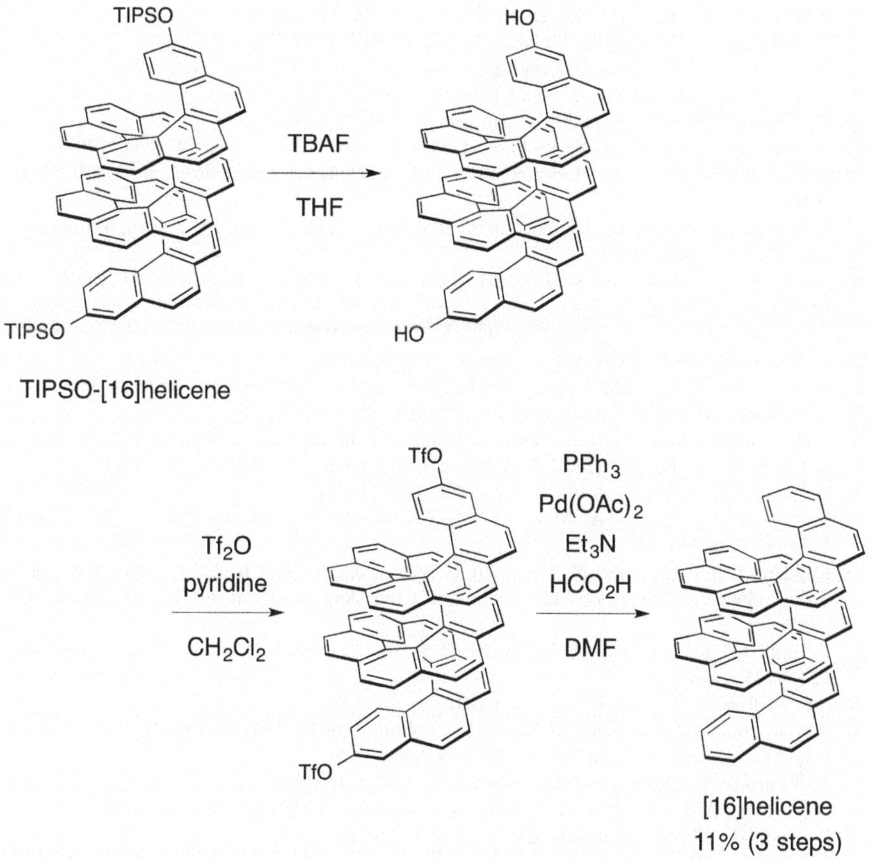

Scheme 4.10 Conversion from TIPSO-[16]helicene to unsubstituted [16]helicene

References

1. Y. Shen, C.-F. Chen, Chem. Rev. **112**, 1463–1535 (2012)
2. M. Gingras, Chem. Soc. Rev. **42**, 968–1006 (2013)
3. M. Gingras, G. Félix, R. Peresutti, Chem. Soc. Rev. **42**, 1007–1050 (2013)
4. M. Gingras, Chem. Soc. Rev. **42**, 1051–1095 (2013)
5. N. Hoffmann, J. Photochem. Photobiol. C: Photochem. Rev. **19**, 1–19 (2014)
6. A. Urbano, M.C. Carreño, Org. Biomol. Chem. **11**, 699–708 (2013)
7. A. Urbano, Angew. Chem. Int. Ed. **42**, 3986–3989 (2003)
8. T.J. Katz, Angew. Chem. Int. Ed. **39**, 1921–1923 (2000)
9. K.P. Meurer, F. Vögtle, Top. Curr. Chem. **127**, 1–76 (1985)
10. W.H. Laarhoven, W.C. Prinsen, Curr. Chem. **125**, 63–130 (1984)
11. H. Martin, Angew. Chem. Int. Ed. Engl. **13**, 649–660 (1974)
12. H. Wynberg, Acc. Chem. Res. **4**, 65–73 (1971)
13. Chen C-F, Shen Y (2017) Helicene chemistry, Springer, Berlin/Heidelberg
14. K. Mori, T. Murase, M. Fujita, Angew. Chem. Int. Ed. **54**, 6847–6851 (2015)
15. R. Weitzenböck, H. Lieb, Monatsh. Chem. **33**, 549–565 (1913)

16. J. Meisenheimer, K. Witte, Ber. Dtsch. Chem. Ges. **36**, 4153–4164 (1903)
17. R. Weitzenböck, A. Klingler, Monatsh. Chem. **39**, 315–323 (1918)
18. M.S. Newman, W.B. Lutz, D. Lednicer, J. Am. Chem. Soc. **77**, 3420–3421 (1955)
19. M.S. Newman, D. Lednicer, J. Am. Chem. Soc. **78**, 4765–4770 (1956)
20. C.S. Wood, F.B. Mallory, J. Org. Chem. **29**, 3373–3377 (1964)
21. M. Flammang-Barbieux, J. Nasielski, R.H. Martin, Tetrahedron Lett. **8**, 743–744 (1967)
22. R.H. Martin, M. Flammang-Barbieux, J.P. Cosyn, M. Gelbcke, Tetrahedron Lett. **9**, 3507–3510 (1968)
23. H. Kagan, A. Moradpour, J.F. Nicoud, G. Balavoine, R.H. Martin, J.P. Cosyn, Tetrahedron Lett. **12**, 2479–2482 (1971)
24. A. Moradpour, H. Kagan, M. Baes, G. Morren, R.H. Martin, Tetrahedron **31**, 2139–2143 (1975)
25. R.H. Martin, J.P. Cosyn, Synth. Commun. **1**, 257–265 (1971)
26. R. Martin, V. Libert, J. Chem. Res. Synop. 130–131 (1960)
27. W.H. Laarhoven, T.J.H.M. Cuppen, R.J.F. Nivard, Tetrahedron **26**, 4865–4881 (1970)
28. R.H. Martin, M. Baes, Tetrahedron **31**, 2135–2137 (1975)
29. R.H. Martin, G. Morren, J.J. Schurter, Tetrahedron Lett. **10**, 3683–3688 (1969)
30. L. Liu, B. Yang, T.J. Katz, M.K. Poindexter, J. Org. Chem. **56**, 3769–3775 (1991)
31. L. Liu, T.J. Katz, Tetrahedron Lett. **31**, 3983–3986 (1990)
32. L. Minuti, A. Taticchi, A. Marrocchi, E. Gacs-Baitz, R. Galeazzi, Eur. J. Org. Chem., 3155–3163 (1999)
33. K. Kamikawa, I. Takemoto, S. Takemoto, H. Matsuzaka, J. Org. Chem. **72**, 7406–7408 (2007)
34. S.K. Collins, A. Grandbois, M.P. Vachon, J. Côté, Angew. Chem. Int. Ed. **45**, 2923–2926 (2006)
35. F. Teplý, I.G. Stará, I. Starý, A. Kollárovič, D. Šaman, L. Rulíšek, P. Fiedler, J. Am. Chem. Soc. **124**, 9175–9180 (2002)
36. F.B. Mallory, C.W. Mallory, Org. React. **30**, 48–49 (1984)
37. R.H. Martin, M.-J. Marchant, M. Baes, Helv. Chim. Acta **54**, 358–360 (1971)
38. R.H. Martin, Angew. Chem. Int. Ed. **13**, 649–660 (1974)
39. A. Sudhakar, T.J. Katz, Tetrahedron Lett. **27**, 2231–2234 (1986)
40. H.R. Talele, A.R. Chaudhary, P.R. Patel, A.V. Bedekar, ARKIVOC **ix**, 15–37 (2011)
41. F. Dietz, M. Scholz, Tetrahedron **24**, 6845–6849 (1968)
42. T. Wismonski-Knittel, G. Fischer, E. Fischer, J. Chem. Soc. Perkin Trans. **2**, 1930–1940 (1974)
43. J. Roose, S. Achermann, O. Dumele, F. Diederich, Eur. J. Org. Chem. 3223–3231 (2013)
44. G. Le Bas, A. Navaza, M. Knossow, C. de Rango, Cryst. Struct. Commun. **5**, 713–718 (1976)
45. Q. Lefebvre, M. Jentsch, M. Rueping, Beilstein J. Org. Chem. **9**, 1883–1890 (2013)
46. A.C. Hernandez-Perez, A. Vlassova, S.K. Collins, Org. Lett. **14**, 2988–2991 (2012)
47. A.-C. Bédard, A. Vlassova, A.C. Hernandez-Perez, A. Bessette, G.S. Hanan, M.A. Heuft, S.K. Collins, Chem. Eur. J. **19**, 16295–16302 (2013)

Chapter 5
Design of the Chiral Environment for Asymmetric Acid-Base Catalysis

Yoshihiro Sohtome, Kazuo Nagasawa, and Mikiko Sodeoka

Abstract Asymmetric catalysis is one of the most powerful methodologies for engineering chiral molecules. But, despite significant advances, there have been few systematic studies aiming to explore and expand the available "catalytic asymmetric space." In general, optimization of the selectivity (enantio-, diastereo-, regio-, and chemoselectivity) of asymmetric catalysis relies on trial-and-error screening using the inherent stereochemistry embedded in "privileged" chiral templates. This review describes our nascent efforts to broaden the conceptual basis of catalyst design, focusing on the development of two different classes of asymmetric catalysts including (1) conformationally flexible guanidine/bisthiourea organocatalysts and (2) centrochiral transition-metal catalysts. We will also discuss physical-organic chemical methodologies available to explore the interplay of structure and selectivity in asymmetric acid-base catalysis.

Keywords Asymmetric catalysts · Acid-base catalysis · Enolate · Hydrogen bonds · Dynamic asymmetric catalysis · Enantioswitching · Entropy · Guanidine · Thiourea · Chirality at metal · Electron density distribution · [3 + 2] Cycloaddition · Nickel · Copper

---

Y. Sohtome (✉) · M. Sodeoka
Synthetic Organic Chemistry lab, RIKEN Cluster for Pioneering Research, Saitama, Japan

Catalysis and Integrated Research Group, RIKEN Center for Sustainable Resource Science, Saitama, Japan
e-mail: sohtome@riken.jp; sodeoka@riken.jp

K. Nagasawa
Department of Biotechnology and Life Science, Tokyo University of Agriculture and Technology, Tokyo, Japan
e-mail: knaga@cc.tuat.ac.jp

S. Shirakawa (ed.), *Designed Molecular Space in Material Science and Catalysis*,
https://doi.org/10.1007/978-981-13-1256-4_5

5.1 Introduction

Chiral molecular catalysts have been often considered as analogous to the catalytic active sites of enzymes [1–4]. However, the stereoselection processes of enzymes and molecular catalysts are regulated by fundamentally different principles. Most enzymes are large proteins, which have considerable conformational flexibility to construct well-defined three-dimensional catalytic active-site architectures in the folded state, facilitating the highly selective molecular recognition of substrates [5, 6]. A remarkable example of the link between structural dynamics and functional changes is provided by signal transduction pathways in the living cell, in which enzymes play a pivotal role in regulating the posttranslational modifications of proteins. In contrast, chiral small molecular catalysts utilize the chiral environments around their catalytic active sites; most organocatalysts contain an active site (s) covalently linked to a chiral template(s), while transition-metal catalysts consist of a central metal(s) with chiral ligand(s). Chemists are generally required to select a suitable combination of the active site and the chiral template/ligand during the optimization process in the development of asymmetric catalysis. With regard to reactivity, the repertoire of reaction patterns susceptible to asymmetric molecular catalysis has rapidly expanded in recent decades, with the discovery of "activation modes" [7, 8]. An important challenge in the field of asymmetric catalysis is to further explore the "chiral catalytic space," because conventional catalysts, including organocatalysts and metallocatalysts, have mainly been predicated on the use of the inherent stereochemistry embedded in "privileged" chiral templates [9, 10].

A primary objective of our work described here is the identification of effective catalyst core frameworks that enable us to manipulate the chiral space. In particular, we have focused on the development of cooperative (bifunctional or multifunctional) catalysis, in which both nucleophilic and electrophilic reactants are activated in the bond-forming process, to attain high selectivity and rate acceleration [11, 12]. We will also discuss physical-organic approaches to gain mechanistic insights into the stereo-discrimination processes involved in catalyst–substrate interactions.

5.2 Conformationally Flexible Guanidine/Bisthiourea Organocatalysts

Hydrogen-bonding (H-bonding) interactions between hydrogen atoms (X–H) and Lewis basic acceptors (A) are critical for controlling the reactivity and stereoselectivity in enzymatic chemical reactions [13]. The energy of individual H-bonds is relatively weak, ranging between those of covalent bonds and van der Waals interactions, but enzymes construct beautifully defined catalytic active sites by exploiting multiple H-bonding interactions. The weakness of H-bonding interactions in enzymes is also the key to efficient product release after the bond-forming reaction, allowing high turnover. Such multiple roles of H-bonds, as well as the

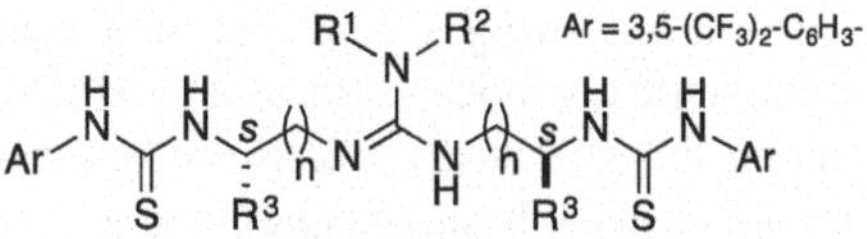

Fig. 5.1 The structures of guanidine/bisthiourea (*S*,*S*)-**1** and (*S*,*S*)-**2**

inertness of H-bond donor catalysts to water and oxygen, have stimulated a wide variety of synthetic applications of asymmetric organocatalysis [14–17].

Beginning with foundational disclosure of the availability of (thio)urea as a double H-bond donor by Sigman and Jacobsen in 1998 [18], the use of chiral double H-bond donors has been recognized as a useful strategy for designing asymmetric organocatalysts. The development of Takemoto's bifunctional thiourea [19, 20] incorporating a dimethylamino group, which acts as a Brønsted base, was another important keystone. By applying the bifunctionality concept [11, 12] to design chiral H-bond donor catalysts, the scope of substrates has been greatly extended. Furthermore, kinetic, computational and structure-catalytic activity relationship studies have provided important guidelines for the development of asymmetric organocatalysis [21–24].

Since Nagasawa's group reported an early example of the development of chiral pentacyclic guanidine catalysts [25–28], we (KN and YS) have continued to explore conformationally flexible guanidine/bisthiourea organocatalysts **1** and **2** (Fig. 5.1) [29–31]. We found great inspiration from the diverse but well-defined kinetic reaction pathways controlled by enzymes. Although several peptide-based organocatalysts have been reported [32, 33], development of organocatalysts using acyclic chiral templates has generally been neglected, likely because acyclic molecules can generate many conformers in situ. In general, catalytic stereo-discrimination processes are discussed in terms of the active complex in the transition state, and structurally defined chiral templates are often used with the aim of reducing the number of transition states. On the other hand, we have explored conformationally flexible non-peptide organocatalysts, which can be easily synthesized from readily available amino acids. In the course of our research program to develop guanidine/bisthiourea catalysts **1** and **2**, our aim was to develop highly tunable catalytic active sites involving H-bond donor and Brønsted base moieties. Basically, we hypothesized that kinetic reaction control through the bifunctional transition state constructed by the guanidine/bisthiourea catalyst would promote catalytic stereoselective bond-forming reactions, regardless of the existence of other unwanted conformers of the catalyst in the reaction media. Building on this basic idea, we have applied conformationally flexible guanidine/bisthiourea organocatalysts **1** and **2** to several classes of catalytic asymmetric reactions, including the nitro-aldol (Henry) reaction [29, 33–37], nitro-Mannich reaction [38], Mannich-type reaction of malonates [39–41], Friedel–Crafts (F–C) reaction of phenols [42–44], Michael reaction of β-dicarbonyl compounds [45], phospha-Michael reaction [46], epoxidation [47], α-hydroxylation [48–51], and α-amination [52].

In the following section, we will mainly discuss the development of dynamic asymmetric organocatalysis as a tool to output dual catalytic functions using a single chiral catalyst. These efforts were based on the working hypothesis that the relative arrangements of basic (guanidine functional group) and H-bond donor sites (thiourea functional groups), which are linked by a flexible chiral spacer in the catalyst, can be altered as required depending on the substrate and reaction conditions. Here, we will place particular emphasis on how to attain switchable properties with a common mechanistic paradigm that utilizes only H-bond donor and Brønsted base moieties.

5.2.1 Enantiodivergent Mannich-Type Reactions

We will begin with enantiodivergent organocatalysis, in which each enantiomer can be catalytically synthesized depending upon the reaction conditions. Stereoswitchable phenotypes to attain high dual enantioselectivities using a single chiral catalyst have long received little attention [53], but more recently, there have been some systematic investigations for the development of chiral switchable catalysts to attain high dual selectivity in the field of asymmetric catalysis and supramolecular catalysis [3, 4, 54–57].

Initially, we selected the Mannich-type reaction of *N*-Boc imines **3** and malonates **4** [58–60] to explore organocatalytic enantioswitching. Because this reaction is irreversible under mild basic conditions, we envisioned that the enantioselectivity would be kinetically controlled in response to conformational change of the guanidine/bisthiourea. After systematic catalyst screening with our working model, we identified the stereoswitching potential of (*S*,*S*)-**1a** in the Mannich-type reaction of **3** with **4** [40]. As shown in Table 5.1, the Mannich-type reactions promoted by (*S*,*S*)-**1a** in *m*-xylene or toluene generate (*S*)-selective adduct (87–97% ee), while the reactions proceeded in an (*R*)-selective manner when acetonitrile was used as a solvent (80–89% ee). Under optimized conditions, we obtained relatively high catalyst turnover frequencies (TOF) in the Mannich-type reaction of **3a** and **4a**: 66 h^{-1} in (*S*)-selective reaction and 25 h^{-1} in (*R*)-selective reaction.

Sequential enantiodivergent catalysis, wherein dual catalytic functions can be exhibited with single-flask operation, emerged as an attractive yet challenging next step [61]. To achieve such a catalytic sequence, the transition states that show different functional outcomes should be constructed with a single chiral source and in a single flask. Before our work [41], only two examples of sequential diastereodivergent catalysis depending on the metal and acidic additive had been reported [62, 63]. Although these examples showed that two diastereomerically different transition states can be constructed in the presence of a single chiral source with single-flask operation, reversible switching of catalytic functions with single-flask operation still remained a formidable task; primarily because exchanging the diastereo-switching trigger, such as metal [62] or acidic additive [63], is inherently difficult without a purification process.

Table 5.1 Solvent-dependent enantiodivergent Mannich-type reaction using (*S*,*S*)-**1a**

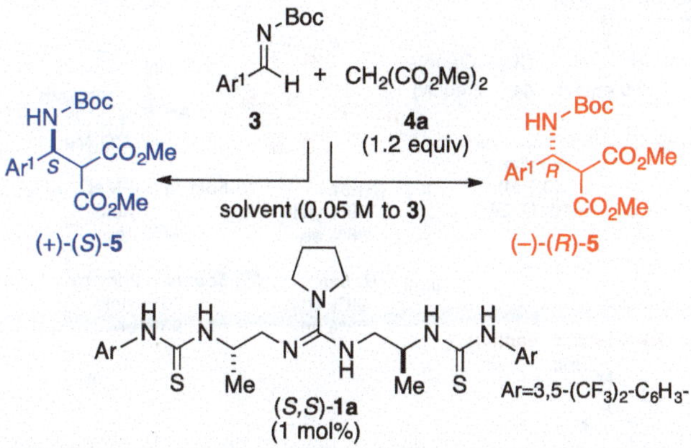

Imine **3** Ar	Solvent	Temp. (°C)	Time (h)	Yield (%)	Ee (%)[a]
3a: Ph	*m*-xylene	0	1.5	99	+92
3b[b]: 4-Me-C_6H_4	*m*-xylene	0	1.5	99	+92
3c: 3-Me-C_6H_4	*m*-xylene	0	2	97	+91
3d: 2-Me-C_6H_4	*m*-xylene	0	3	98	+93
3e[b]: 4-Cl-C_6H_4	*m*-xylene	0	2	99	+90
3f: 4-MeO-C_6H_4	*m*-xylene	0	8	99	+97
3 g: 2-naphthyl	*m*-xylene	0	1.5	99	+94
3h[c]: 2-furyl	*m*-xylene	−10	1	97	+89
3i[c]: 2-thienyl	toluene	−10	1	98	+97
3a: Ph	MeCN	−30	4	99	−88
3b: 4-Me-C_6H_4	MeCN	−40	18	98	−89
3c: 3-Me-C_6H_4	MeCN	−40	15	96	−89
3d[d]: 2-Me-C_6H_4	MeCN	−30	14	90	−80
3e: 4-Cl-C_6H_4	MeCN	−40	6	99	−82
3f: 4-MeO-C_6H_4	MeCN	−40	20	88	−84
3 g: 2-naphthyl	MeCN	−40	12	97	−80
3 h: 2-furyl	MeCN	−30	10	92	−84
3i[d]: 2-thienyl	MeCN	−40	12	97	−86

[a]The *ee* of (*S*)-**5** is defined as plus and that of (*R*)-**5** as minus
[b]The reaction was carried out at 0.1 M concentration of **3**
[c]2 mol% of (*S*,*S*)-**1a** was used
[d]3 mol% of (*S*,*S*)-**1a** was used

Inspired by these excellent sequential diastereodivergent reactions, we developed sequential enantiodivergent reactions using (*S*,*S*)-**1a**. Because we can easily switch the reaction solvent as required with single-flask operation, we designed the protocol shown in Scheme 5.1. After the imine **3a** was consumed in the first Mannich-type reaction with **4a**, the solvent was removed, and then the second Mannich-type reaction with a different imine was carried out in a different solvent. In the upper

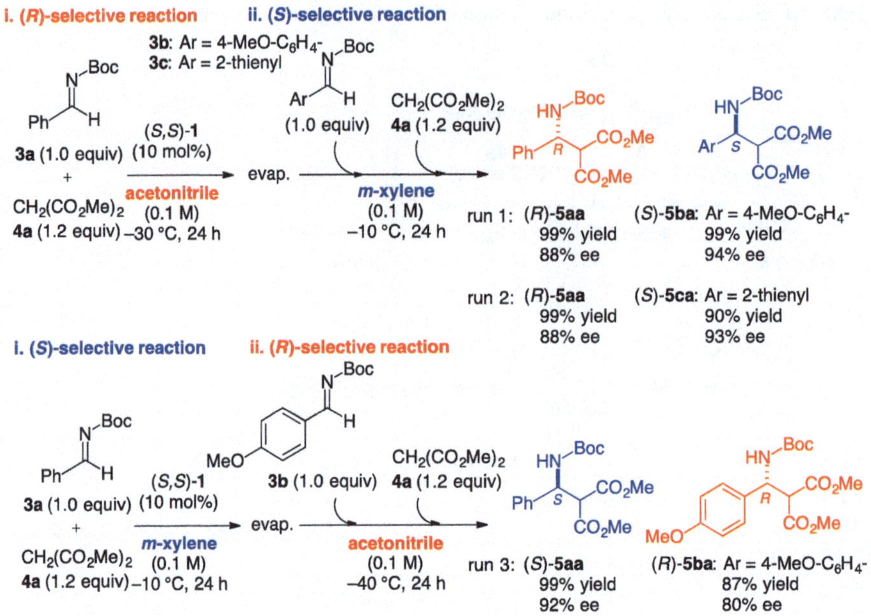

Scheme 5.1 Bidirectional sequential enantiodivergent Mannich-type reactions

scheme, (*R*)-**5aa** (99% yield, 88% ee) was obtained in acetonitrile, while (*S*)-**5ba** (run 1: 99% yield, 94% ee) and (*S*)-**5ca** (run 2: 90% yield with 93% ee) were obtained in the second Mannich-type reaction using *m*-xylene. Notably, this catalytic strategy can be applied to bidirectional switching. As shown in the bottom part of the scheme, (*S*)-**5aa** (99% yield, 92% ee) in *m*-xylene was produced in the first reaction, followed by generation of (*R*)-**5ba** (run 3: 87% yield, 80% ee) in the second reaction using acetonitrile.

Thus, our work illustrated in Scheme 5.1 showcased that the catalytic activity of (*S,S*)-**1a** is not inhibited even in the presence of an excess amount of the chiral Mannich adduct produced in the first reaction. This is a remarkable example of artificially switchable catalysis, which enables ON/ON functional switching with relatively high dual enantioselectivity.

Next, we will briefly discuss the mechanism of the characteristic enantioswitching [40, 41]. First, the critical role of both the guanidine and thiourea functional groups on (*S,S*)-**1a** was confirmed by comparison of the reactions in toluene and acetonitrile promoted by (*S,S*)-**1** versus its structural variants **6** and **7** (Scheme 5.2). When the same conditions used in the (*S,S*)-**1a**-catalyzed reactions were employed for **6**, no reaction occurred. Guanidine **7**, in which the side arms were protected with Boc, can catalyze the Mannich-type reaction in high yield, albeit with low enantioselectivity (toluene: 99% yield, 0% ee and acetonitrile: 97% yield, −1% ee). These control experiments indicate that both guanidine and thioureas are crucial for attaining the high reactivity and selectivity in the (*S*)-, and (*R*)-selective

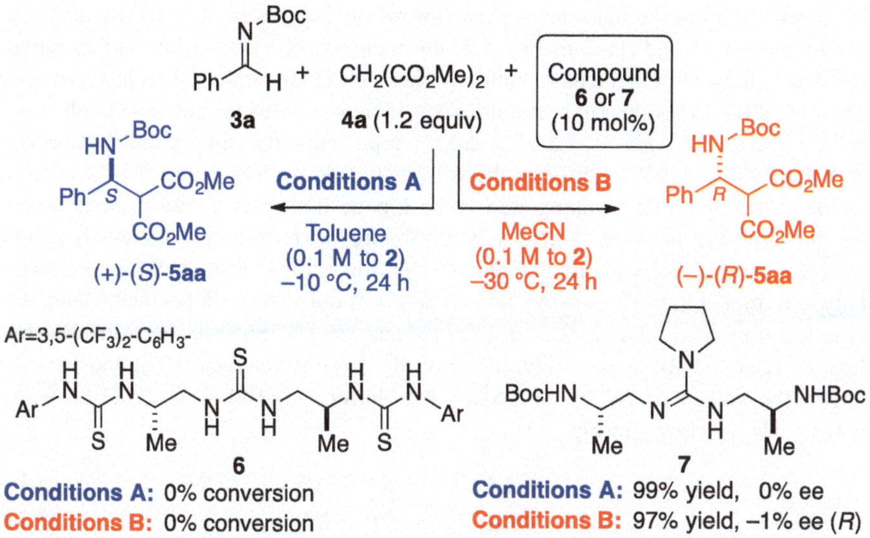

Scheme 5.2 Structure and catalytic activity relationships in the enantiodivergent Mannich-type reactions of **3a** with **4a**

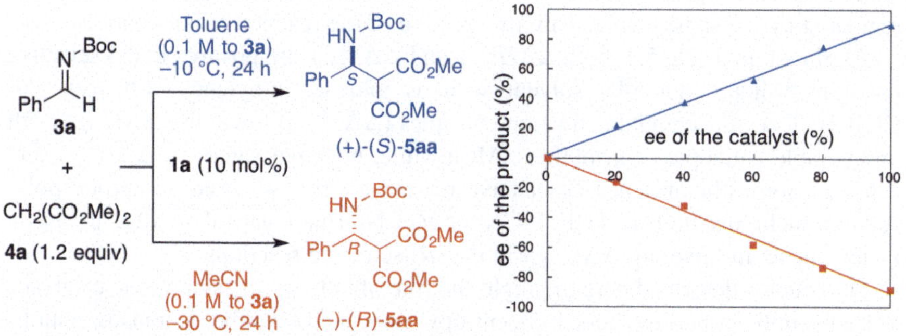

Fig. 5.2 Relationships between the % ee of catalyst **1a** and that of Mannich adduct **5aa**

Mannich-type reactions. The fact that asymmetric induction did not occur with the chiral guanidine **7** tends to rule out a mechanistic scenario based on functional masking with interactions between thiourea and acetonitrile.

Second, in order to examine whether the active species is monomeric or oligomeric, we investigated the relationships between the % ee of catalyst **1a** and that of Mannich adduct **5aa** in both toluene and acetonitrile. The resulting linear relationships (Fig. 5.2) suggest that the monomeric species of **1a**, not oligomers [36], controls the stereo-discrimination process in the Mannich-type reaction.

Most importantly, we identified a characteristic temperature profile of the solvent-dependent Mannich-type reaction. In particular, as the reaction temperature was raised, the (*S*)-selectivity in *m*-xylene increased. Considering that the stereoselectivity in the (*S*)-selective reaction would be governed by the monomeric active species, as indicated in Fig. 5.2, the temperature effect in the (*S*)-selective reaction is unusual. On the other hand, as the reaction temperature was lowered, the ee value of the (*R*)-adduct in acetonitrile was increased as usual. Because Gibbs free energy $\Delta\Delta G^{\ddagger}$ ($\Delta\Delta H^{\ddagger}$ – $T\Delta\Delta S^{\ddagger}$; $\Delta\Delta H^{\ddagger}$ represents the differential activation enthalpy, while $\Delta\Delta S^{\ddagger}$ represents the differential activation entropy) is a crucial factor in the stereo-determining step under kinetic domination, and it is regulated by the interplay between $\Delta\Delta H^{\ddagger}$ and $\Delta\Delta S^{\ddagger}$, we next examined the relationship between the enantioswitching and enthalpy-entropy compensation. We experimentally determined the values of $\Delta\Delta H^{\ddagger}$ and $\Delta\Delta S^{\ddagger}$ in our solvent-dependent Mannich-type reaction. The relative rates of (*S*)-(+)- and (*R*)-(−)-**5aa** can be expressed by Eqs. (5.1) and (5.2), respectively, based on the Eyring treatment [64], and we can obtain the values of $\Delta\Delta H^{\ddagger}$ and $\Delta\Delta S^{\ddagger}$ by plotting ln[(100 + % ee)/(100 – % ee)] vs. 1/T, as shown in Fig. 5.1.

$$\ln(k_S/k_R) = -\Delta\Delta H^{\ddagger}{}_{S-R}/RT + \Delta\Delta S^{\ddagger}{}_{S-R}/R \quad (5.1)$$

$$\ln(k_R/k_S) = -\Delta\Delta H^{\ddagger}{}_{R-S}/RT + \Delta\Delta\ S^{\ddagger}{}_{R-S}/R \quad (5.2)$$

Importantly, the plots gave reasonably straight lines with good correlation coefficients over the temperature range (from 0 to −30 °C) we explored (Fig. 5.3). These observations strongly suggest that the stereo-discriminating step in each solvent was regulated by the same mechanism, irrespective of the reaction temperature.

As shown in Table 5.2, both $\Delta\Delta H^{\ddagger}{}_{S\text{–}R}$ and $\Delta\Delta S^{\ddagger}{}_{S\text{–}R}$ are positive in (*S*)-selective reactions using a nonpolar solvent such as toluene, *m*-xylene, Cl-benzene, or CH_2Cl_2. Thus, differential activation entropies ($\Delta\Delta S^{\ddagger}{}_{S\text{–}R}$) lower the $\Delta\Delta G^{\ddagger}{}_{S\text{–}R}$ with unfavorable enthalpic contributions. Meanwhile, negative values of $\Delta\Delta H^{\ddagger}{}_{R\text{–}S}$ and $\Delta\Delta S^{\ddagger}{}_{R\text{–}S}$ were obtained in (*R*)-selective reactions when we used an aprotic polar solvent, including EtOAc, THF, EtCN, or MeCN. This means that $\Delta\Delta H^{\ddagger}{}_{R\text{–}S}$ has a major impact in lowering $\Delta\Delta G^{\ddagger}{}_{R\text{–}S}$ in the (*R*)-selective reactions.

Our studies described here highlight the role of entropy in asymmetric catalysis. Since the initial report by Inoue [65], entropy-associated asymmetric transformations have been sporadically reported, but there are still only a few examples where high enantioselectivity has been achieved. Furthermore, functional switching based on enthalpy-entropy compensation mode switching is a new idea, which is different from the typical stereoselectivity principle based on catalyst–substrate binding.

5.2.2 *Entropy-Controlled Organocatalytic Friedel–Crafts Reaction of Phenols*

In this section, we will describe our work to apply the 1,3-diamine-tethered guanidine/bisthiourea organocatalyst **2** for 1,4-addition (Michael) reactions of nitroolefins. Because the Michael adduct of nitroolefins involves the reactive site at the α-position of the nitro group, we also selected this reaction as a starting point to

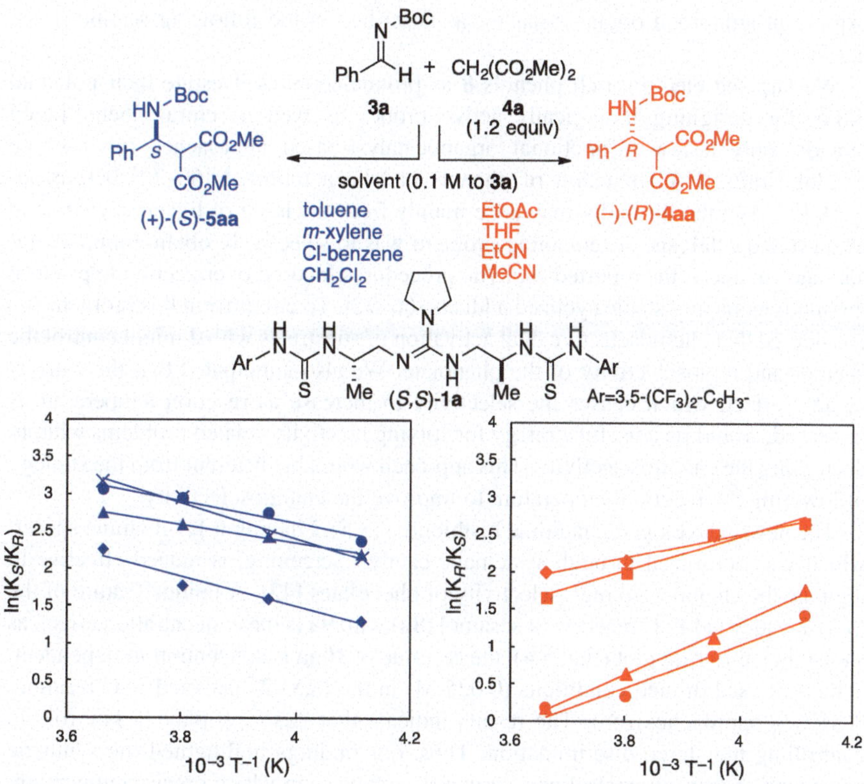

Fig. 5.3 Eyring plots of ln[(100 + % ee)/(100 − % ee)] vs. 1/T for (*S*,*S*)-**1**-catalyzed Mannich-type reactions in various solvents

[a]Blue color: (*S*)-Selective reactions in toluene (circle: $R^2 = 0.949$), *m*-xylene (cross: $R^2 = 0.983$), chlorobenzene (triangle: $R^2 = 0.992$), and CH_2Cl_2 (diamond: $R^2 = 0.956$)

[b]Red color: (*R*)-Selective reactions in EtOAc (circle: $R^2 = 0.967$), THF (triangle: $R^2 = 0.998$), EtCN (square: $R^2 = 0.955$), and MeCN (diamond: $R^2 = 0.981$)

Table 5.2 Differential activation parameters obtained in (*S*,*S*)-**1a** catalyzed Mannich-type reactions using various solvents

Solvent	$\Delta\Delta H^\ddagger$ (kJ mol^{-1})	$\Delta\Delta S^\ddagger$ (J mol^{-1} K^{-1})
Toluene	+15.1	+ 82.3
m-xylene	+19.9	+ 99.6
Cl-benzene	+ 9.99	+ 59.6
CH_2Cl_2	+16.8	+ 79.7
EtOAc	−24.0	− 87.3
THF	−29.4	−106
EtCN	−19.4	− 57.2
MeCN	−14.6	− 37.2

explore programmed organocascades, as described in the following section (Sect. 5.2.3).

We targeted electron-rich phenols **8** as pronucleophiles. Despite their potential utility for designing biologically active probes as well as chiral phenol-based ligands, only a few bifunctional organocatalyst-based approaches for 1,4-type Friedel–Crafts (F–C) reaction of phenols had been reported [66–72] before our work [42–44]; the difficulty may arise mainly from the issue of ligand exchange in metal-based catalysis. A remaining problem was to selectively obtain mono-*ortho*-alkylated adducts; the reported catalytic procedures favored overreaction to produce thermodynamically stable cyclized adducts [66–72]. To circumvent this problem, we considered that chemoselective dual activation of substrates with **2** might control the chemo- and regioselectivity of the phenolate. We also anticipated that the entropy ($\Delta\Delta S^{\ddagger}$) effect, which causes the selectivity to increase as reaction temperature is increased, would be a useful strategy for solving reactivity-related problems without decreasing the enantioselectivity. This approach would be different from the strategy of lowering the reaction temperature to improve the enantioselectivity.

The newly developed guanidine/bisthiourea (*S*,*S*)-**2** having a 1,3-diamine spacer, which was optimized through systematic catalyst screening, is indeed effective to regulate the chemo- and regioselectivity of phenolates [42]. A unique feature of the (*S*,*S*)-**2**-catalyzed F–C reaction of sesamol (**8a**) with **9a** is the concentration effect, as shown in the Eyring plot (Fig. 5.4); the ee value of **10aa** is concentration-dependent. When we used diluted conditions (0.025 M) in the (*S*,*S*)-**2**-catalyzed F–C reaction, $\Delta\Delta H^{\ddagger}_{S-R}$ approaches zero. The results indicate that $\Delta\Delta S^{\ddagger}_{S-R}$ plays a key role in controlling the stereo-discrimination. Thus, our findings highlighted the synthetic advantage of entropy-controlled asymmetric catalysis, in which precise temperature control is not required to attain maximum enantioselectivity.

Structure-catalytic activity relationship studies using structural variants of (*S*,*S*)-**2**, shown in Scheme 5.3, demonstrated the critical roles of both catalytic active sites and the chiral spacer in (*S*,*S*)-**2** [43]. First, control experiments using tris-thiourea **11** and guanidine **12** indicated that guanidine and thiourea in (*S*,*S*)-**2** are essential for the reaction. Second, guanidine/bisthiourea **1b** having a 1,2-diamine spacer showed different characteristics from catalyst **2** bearing a 1,3-diamine spacer. In (*S*,*S*)-**1b**-catalyzed reaction (*R*)-**10aa** was a major product, while (*S*,*S*)-**2** afforded (*S*)-**10aa**. Furthermore, when we used **1b** as a catalyst, the yield and ee were both drastically decreased. In the case of (*S*,*S*)-**1b**, the enantioselectivity increased as the reaction temperature was lowered. These results provide substantial evidence that the 1,3-diamine spacer is critical for reactivity, selectivity, and the unique entropy effect in the catalytic F–C reaction of **8a** with **9a**.

The observed linear relationship between the ee of the catalyst **2** and the ee of F–C adduct **10aa**, regardless of the concentration, also suggested that the stereoselectivity in the F–C reaction using 1,3-diamine tethered **2** is controlled by the monomeric species [43].

The entropy effect observed in the F–C reaction of electron-rich phenols **8** and nitroolefin **9a** is fairly general. As shown in Scheme 5.4, (*S*,*S*)-**2** promotes the 1,4-type F–C reactions of sesamol (**8a**), 2-naphthol (**8b**), and 1-naphthol (**8c**) with

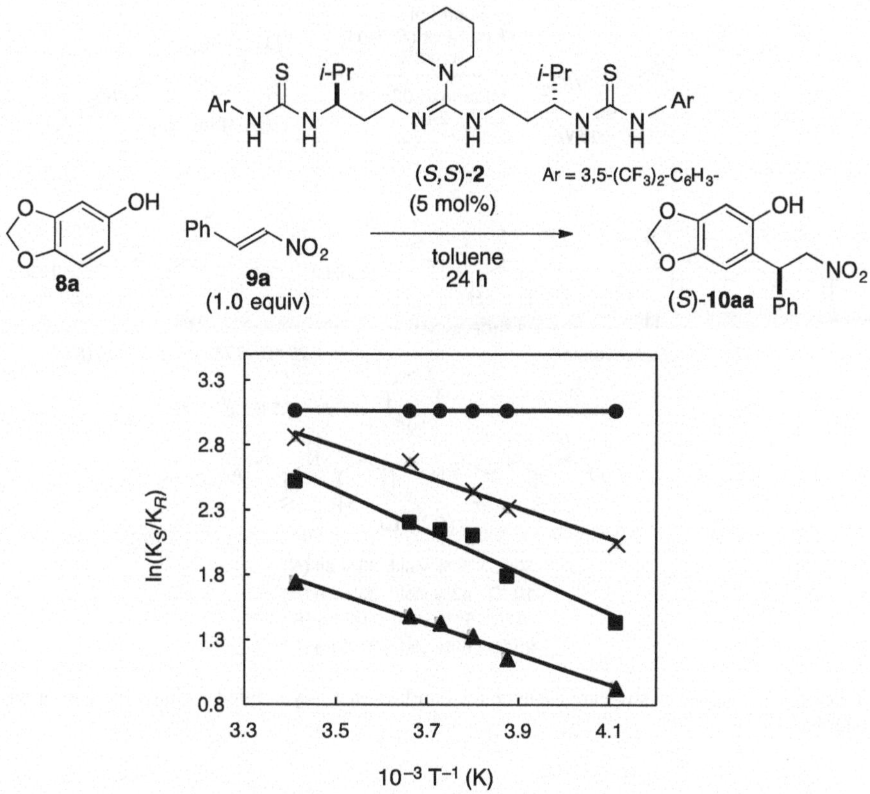

Fig. 5.4 Eyring plots of ln[(100 + % ee)/(100 − % ee)] vs. 1/T for (*S*,*S*)-**2**-catalyzed F–C reactions at a variety of concentrations; 0.025 M (circle), 0.05 M (cross), 0.1 M (square) and 0.2 M (triangle)

a range of nitroolefins **9** with 82–94% ee. Most importantly, these reactions take place at 20 °C, and only trace amounts of by-products were detected [42].

5.2.3 Programmed Organocascades

Programmed cascade catalysis (cycle-specific catalysis) is an attractive strategy to selectively access distinct classes of chiral product, depending on the catalytic system employed [72, 73]. In this section we will describe our efforts to use (*S*,*S*)-**2** in programmed organocascades by employing two distinct reaction conditions [74].

The idea for the development of programmed organocascades came from an early example of the catalytic asymmetric 1,4-type F–C reaction of **8** and **9** reported by Chen [66], who found that overreaction occurred to generate the dimeric dihydronaphthofurans **13** in the presence of cinchona-based bifunctional thiourea

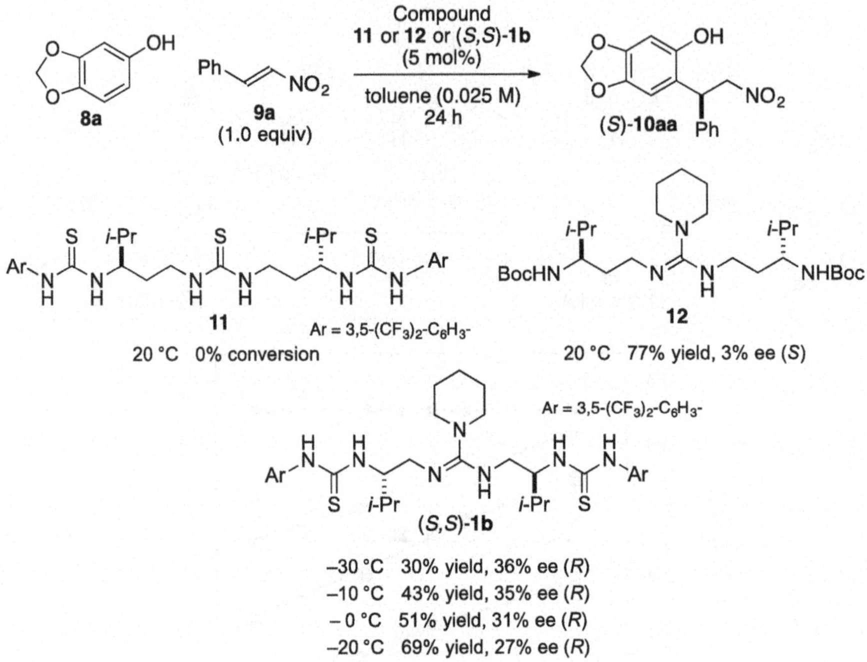

Scheme 5.3 Studies on structure-catalytic relationships in 1,4-type Friedel–Crafts reaction of **8a** with **9a**

catalyst, even when the reaction temperature was decreased to −50 °C. We postulated that if F–C adducts **10** could be catalytically stimulated in the chiral environment upon completion of the first 1,4-type F–C reaction using (*S*,*S*)-**2**, subsequent dimer formation (**10** to **13**) would take place. Encouraged by the developed solvent-dependent enantioswitching, as described in Sect. 5.2.1, we initially examined whether the solvent effect could be used to develop the 1,4-type F–C reaction/dimer formation sequence in the presence of (*S*,*S*)-**2**. However, no rate acceleration for dimer formation starting from **10ba** (88% ee) was observed, although we used various solvents. Instead, we found that potassium carbonate (50 mol%) in the presence of (*S*,*S*)-**2** can trigger formation of the corresponding dihydrofuranyl-hydroxyamines **13ba** with single-flask operation (Scheme 5.5). Most importantly, (*S*,*S*)-**2** induces kinetic resolution [(*S*,*S*)-**13ba** vs. *meso*-**13ba**]; the ee value of (*S*,*S*)-**13ba** is higher than that of (*S*)-**10ba**. On the other hand, the reactivity was drastically reduced in the absence of (*S*,*S*)-**2**, resulting in recovery of (*S*)-**10ba** (65%). Furthermore, in the absence of (*S*,*S*)-**2**, the ee value of both (*S*,*S*)-**13ba** and the recovered-(*S*)-**10ba** showed no marked change, suggesting that (*S*,*S*)-**2** participates in the catalyst-controlled diastereoselective bond formation. This is a remarkable example of single chiral catalyst-based sequential asymmetric catalysis in a single flask. The observed kinetic resolution in the dimer formation is significantly different from typical substrate-controlled diastereoselective reactions that rely on the chirality in the first adduct.

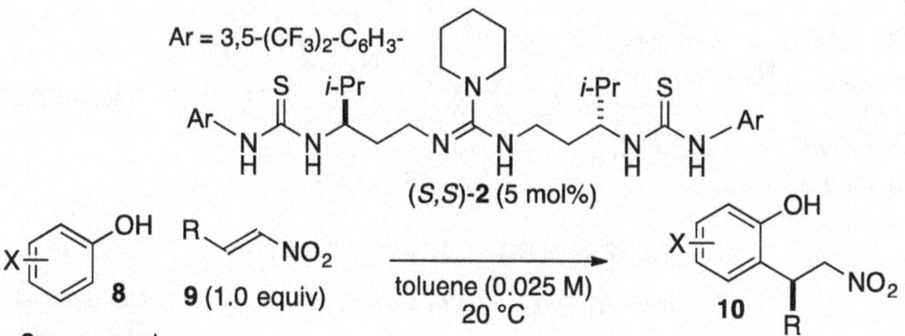

8a: sesamol
8b: 2-naphthol
8c: 1-naphthol

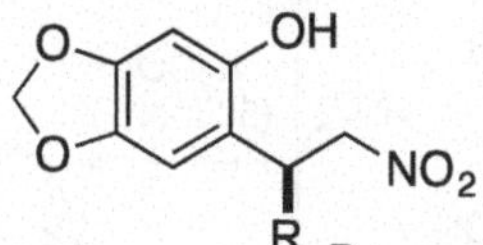

R			
Ph	9 h	97%	91% ee
4-Cl-Ph	12 h	99%	90% ee
4-MeO-Ph	12 h	91%	87% ee
2-naphthyl	12 h	99%	94% ee
2-thienyl	9 h	98%	90% ee

R			
2-furyl	12 h	99%	88% ee
c-C_6H_{11}-	48 h	84%	91% ee
$CH_3(CH_2)_5$-	24 h	66%	90% ee
$Ph(CH_2)_2$-	24 h	88%	89% ee

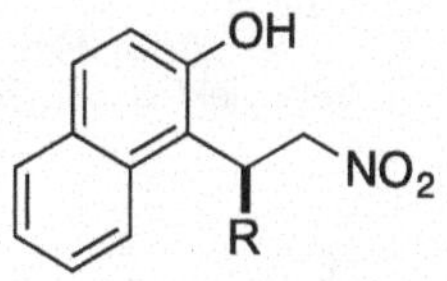

R			
Ph	12 h	95%	88% ee
2-naphthyl	12 h	84%	84% ee
2-thienyl	12 h	99%	84% ee
$CH_3(CH_2)_2$-	24 h	90%	93% ee

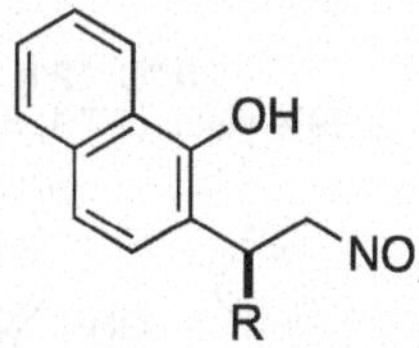

R			
Ph	24 h	77%	82% ee
4-Cl-Ph	24 h	78%	83% ee

Scheme 5.4 Catalytic *ortho*selective 1,4-type F–C alkylation of **8** with **9** utilizing (*S*,*S*)-**2**

As shown in Scheme 5.6, the protocol using potassium carbonate (50 mol%) as an additive is useful for the one-flask synthesis of (*S*,*S*)-**13** starting from phenol **8** and nitroolefins **9**. These results proved that both chiral phenols **10** (Sect. 5.2.2) and *N*,*N*-bis-dihydrofuranyl-hydroxyamines **13** (Sect. 5.2.3) can be synthesized in the presence of (*S*,*S*)-**2** by suitably selecting the reaction conditions.

Before closing this section, we would like to mention that the developed organocatalytic methodologies have been applied to total synthesis of several natural products, such as (+)-rishirilide B [75], (+)-*trans*-dihydrolycoridine [76], (+)-gracilamine [77], and (+)-linoxepin [51]. In addition, our recent density functional theory (DFT) calculations [78] have identified a key *S*-shaped transition structure

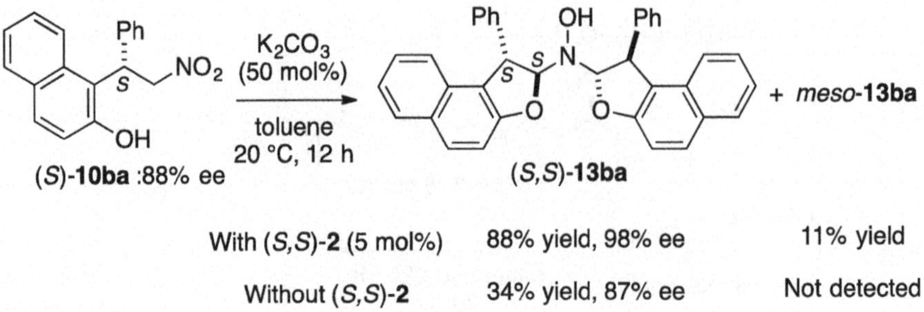

Scheme 5.5 Catalytic synthesis of (*S*)-**13ba** in the presence and absence of (*S*,*S*)-**2**

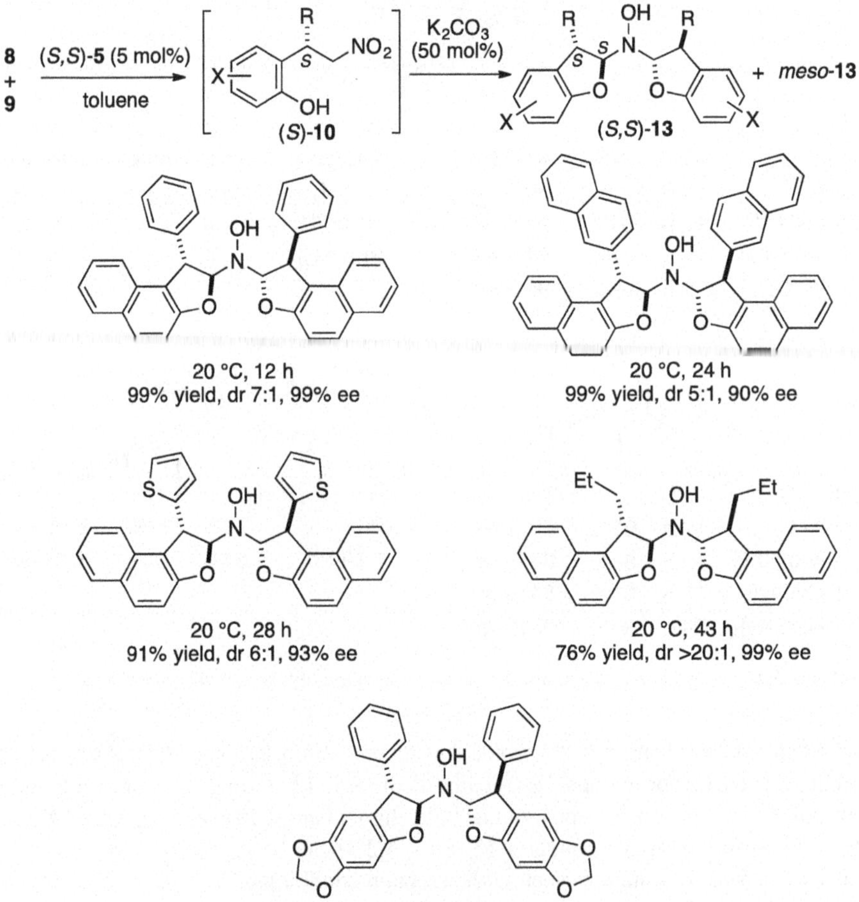

Scheme 5.6 One-flask synthesis of (*S*)-**13** from **8** and **9**

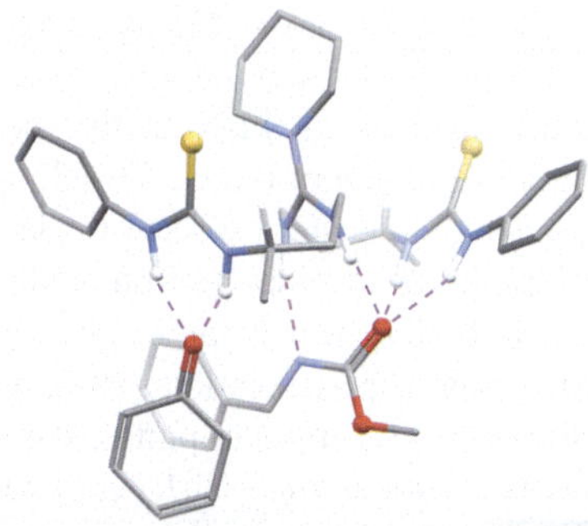

Fig. 5.5 Simplified transition-state model for 1,2-type Friedel–Crafts reaction of phenol with imine, calculated at the level of B3LYP/6-31G(d) [44]

that can explain how the conformationally flexible guanidine-bis(thiourea) controls the highly enantioselective bond-forming reactions [44, 51]. In this *S*-shaped model, shown in Fig. 5.5, all of H-bond donors (i.e., guanidinium and ureas) participate in interacting with the reaction components. Considering that linear conformation of the guanidine/bisthiourea is predominant in the ground state, the dynamic structural change to form the *S*-shaped transition state in response to substrate binding may be the key to attaining a positive value of $\Delta\Delta S^{\ddagger}$.

5.3 Centrochiral Transition-Metal Catalysts

Transition-metal catalysts having *d*-orbitals (*dxy*, *dyz*, *dzx*, dx^2–y^2, and dz^2) can adopt many geometries owing to splitting of the *d*-orbitals. Because the *d*-orbitals can participate in activation of functionalities in the reaction components, it is clear that transition-metal catalysts work according to a different mechanistic principle from organocatalytic activation modes that utilize *s*- and *p*-orbitals [79]. The geometrical arrangement of the metal complex can also vary depending upon the number and type of ligands bonded to the metal center and the coordination preference of the central atom. Thus, selection of suitable combinations of metal (s) and chiral ligand(s) is a general strategy in the development and optimization of asymmetric transition-metal catalysis.

Metal enolates are important active species for the development of asymmetric catalysis [80]. Historically, metal enolates have often been prepared by the treatment of carbonyl compounds with a relatively strong Brønsted base. On the other hand, Sodeoka's group has reported that late transition-metal complexes act as acid-base catalysts to generate metal enolates even under acidic reaction conditions, where the electrophile can be also activated by the proton source [81]. Catalytic transformations developed on this mechanistic basis [82] include Michael [81], halogenation [83–85], Mannich-type [86, 87], and aldol-type reactions [88] of 1,3-dicarbonyl compounds, as well as Michael additions [89, 90] and halogenations [91] using α-ketoesters (1,2-dicarbonyl compounds).

In this section, we (MS and YS) will introduce our recent efforts to integrate metal enolates and 1,3-dipoles as unconventional electrophiles [92, 93] under these mechanistic paradigms. Importantly, the regioselectivity of these transformations is different from that of typical [3 + 2] cycloadditions; most catalytic asymmetric [3 + 2] cycloadditions are restricted to electron-deficient olefins [94]. There are only a few examples of inverse-electron-demand (IED) cycloadditions using *electron-rich olefins* [94–103], which can be also categorized as type III cycloadditions according to Sustmann's classification [104]. In those cases, enol ethers or silyl ethers were utilized as 1,3-dipolarophiles. On the other hand, we focused on the use of metal enolates including Ni(II)-enolates [92] and Cu(II)-enolates [93] as formal 1,3-dipolarophiles, leading to the development of the catalytic asymmetric IED [3 + 2] cycloadditions. These catalytic IED [3 + 2] cycloadditions provided efficient access to stereochemically complex 5-membered heterocycles having adjacent stereocenters, in which a unique hemiketal structure is involved. Here we will also present structural characterization of a Ni(II) complex that involves metal centrochirality [105–110], in order to illustrate the relationship between the electronic structure of the Ni(II) complex and its catalytic activity. A plausible synergistic mode of action for ligand-induced metal centrochirality will be also discussed.

5.3.1 [3 + 2] Cycloaddition of α-Ketoester Enolates and (E)-Nitrones

Following the disclosure in 1926 of the X-ray structure of urease [111], which ultimately led to the confirmation that enzymes are pure proteins, numerous structures of Ni-enzymes have been reported. Intriguingly, combinations of natural ligands (i.e., amino acid residues and/or modified amino acids) enable construction of quite distinct chiral environments that involve different coordination geometry and assembly states, thereby affording different functional outcomes [112].

Since the seminal work in 1976 [113], the field of asymmetric nickel catalysis has been extended to include acid-base and redox catalysis [114–117]. Nevertheless, it is still difficult to design asymmetric nickel catalysts due to the complex nature of nickel, which, for example, has multiple oxidation states including Ni(0), Ni(I), Ni (II), Ni(III), and Ni(IV). Depending on the oxidation state of Ni and the ligands used, the electronic configurations of Ni complexes can vary from 16 to 20 electrons.

Focusing on the azophilicity of Ni, we began by exploring new Ni(II)–diamine complexes on the basis of our early work on Ni(II)–enolate chemistry [89]. After optimizing the reaction conditions, we found that the catalytic triad including Ni $(OAc)_2{\cdot}4H_2O$, chiral diamine (*R,R*)-**14**, and $^{i}Pr_2NH$ is effective to selectively promote the [3 + 2] cycloaddition between **15** and **16** with reasonably high regio-, diastereo-, and enantioselectivity (Scheme 5.7) [92].

As shown in Fig. 5.6a, b, the structures of key Ni(II)–diamine complexes **I** (CCDC 1482739) and **II** (CCDC 1482740) were unambiguously identified by

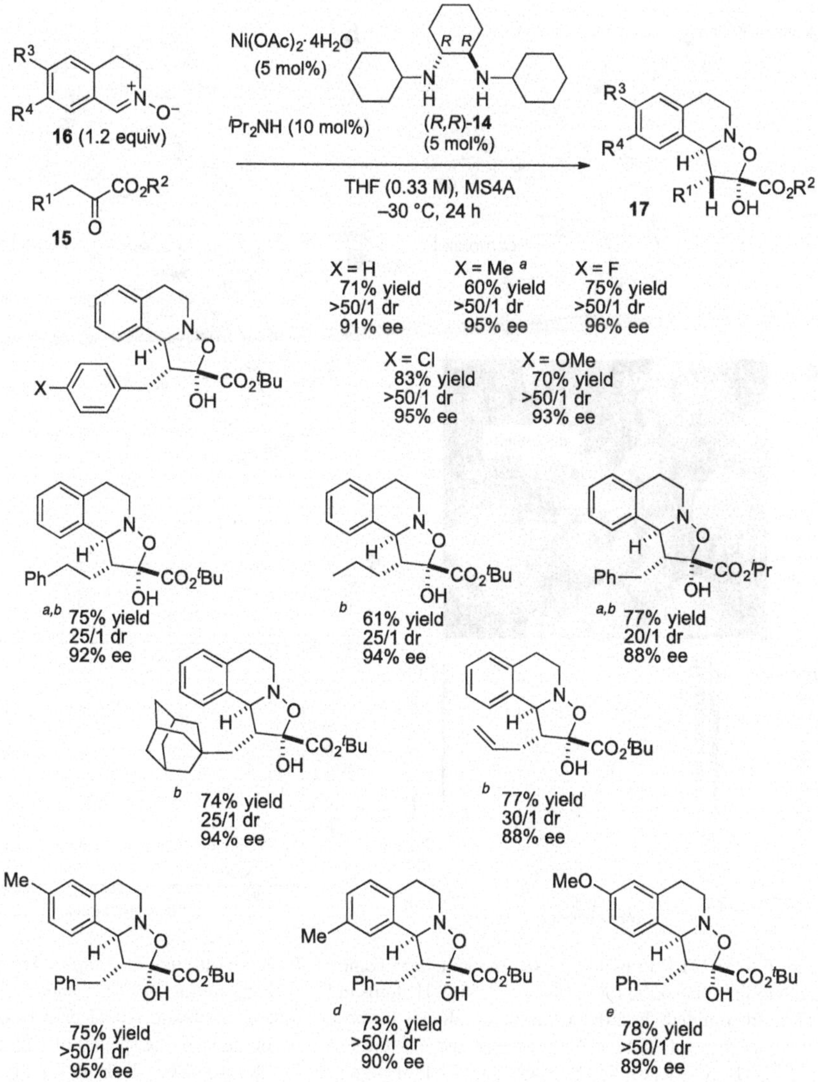

[a]Run for 48 h.
[b] 2.0 equiv. of **16** was used.
[c] Run at a concentration of 1.0 M.
[d] 10 mol% of metal complex and 3.0 equiv of **16** and Et_3N (10 mol%) were used.
[e] 10 mol% of metal complex and 3.0 equiv of **16** were used.

Scheme 5.7 Ni(II)-catalyzed formal [3 + 2] cycloaddition using a variety of **15** and **16**
[a]Run for 48 h
[b]2.0 equiv. of **16** was used
[c]Run at a concentration of 1.0 M
[d]10 mol% of metal complex and 3.0 equiv. of **16** and Et_3N (10 mol%) were used
[e]10 mol% of metal complex and 3.0 equiv. of **16** were used

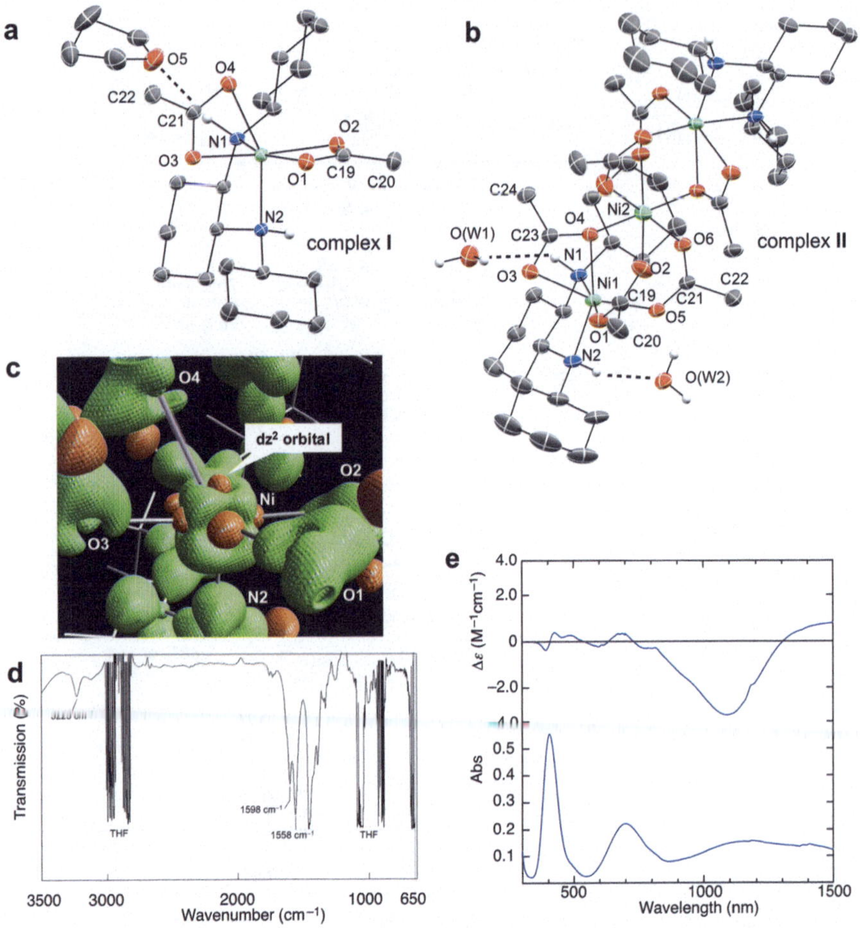

Fig. 5.6 ORTEP drawings of (**a**) mononuclear complex **I** and (**b**) trinuclear complex **II** (50% probability ellipsoids; for the sake of clarity, H atoms in **I** and **II** and minor disorder components of THF are omitted; hydrogen bonds are shown by broken lines). (**c**) Static model density of **I**; contours drawn at +0.2 e Å^{-3} in green and at −0.2 e Å^{-3} in orange. (**d**) IR spectra of **I** in THF (0.017 M). (**e**) Electronic absorption and ECD spectra (300–1500 nm) of **I** in THF (0.017 M)

means of X-ray analyses [92]. We would like to emphasize that the simple diamine ligand **14** plays a key role in providing a well-defined catalytic active site, featuring distorted octahedral architecture and (*Λ*)-chirality at the Ni(II) center, even though the diamine **14** has generally been overlooked as a chiral ligand [118]. Complex **I**, which can be obtained by crystallization from CH_2Cl_2, is a monomer (*Λ*)-Ni$(OAc)_2$**14**. On the other hand, complex **II**, which can be obtained by crystallization from *n*-hexane, is a μ-acetate-bridged trinuclear Ni(II) complex: (*Λ*,*Λ*)-[$Ni_3(\mu\text{-}OAc)_4(OAc)_2$**14**$_2$].

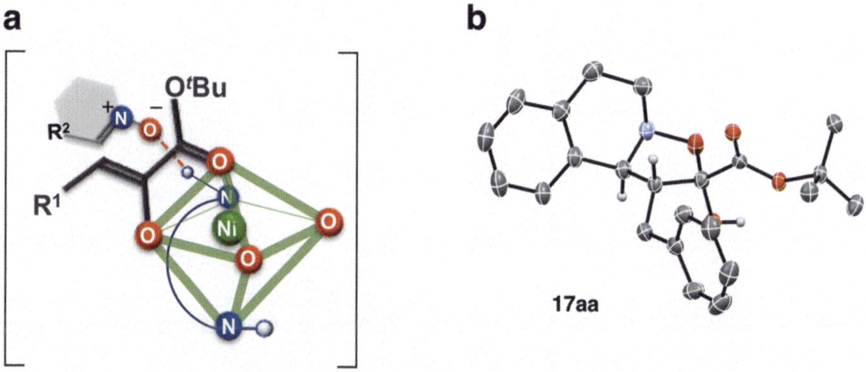

Fig. 5.7 (**a**) Stereochemical model showing enolate formation and outer-sphere hydrogen-bonding. (**b**) ORTEP structure of **17aa** (50% probability ellipsoids; H atoms except for those on chiral carbons and alcohol are omitted for the sake of clarity)

As we can see in the structure of mononuclear Ni(II) complex **I** (Fig. 5.6a), the initial C_2-symmetry of **14** is desymmetrized through the formation of Ni(II) complex **I**. For example, the Ni(II)–N(2) bond [2.109(1) Å] is significantly longer than the Ni (II)–N(1) bond [2.091(1) Å]; in the figure, Ni(II)–N(2) lies on the z-axis (vertical direction). An important finding is the existence of the intermolecular H-bonding interaction of N(1)–H with the remaining THF in the crystal, showing that the N(1)–H can act as a H-bond donor; this finding can explain how the outer-sphere hydrogen-bonding serves to activate (*E*)-nitrone **16** [103, 119–121] in this catalysis (See, Fig. 5.7a). Another striking feature in the distorted octahedral Ni(II) complex **I** is the Ni(II)–O(4) distance [2.302(2) Å] at the pseudoapical position. The angle of N(2)–Ni(II)–O(4) [155.99(5)°], which is far from 180°, highlights the unique distorted architecture. Considering other Ni(II)–O and Ni(II)–N distances [2.045 (1)~2.140(1) Å] in this complex, the distortion and elongated Ni(II)–O(4) bond suggest that one acetate that binds to the Ni(II) center in the apical-equatorial coordination mode is more labile than the other acetate that coordinates to the Ni (II) center in the equatorial-equatorial coordination mode (Fig. 5.6a). Dissociation of the labile acetate from the Ni(II) center in mononuclear complex **I** is also suggested by the analogy to mononuclear Ni(II) complex in the trinuclear Ni(II) complex **II** (Fig. 5.6 b), wherein acetate bridges link the distorted units at the terminal position with symmetric octahedral Ni(II) at the central part in **II**.

Catalytic activity tests using complexes **I** and **II** under optimized reaction conditions for [3 + 2] cycloaddition of **15a** and **16a** revealed that the mononuclear Ni (II) complex **I** provided superior ee (Table 5.3, entry 1: 90% ee), compared with the trinuclear **II** (Table 5.3, entries 2 and 3: 60% ee). We also found that ^{i}Pr$_2$NH (10 mol %) is an effective additive to improve the moderate ee values obtained with trinuclear catalyst **II** [Table 2, entry 2, without ^{i}Pr$_2$NH: 60% ee vs. entry 4, with ^{i}Pr$_2$NH: 90% ee]. Although the mechanism through which ^{i}Pr$_2$NH improves the

Table 5.3 Catalytic activities of complex **I** and **II**

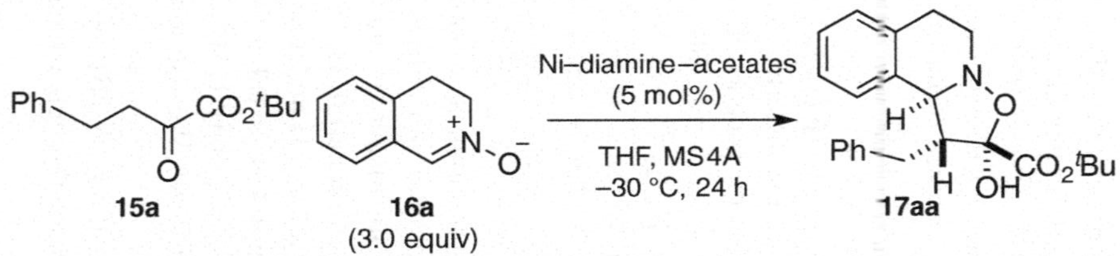

Entry	Ni complex	Additive (mol%)	Yield (%)	dr	Ee (%)
1	Mononuclear **I**	None	62	>50/1	90
2[a]	Trinuclear **II**	None	52	>50/1	60
3[b]	Trinuclear **II**	None	78	>50/1	60
4	Trinuclear **II**	$^{i}Pr_2NH$	83	>50/1	90

[a] 5 mol% of **II** (Ni-based) was used
[b] 5 mol% of **II** (MW-based) was used

enantioselectivity is unclear at this stage, this finding suggests that the reaction pathway and/or the chiral environment constructed by Ni(II) acetates with chiral diamine can be tuned by the addition of *achiral* amine [122, 123].

Considering the unique distorted octahedral structure and catalytic activity of mononuclear complex **I**, we further characterized the bonding nature around the Ni (II) center, a plausible catalytic active site. We used electron density distribution (EDD) analysis, enabling us to visualize the valence electron density topology by employing an aspherical atomic model [124–126]. As shown in Fig. 5.6c, a three-dimensional plot of the static deformation density of **I** can visualize the shapes of *d*-orbitals at the Ni(II) center, wherein the electron-rich region is shown in green, while the electron-poor region is shown in orange. As regards the Ni(II) and ligand interactions, we can see that the green regions that show the lone pairs [O(1), O (2), and O(3)] and N [N(1) and N(2)] are directed toward the electron-deficient *d*-orbitals shown in orange; these are typical interactions that can be seen in coordination bonds. In marked contrast, the lone pair of O(4) (green) is directed toward the electron-rich region at Ni(II) (green) as highlighted in pink in Fig. 5.6c. These observations indicate that the distortion around the Ni(II) center, induced by the sterically demanding ligand **14**, makes the dz^2 orbital partially "naked"; the labile acetate ligand at the pseudoapical position is weakly coordinated to the Ni(II) center via electrostatic interaction, in accordance with the experimental findings. Furthermore, the density at the bond critical point (BCP) obtained by the EDD analysis provides quantitative insight into the bond strength. The smaller value of the density of the Ni(II)–O(4) bond [0.260(2) e $Å^{-3}$] compared with the other coordination bonds [Ni(II)–O(1); 0.431(2), Ni(II)–O(2); 0.480(2), Ni(II)–O(3); 0.494(2), Ni(II)–N(1); 0.559(3), and Ni(II)–N(2); 0.550(3) e $Å^{-3}$] shows that O(4) binds weakly to the Ni(II) center. The density at the BCP of the N(1)–H···O(5) [0.084(10) e $Å^{-3}$] is also consistent with the topological parameters [d(H···O): 2.200, d(N···O): 3.155 (2) and α(N–H···O): 159.20] of typical H-bonds [125]. Thus, EDD analysis of the chiral metal complex in the ground state can provide a starting point for discussing how the catalyst initiates the reaction, based on the electron density topology of the metal complex, with quantitative insights from the BCP.

In order to further link the electronic structure of **I**, obtained by crystallographic analyses, with the catalytic activities, it is also important to characterize the structure in the solution state at the same concentration as used in the developed catalysis (Fig. 5.6d, e). From a series of structural analyses, we concluded that the Ni(II)–diamine–acetate **I** retains pseudo-octahedral structure in the solution state. For example, the unique carboxylate shifts (1598 and 1558 cm^{-1}) in the infrared (IR) spectrum indicate the existence of two different acetates, which bidentately interact with the Ni(II) center (Fig. 5.6d) [127]. The N–H stretching vibration on the ligand was also observed at 3225 cm^{-1} [128]. As shown in Fig. 5.6e, bottom, the electronic absorption spectrum of **I** showed typical broad bands of pseudo-octahedral structure of Ni(d^8) (408, 753 and 1182 nm). In the electronic circular dichroism (ECD) analysis of **I**, the intensities in the near-IR region are stronger than in the UV-Vis region (Fig. 5.6e, top). The simulated IR and ECD spectra, obtained by DFT calculations [129] on the distorted Ni(II) complex **I** at the level of UM06/6-

311G(d,p) (SDD for Ni), showed good agreement with the experimental results [130].

In Fig. 5.7, a plausible stereochemical model that accounts for the synergy of the H-bonding activation and enolate formation is shown. In the initial step of enolate generation, we assumed that the labile acetate anion acts as a Brønsted base to cooperatively deprotonate the α-ketoester with the electron-deficient dz^2 orbital at the Ni(II) center, giving the (*Λ*)-Ni(II)–enolate. On the basis of X-ray (Fig. 5.6a) and EDD analyses of **I** (Fig. 5.6c), we proposed the involvement of a transient Ni(II)–enolate, wherein the ester carbonyl in **15** coordinates to Ni(II) at the pseudoapical position. Notably, the coordination pattern in the octahedral Ni(II)–enolate complex illustrated here is different from that of the Ni(II)–enolate reported by Evans [131, 132], in which the enolate interacts with the Ni(II) center in the equatorial-equatorial coordination mode relative to the diamine ligand. The observed H-bonding ability, supported by a series of spectroscopic analyses, appears to have a key role in fixing the nitrone to enhance the bond-forming reaction with high diastereo- and enantioselectivity. The model shown in Fig. 5.7a can explain the obtained absolute stereochemistry of the [3 + 2] adduct (CCDC 1482737).

5.3.2 [3 + 2] Cycloaddition of α-Ketoester Enolates and Nitrile Oxides

To broaden the substrate scope with respect to 1,3-dipoles of our catalytic strategy using the metal–diamine–acetate complex, we planned to employ the nitrile oxide **18**, which can be prepared in situ from the *N*-hydroxyimidoyl chloride **19**. We became interested in the catalytic synthesis of new isoxazolines **20** using a transient metal enolate and nitrile oxide as an attractive yet challenging next step. As illustrated in Scheme 5.8, the difficulty in selectively synthesizing **20** using nitrile oxide and α-ketoester enolate arises from the multiple reaction pathways that are available, including dimerization of nitrile oxide (path A) [133], homo-aldol reaction of the α-keto ester **15** (path B) [134], hetero-[3 + 2] cycloaddition on the ketone in **15** (path D) [135], and *O*-trapping of the metal enolate (path E). Because the HOMO–LUMO gaps between (*E*)-nitrone **16a** (HOMO: −7.09 eV and LUMO: −0.831 eV) and nitrile oxide **18a** (R^2 = Ph: HOMO: −8.12 eV and LUMO: −0.816 eV) are not so large, judging from our preliminary DFT calculations at the level of M06-2X/6–311 + G(d) [130], we envisioned that tuning the catalytic activity by changing the central metal and ligand on the basis of the stereochemical model discussed in Sect. 5.3.1 (shown in square brackets in Scheme 5.8) would be an effective strategy to optimize the reaction conditions for selectively forming the desired isoxazolines **20**.

After thorough optimization studies, we identified two key factors in this catalytic transformation. First, the use of $Cu(OAc)_2{\cdot}H_2O$ in ^{i}PrOH [90] can eliminate the *O*-addition adduct (path E), the formation of which was a major problem when we used $Ni(OAc)_2{\cdot}4H_2O$. Second, the diamine **21a** having a mesityl group or **21b** bearing

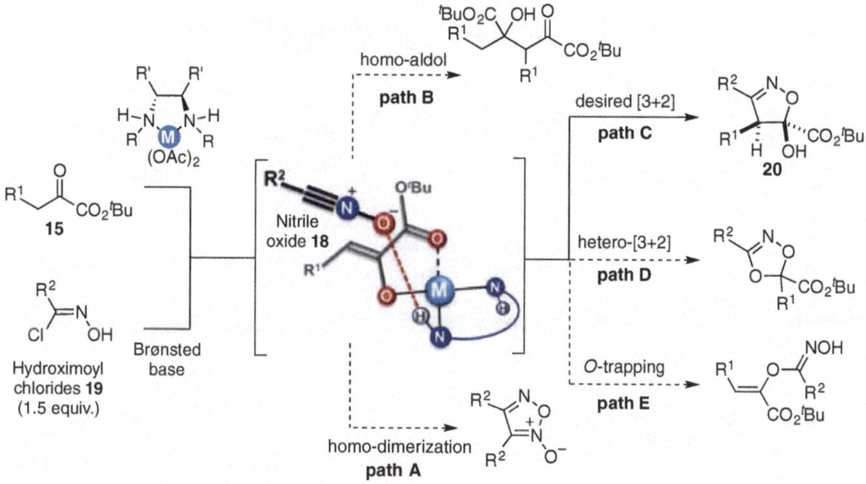

Scheme 5.8 Possible reaction pathways in the reaction of α-ketoesters **15** and nitrile oxides **18** prepared from *N*-hydroxyimidoyl chlorides **19**

2-CF_3 is best to attain high enantioselectivity in the formal [3 + 2] cycloaddition starting from **15** and **19** in the presence of a stoichiometric amount of Et_3N [93]. Mesityl ligand **21a** displayed lower levels of enantioselection when 2-substituted aryl nitrile oxides were used, while the 2-CF_3 analogue **21b** provided high selectivity with these substrates. By exploiting such fine ligand tuning, a versatile array of nitrile oxides and α-ketoester enolates could participate in the cycloaddition, providing novel 5-hydroxy-2-isoxazolines in high chemical yield with high levels of diastereo- and enantioselectivity (Scheme 5.9).

During the optimization process, we also observed unique metal-dependent enantioswitching. The Cu(II)–(*R,R*)-diamine complex mainly afforded (*S,S*)-**20**, while the Ni(II)–(*R,R*)-diamine complex gave (*R,R*)-**20** as the major product [3, 4, 54–57]. The enantioselectivity trend [Cu(II): (*S,S*)-selective vs. Ni(II): (*R,R*)-selective] is general, regardless of the substituents (R and R') on the (*R,R*)-diamine and the solvent used. Database search (using Cambridge Structural Database) suggested that the Cu(II)–diamine–acetate complexes would have distorted square-pyramidal geometry (ex. CCDC 299401 [136], CCDC 665454 [137] and CCDC 894524 [138]). As can be seen in the example (CCDC 299401) in Fig. 5.8 a, the Cu(II)–O (2) [2.78(1) Å] is elongated at the pseudoapical position. This distance is markedly longer than other Cu(II)–O and Cu(II)–N bonds [1.907(8)–2.054(8) Å]. By analogy with the Ni(II)–enolate described in Sect. 5.3.1 [92], these results suggest that the apical acetate can act as a Brønsted base to form a Cu(II)–enolate, in which the ester carbonyl in **15** coordinates to Cu(II) at the top of the pyramidal architecture. Considering that the N–H on the diamine ligand in the Cu(II)–diamine complex can also act as an H-bond donor [139], the existence of outer-sphere H-bonding to hold the nitrile oxide close to the Cu(II)–enolate can reasonably explain the obtained absolute stereochemistry of **20A** (CCDC 147644).

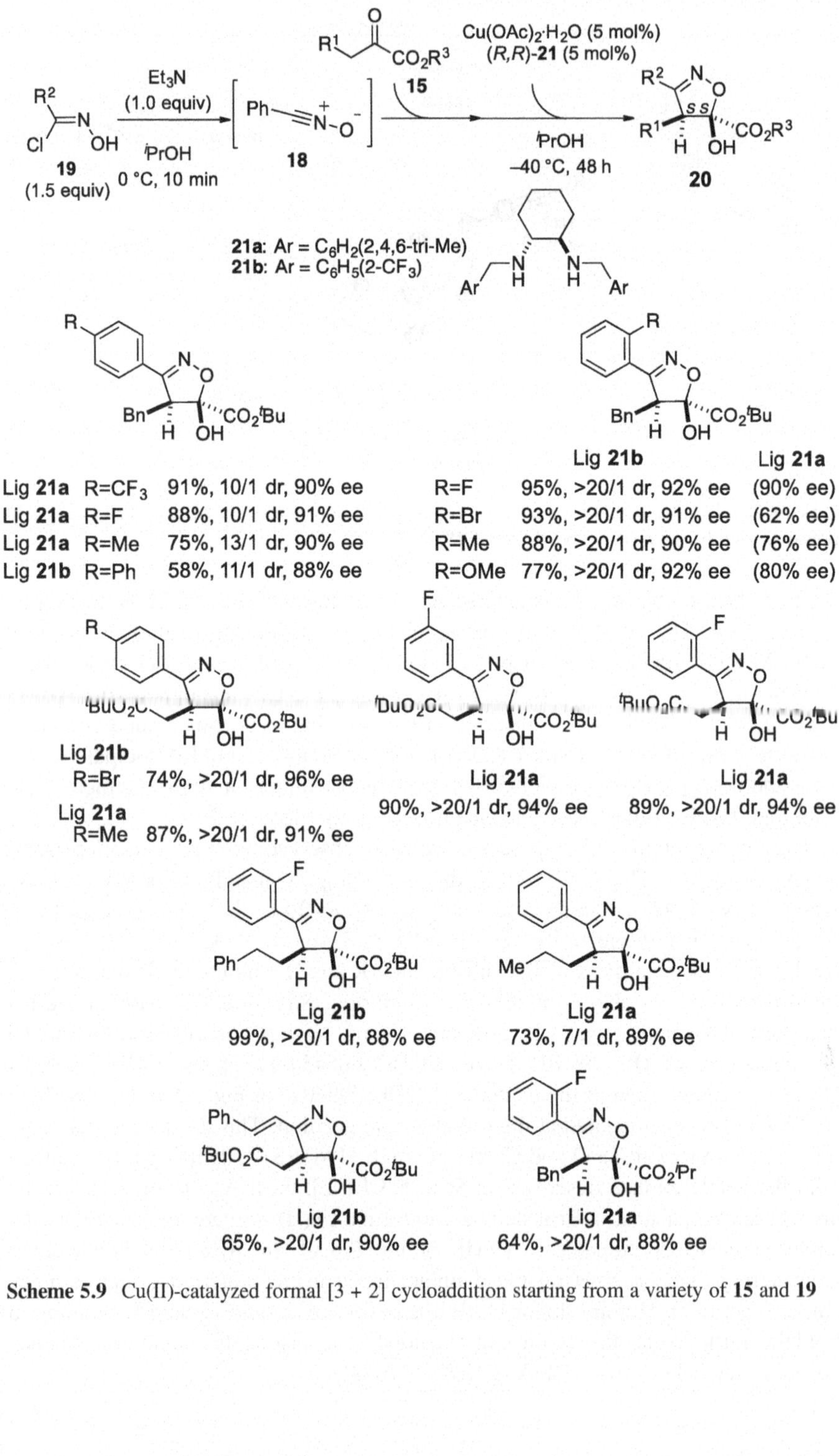

Scheme 5.9 Cu(II)-catalyzed formal [3 + 2] cycloaddition starting from a variety of **15** and **19**

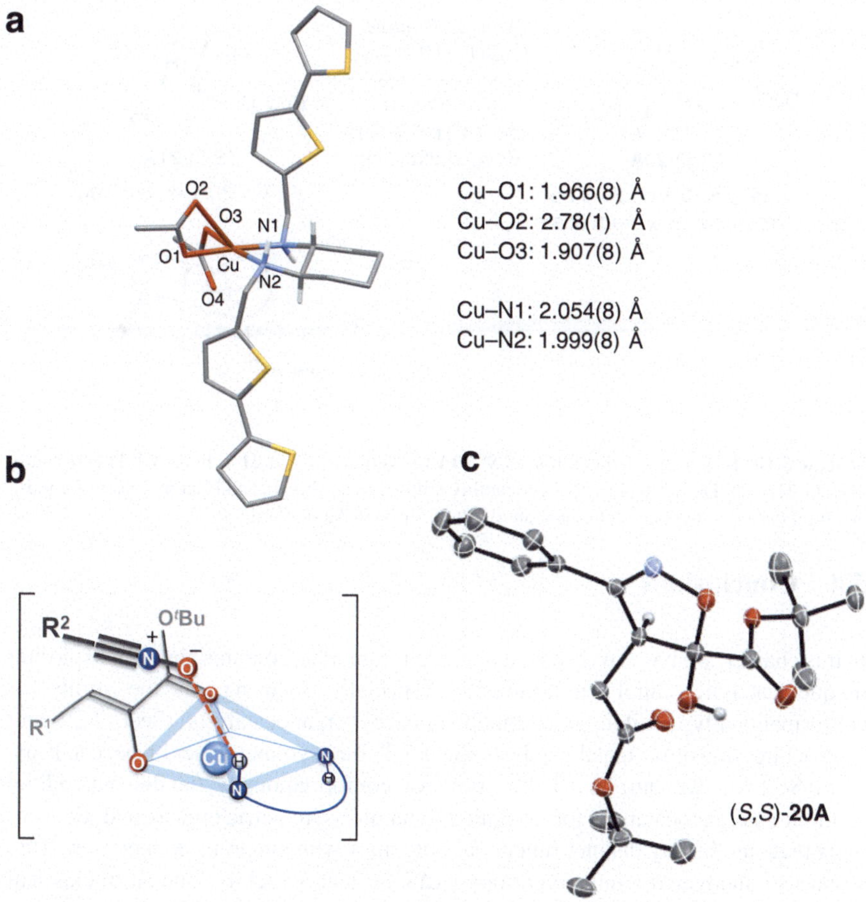

Fig. 5.8 (**a**) Mirror-inverted structure of CCDC 299401 (made by GaussView). (**b**) The stereochemical model involving Cu(II)–enolate formation and outer-sphere hydrogen-bonding. (**c**) ORTEP structure of (*S*,*S*)-**20A** (50% probability ellipsoids; H atoms except for those on chiral carbons and alcohol are omitted for the sake of clarity)

The developed catalytic protocol can apply to synthesize (*S*,*S*)-**20A** on a gram scale using 2.5 mol% Cu(II) catalyst loading (Scheme 5.10). The C=N bond in (*S*,*S*)-**20A** (90% ee) was reduced by treatment with $NiCl_2/NaBH_4$ [140], facilitating construction of the γ-lactam **21A** containing three vicinal stereocenters (2.5 mmol scale). The lactam **21A** was slightly enriched by recrystallization, and the structure of this product was confirmed by X-ray crystallography.

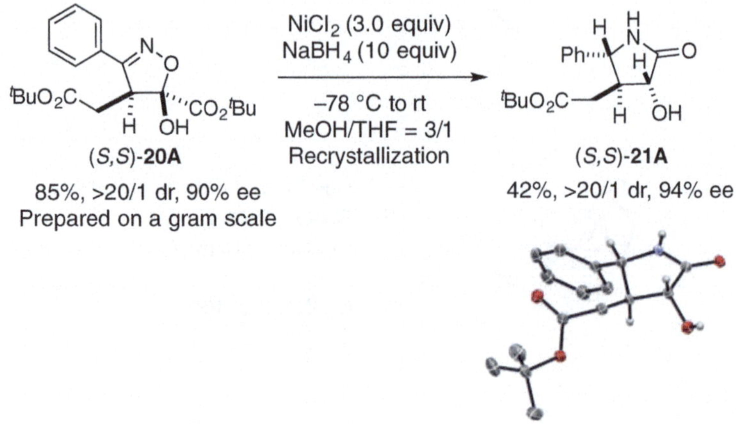

Scheme 5.10 Isoxazoline reduction in (*S*,*S*)-**20A** to synthesize (*S*,*S*)-**21A**. In the ORTEP structure of (*S*,*S*)-**21A** (CCDC 1476445), 50% probability ellipsoids are shown, and H atoms except for those on chiral carbons, alcohol, and amide are omitted for the sake of clarity

5.4 Conclusions

In this chapter, we have overviewed our recent research programs aimed at exploring unique catalytic chiral environments, especially focusing on the utility of conformationally flexible guanidine/bisthiourea organocatalysts (Sect. 5.2) and centrochiral transition-metal catalysts (Sect.5.3) for asymmetric acid-base catalysis.

In Sect.5.2, we introduced the utility of conformationally flexible guanidine/bisthiourea organocatalysts for designing dynamic asymmetric organocatalytic reactions that can display distinct functional outcomes with single-flask operation. That work also uncovered a unique entropy ($\Delta\Delta S^{\ddagger}$) effect, which is contrast to classical asymmetric catalysis that relies on enthalpic ($\Delta\Delta H^{\ddagger}$) binding contributions to regulate the reactivity and selectivity. Although the available activation modes using *s*- and *p*-orbitals in our organocatalysts are limited, the tolerance of the guanidine and bisthiourea units, linked by a carbon spacer, to various conditions provides broad opportunities to tune the reaction conditions. Indeed, we used solvent and potassium carbonate as switching triggers for enantiodivergent reactions and programmed organocascade reactions, respectively.

In Sect. 5.3, we describe our recent efforts to achieve mechanistic insights into the developed IED [3 + 2] cycloadditions of α-ketoester enolates though characterization of the well-defined catalytic active site of Ni(II)–diamine–acetates by means of a series of structural analyses and DFT calculations in the ground state (both in the solid and solution states). The stereochemical model that enables the merging of enolate formation and H-bonding activation is applicable not only to the octahedral framework constructed by Ni(II)–diamine–acetates but also to the pyramidal geometry of Cu(II)–diamine–acetates. These two different catalytic IED [3 + 2] cycloadditions provide efficient access to stereochemically complex 5-membered heterocycles having adjacent stereocenters with high regio-, diastereo-, and enantioselectivity.

Overall, we believe that the continued exploration of less conventional catalyst frameworks based on conformationally flexible chiral spacers and metal centrochirality has vast potential for accessing stereochemically complex architectures. Because fully visualizing the dynamic structural changes of the catalyst and/or the transition state in the reaction media is still difficult, physical organic chemistry approaches, supported by DFT calculations, may be the key to understanding the interplay of structure and selectivity, enabling rational design of even more sophisticated asymmetric catalysts.

It is a great pleasure to thank the many wonderful co-workers who have greatly contributed to our scientific research program. Their names appear in the cited references, but we should especially like to mention Nobuko Takemura, Toshitsugu Iguchi, Kaori Cho, Rika Takagi, Keisuke Takada, Bongki Shin, Shinji Tanaka, Natsuko Horitsugi, Takahisa Yamaguchi, Kohei Kojima, Dr. Koji Yasui, Dr. Masaru Kato, Shogo Hirao, Katsuhiro Nakano, Makoto Sato, Prof. Dr. Masahiro Yamanaka, Genta Nakamura, Dr. Atsuya Muranaka, Dr. Daisuke Hashizume, Prof. Dr. Teruhisa Tsuchimoto, Prof. Dr. Masanobu Uchiyama, Dr. Samuel L. Bartlett, Dr. Peter White, Miki Sawamura, and Prof. Dr. Jeffrey S. Johnson. We also thank Grant-in-Aid for Scientific Research (B)17H02213 and Japan Society for the Promotion of Science (JSPS) KAKENHI Grant number JP18H04277 in Precisely Designed Catalysts with Customized Scaffolding.

References

1. J. Meeuwissen, N.H. Reek, Nat. Chem. **2**, 615–621 (2010)
2. L. Marchetti, M. Leivine, ACS Cat **1**, 1090–1118 (2011)
3. N. Kumagai, M. Shibasaki, Cat. Sci. Technol. **3**, 41–57 (2013)
4. T. Imahori, S. Kurihara, Chem. Lett. **43**, 1524–1431 (2014)
5. A. Fersht, *Structure and Mechanism in Protein Science: A Guide to Enzyme Catalysis and Protein Folding* (Freeman, New York, 1999)
6. S. Hammes-Schiffer, J. Klinman, Acc. Chem. Res. **48**, 899–899 (2015)
7. D.W.C. MacMillan, Nature **455**, 304–308 (2008)
8. E. M. Carreira, H. Yamamoto (eds.), *Comprehensive Chirality* (Elsevier, Oxford, 2012)
9. T.P. Yoon, E.N. Jacobsen, Science **299**, 1691–1693 (2003)
10. Q. L. Zhou (ed.), *Privileged Chiral Ligands and Catalysts* (Wiley-VCH, Weinheim, 2011)
11. A.E. Allen, D.W.C. MacMillan, Chem. Sci. **3**, 633–658 (2012)
12. M. Shibasaki, M. Kanai, S. Matsunaga, N. Kumagai, Acc. Chem. Res. **42**, 1117–1127 (2009)
13. G.R. Desiraju, T. Steiner, *The Weak Hydrogen Bond in Structural Chemistry and Biology* (Oxford University Press, Oxford, 1999)
14. P. M. Pihko (ed.), *Hydrogen Bonding in Organic Synthesis* (Wiley-VCH, Weinheim, 2009)
15. A.G. Doyle, E.N. Jacobsen, Chem. Rev. **107**, 5713–5743 (2007)
16. T. Akiyama, K. Mori, Chem. Rev. **115**, 9277–9306 (2015)
17. Z. Zhang, P.R. Schreiner, Chem. Soc. Rev. **38**, 1187–1198 (2009)
18. M.S. Sigman, E.N. Jacobsen, J. Am. Chem. Soc. **120**, 4901–4902 (1998)
19. T. Okino, Y. Hoashi, Y. Takemoto, J. Am. Chem. Soc. **125**, 12672–11263 (2003)

20. Y. Takemoto, Chem. Pharm. Bull. **58**, 593–601 (2010)
21. P.H.Y. Cheong, C.Y. Legault, J.M. Um, N. Çelebi-Ölçüm, K.N. Houk, Chem. Rev. **111**, 5042–5137 (2011)
22. S.E. Wheeler, T.J. Seguin, Y. Guan, A.C. Doney, Acc. Chem. Res. **49**, 1061–1069 (2016)
23. G. Jinadal, H.K. Kisan, R.B. Sunoj, ACS Catal. **5**, 480–503 (2015)
24. R.R. Knowles, E.N. Jacobsen, Proc. Natl. Acad. Sci. USA. **107**, 20678–20685 (2010)
25. T. Kita, A. Georgieva, Y. Hashimoto, T. Nakata, K. Nagasawa, Angew. Chem. Int. Ed. **41**, 2832–2834 (2002)
26. P. Selig, Synthesis **45**, 703–718 (2013)
27. P. Selig (ed.), *Guanidinees as Reagents and Catalysts I & II* (Springer, Cham, 2017)
28. K. Hosoya, M. Odagi, K. Nagasawa, Tetrahedron Lett. **59**, 687–696 (2018)
29. Y. Sohtome, Y. Hashimoto, K. Nagasawa, Adv. Synth. Catal. **347**, 1643–1648 (2005)
30. Y. Sohtome, K. Nagasawa, Synlett, 1–22 (2010)
31. Y. Sohtome, K. Nagasawa, Chem. Commun. **48**, 7777–7789 (2012)
32. S.J. Miller, G.T. Copeland, N. Papaioannou, T.E. Horstmann, E.M. Ruel, J. Am. Chem. Soc. **120**, 1629–1630 (1998)
33. E.A.C. Davie, S.M. Mennen, Y. Xu, S.J. Miller, Chem. Rev. **107**, 5759–5812 (2007)
34. Y. Sohtome, N. Takemura, T. Iguchi, Y. Hashimoto, K. Nagasawa, Synlett, 144–146 (2006)
35. Y. Sohtome, Y. Hashimoto, K. Nagasawa, Eur. J. Org. Chem., 2894–2897 (2006)
36. Y. Sohtome, N. Takemura, K. Takada, R. Takagi, T. Iguchi, K. Nagasawa, Chem. Asian J. **2**, 1150–1160 (2007)
37. K. Takada, N. Takemura, K. Cho, Y. Sohtome, K. Nagasawa, Tetrahedron Lett. **49**, 1623–1626 (2008)
38. K. Takada, K. Nagasawa, Adv. Synth. Catal. **351**, 345–347 (2009)
39. K. Takada, S. Tanaka, K. Nagasawa, Synlett, 1643–1646 (2009)
40. Y. Sohtome, S. Tanaka, K. Takada, T. Yamaguchi, K. Nagasawa, Angew. Chem. Int. Ed. **49**, 9254–9257 (2010)
41. Y. Sohtome, T. Yamaguchi, S. Tanaka, K. Nagasawa, Org. Biomol. Chem. **11**, 2780–2786 (2013)
42. Y. Sohtome, B. Shin, N. Horitsugi, R. Takagi, K. Noguchi, K. Nagasawa, Angew. Chem. Int. Ed. **49**, 7299–7303 (2010)
43. Y. Sohtome, B. Shin, N. Horitsugi, R. Takagi, K. Noguchi, K. Nagasawa, Chem. Asian J. **6**, 2463–2470 (2011)
44. M. Kato, S. Hirao, K. Nakano, M. Sato, M. Yamanaka, Y. Sohtome, K. Nagasawa, Chem. Eur. J. **21**, 18606–18612 (2015)
45. N. Horitsugi, K. Kojima, K. Yasui, Y. Sohtome, K. Nagasawa, Asian J. Org. Chem. **3**, 445–448 (2014)
46. Y. Sohtome, N. Horitsugi, R. Takagi, K. Nagasawa, Adv. Synth. Catal. **353**, 2631–2636 (2011)
47. S. Tanaka, K. Nagasawa, Synlett, 667–670 (2009)
48. M. Odagi, K. Furugori, T. Watanabe, K. Nagasawa, Chem. Eur. J. **19**, 16740–16745 (2013)
49. T. Watanabe, M. Odagi, K. Furukori, K. Nagasawa, Chem. Eur. J. **20**, 591–597 (2014)
50. M. Odagi, K. Takayama, K. Furukori, T. Watanabe, K. Nagasawa, Aust. J. Chem. **67**, 1017–1020 (2014)
51. M. Odagi, K. Furukori, Y. Yamamoto, M. Sato, K. Iida, M. Yamanaka, K. Nagasawa, J. Am. Chem. Soc. **137**, 1909–1915 (2015)
52. M. Odagi, Y. Yamamoto, K. Nagasawa, Beilstein J. Org. Chem. **12**, 198–203 (2016)
53. M. Bartók, Chem. Rev. **110**, 1663–1705 (2010)
54. J. Escorihuela, M.I. Burguete, S.V. Luis, Chem. Soc. Rev. **42**, 5595–5617 (2013)
55. V. Blanco, D.A. Leigh, V. Marcos, Chem. Soc. Rev. **44**, 5341–5370 (2015)
56. G. Romanazzi, L. Degennaro, P. Mastrorilli, R. Luisi, ACS Catal. **7**, 4100–4114 (2017)
57. S. Krautwald, E.M. Carreira, J. Am. Chem. Soc. **139**, 5627–5639 (2017)
58. A. Ting, S.E. Schaus, Eur. J. Org. Chem., 5797–5815 (2007)

59. J.M. Verkade, J.C. van Hemert, P.L.M. Quaedflieg, F.J.T. Rutjes, Chem. Soc. Rev. **37**, 29–41 (2008)
60. R.G. Arrayás, J.C. Carrtero, Chem. Soc. Rev. **38**, 1940–1948 (2009)
61. Y. Sohtome, K. Nagasawa, Org. Bioml. Chem. **12**, 1681–1685 (2014)
62. A. Nojiri, N. Kumagai, M. Shibasaki, J. Am. Chem. Soc. **131**, 3779–3784 (2009)
63. X. Tian, C. Cassani, Y. Liu, A. Moran, A. Urakawa, P. Galzerano, E. Arceo, P. Melchiorre, J. Am. Chem. Soc. **133**, 17934–17941 (2011)
64. H. Eyring, J. Chem. Phys. **3**, 107–115 (1935)
65. Y. Inoue, T. Yokoyama, N. Yamasaki, A. Tai, Nature **341**, 225–226 (1989)
66. T.Y. Liu, H.L. Cui, Q. Chai, J. Long, B.J. Li, Y. Wu, L.S. Ding, Y.C. Chen, Chem. Commun., 2228–2230 (2007)
67. X.S. Wang, G.S. Yang, G. Zhao, Tetrahedron Asymmetry **19**, 709–714 (2008)
68. X.S. Wang, C.S. Zheng, S.L. Zhao, Z. Chai, G. Zhao, G.S. Yang, Tetrahedron Asymmetry **19**, 2699–2704 (2008)
69. L. Hong, L. Wang, W. Sun, K. Wong, R. Wang, J. Org. Chem. **74**, 6881–6884 (2009)
70. H. Zhang, Y.H. Liao, W.C. Yuan, X.M. Zhang, Eur. J. Org. Chem., 3215–3218 (2010)
71. M. Montesinos-Magraner, C. Vila, G. Blay, J. Pedro, Synthesis **48**, 2151–2164 (2016)
72. A.M. Walji, D.W.C. MacMillan, Synlett, 1477–1489 (2007)
73. Y. Wang, H. Lu, P.F. Xu, Acc. Chem. Res. **48**, 1832–1844 (2015)
74. Y. Sohtome, T. Yamaguchi, B. Shin, K. Nagasawa, Chem. Lett. **40**, 843–845 (2011)
75. M. Odagi, K. Furukori, K. Takayama, K. Noguchi, K. Nagasawa, Angew. Chem. Int. Ed. **56**, 6609–6612 (2017)
76. M. Kato, K. Yasui, M. Yamanaka, K. Nagasawa, Asian J. Org. Chem. **5**, 380–388 (2016)
77. M. Odagi, Y. Yamammoto, K. Nagasawa, Angew. Chem. Int. Ed. **57**, 2229–2232 (2018)
78. M.J. Frisch et al., *Gaussian 09, revision B.01* (Gaussian, Wallingford, 2010)
79. J.F. Hartiwig, Organotransition metal chemistry: From bonding to catalysis, in *University Science Books*, (Mill Valley, California, 2010)
80. M. Braun, *Modern Enolate Chemistry* (Wiley-VCH, Weinheim, 2016)
81. Y. Hamashima, D. Hotta, M. Sodeoka, J. Am. Chem. Soc. **124**, 11240–11241 (2002)
82. M. Sodeoka, Y. Hamashima, Chem. Commun., 5787–5798 (2009)
83. Y. Hamashima, K. Yagi, H. Takano, L. Tamás, M. Sodeoka, J. Am. Chem. Soc. **124**, 14530–14531 (2002)
84. Y. Hamashima, T. Suzuki, H. Takano, Y. Shimura, M. Sodeoka, J. Am. Chem. Soc. **127**, 10164–10165 (2005)
85. Y. Hamashima, T. Nagi, R. Shimizu, T. Tsuchimoto, M. Sodeoka, Eur. J. Org. Chem., 3675–3678 (2011)
86. N. Sasamoto, C. Dubs, Y. Hamashima, M. Sodeoka, J. Am. Chem. Soc. **128**, 14010–14011 (2006)
87. C. Dubs, Y. Hamashima, N. Sasamoto, T.M. Seidel, S. Suzuki, D. Hashizume, M. Sodeoka, J. Org. Chem. **73**, 5859–5871 (2008)
88. N. Umebayashi, Y. Hamashima, D. Hashizume, M. Sodeoka, Angew. Chem. Int. Ed. **47**, 4196–4199 (2008)
89. A. Nakamura, S. Lectard, D. Hashizume, Y. Hamashima, M. Sodeoka, J. Am. Chem. Soc. **132**, 4036–4037 (2010)
90. A. Nakamura, S. Lectard, R. Shimizu, Y. Hamashima, M. Sodeoka, Tetrahedron Asymmetry **21**, 1682–1687 (2010)
91. S. Suzuki, Y. Kitamura, S. Lectard, Y. Hamashima, M. Sodeoka, Angew. Chem. Int. Ed. **51**, 4581–4585 (2012)
92. Y. Sohtome, G. Nakamura, A. Muranaka, D. Hashizume, S. Lectard, T. Tsuchimoto, M. Uchiyama, M. Sodeoka, Nat. Commun. **8**, 14875 (2017)
93. S.L. Bartlett, Y. Sohtome, D. Hashizume, P.S. White, M. Sawamura, J.S. Johnson, M. Sodeoka, J. Am. Chem. Soc. **139**, 8661–8666 (2017)
94. T. Hashimoto, K. Maruoka, Chem. Rev. **115**, 5366–5412 (2015)

95. A. Yanagisawa, A. Izumiseki, T. Sugita, N. Kushihara, K. Yoshida, Synlett **23**, 107–112 (2012)
96. J.P.G. Seerden, A.W.A. Scholte op Reimer, H.W. Scheeren, Tetrahedron Lett. **35**, 4419–4422 (1994)
97. J.P.G. Seerden, M.M.M. Kuypers, H.W. Scheeren, Tetrahedron Asymmetry **6**, 1441–1450 (1995)
98. J.P.G. Seerden, M.M.M. Boeren, H.W. Scheeren, Tetrahedron **53**, 11843–11852 (1997)
99. K.B. Simonsen, P. Bayón, R.G. Hazell, K.V. Gothelf, K.A. Jørgensen, J. Am. Chem. Soc. **121**, 3845–3853 (1999)
100. K.B. Jensen, R.G. Hazell, K.A. Jørgensen, J. Org. Chem. **64**, 2353–2360 (1999)
101. K.B. Jensen, M. Roberson, K.A. Jørgensen, J. Org. Chem. **65**, 9080–9084 (2000)
102. T. Ashizawa, N. Ohtsuki, T. Miura, M. Ohya, T. Shinozaki, T. Ikeno, T. Yamada, Heterocycles **68**, 1801–1810 (2006)
103. P. Jiao, D. Nakashima, H. Yamamoto, Angew. Chem. Int. Ed. **47**, 2411–2413 (2008)
104. R. Sustmann, Pure Appl. Chem. **40**, 569–593 (1974)
105. J.P. Abell, H. Yamamoto, Chem. Soc. Rev. **39**, 61–69 (2010)
106. E. Meggers, Eur. J. Inorg. Chem., 2911–2926 (2011)
107. L. Gong, L.A. Chen, E. Meggers, Angew. Chem. Int. Ed. **53**, 10868–10874 (2014)
108. E.C. Constable, Chem. Soc. Rev. **42**, 1637–1651 (2013)
109. A. Ehnbom, S.K. Ghosh, K.G. Lewis, J.A. Gladysz, Chem. Soc. Rev. **45**, 6799–6811 (2016)
110. L. Zhang, E. Meggers, Chem. Asian J. **12**, 2335–2342 (2017)
111. J.B. Sumner, J. Biol. Chem. **69**, 435–441 (1926)
112. M.J. Maroney, S. Ciurli, Chem. Rev. **114**, 4206–4228 (2014)
113. T. Hayashi, M. Tajika, K. Tamao, M. Kumada, J. Am. Chem. Soc. **98**, 3718–3719 (1976)
114. E.A. Standley, S.Z. Tasker, K.L. Jensen, T.F. Jamison, Acc. Chem. Res. **48**, 1503–1514 (2015)
115. S.Z. Tasker, E.A. Standley, T.F. Jamison, Nature **509**, 299–309 (2014)
116. V.P. Ananikov, ACS Catal. **5**, 1964–1971 (2015)
117. H. Pellissier, Adv. Synth. Catal. **357**, 2745–2780 (2015)
118. H. Mimoun, J.Y. de Saing Laumer, L. Giannini, R. Scopelliti, C. Floriani, J. Am. Chem. Soc. **121**, 6158–6166 (1999)
119. B. Zhao, Z. Han, K. Ding, Angew. Chem. Int. Ed. **52**, 4744–4788 (2013)
120. A. Wittkopp, P.R. Schreiner, Chem. Eur. J. **9**, 407–414 (2003)
121. T. Okino, Y. Hoashi, Y. Takemoto, Tetrahedron Lett. **44**, 2817–2821 (2003)
122. E.M. Vogl, H. Gröger, M. Shibasaki, Angew. Chem. Int. Ed. **38**, 1570–1577 (1999)
123. L. Hong, W. Sun, D. Yang, G. Li, R. Wang, Chem. Rev. **116**, 4006–4123 (2016)
124. R.F.W. Bader, *Atoms in Molecules: A Quantum Theory* (Oxford University Press, Oxford, 1990)
125. D. Stalke (ed.), *Structure and Bonding 146: Electron Density and Chemical Bonding I, Experimental Charge Density Studies* (Springer, Heidelberg/New York/Dordrecht/ London, 2012)
126. D. Hashizume, Adv. Mater. **29**, 1605175 (2017)
127. M. Arnold, D.A. Brown, O. Deeg, W. Errington, W. Haase, K. Herlihy, T.J. Kemp, H. Nimir, R. Werner, Inorg. Chem. **37**, 2920–2925 (1998)
128. J.H. Lady, K.B. Whetsel, J. Phys. Chem. **71**, 1421–1429 (1967)
129. Frisch M. J. *et al.* Gaussian 09, revision C.01 (2010), Gaussian, Inc.: Wallingford, CT
130. Y. Zhao, D.G. Truhlar, Acc. Chem. Res. **41**, 157–167 (2008)
131. D.A. Evans, D. Seidel, J. Am. Chem. Soc. **127**, 9958–9959 (2005)
132. D.A. Evans, S. Mito, D. Seidel, J. Am. Chem. Soc. **129**, 11583–11592 (2007)
133. Z.H. Yu, P. Caramella, K.N. Houk, J. Am. Chem. Soc. **125**, 15420–15425 (2003)
134. K. Juhl, N. Gathergood, K.A. Jørgensen, Chem. Commun., 2211–2212 (2000)
135. R. Huisgen, M. Wilhelm, Tetraedern Lett. **2**, 583–586 (1961)

136. M. Bandini, F. Piccinelli, S. Tommasi, A. Umani-Ronchi, C. Ventrici, Chem. Commun., 616–618 (2006)
137. G. Zhang, E. Yashima, W.D. Woggon, Adv. Synth. Catal. **351**, 1255–1262 (2009)
138. K. Tanaka, A. Asakura, T. Muraoka, P. Kalicki, Urbanczyk-Lipkowska. New J. Chem. **37**, 2851–2855 (2013)
139. K. Tanaka, T. Iwashita, E. Yoshida, T. Ishikawa, S. Otuka, Z. Urbanczyk-Lipkowska, H. Takahashi, Chem. Commun. **51**, 7907–7910 (2015)
140. M.K. Zhu, J.F. Zhao, T.P. Loh, J. Am. Chem. Soc. **132**, 6284–6285 (2010)

Chapter 6
Biaryl Amino Acids and Their Surrogates: A Unique Class of Unnatural Amino Acid

Takumi Furuta

Abstract Unnatural amino acids have attracted extensive attention as potential building blocks for bioactive molecules, artificial peptides, and asymmetric catalysts in the context of fine organic synthesis. Biaryl amino acids, a unique class of unnatural amino acids possessing a biaryl moiety, permit the development of a variety of functional molecules by virtue of their structural features, including axial chirality. This chapter discusses typical examples of the synthesis and applications of biaryl amino acids as well as their surrogates, including the author's approaches.

Keywords Biaryl amino acid · Unnatural amino acid · Axial chirality · Chiral building block · Peptidomimetics · Organocatalyst · Chiral ligand

6.1 Introduction

Unnatural amino acids have played a significant role as key building blocks for the production of bioactive compounds in the pharmaceutical industry, artificial peptides in peptidomimetics, and artificial protein production [1, 2]. Unnatural amino acids bearing an asymmetric center can be employed in chiral organocatalysts and chiral ligands in catalytic asymmetric synthesis. A variety of unnatural amino acids possessing unique structural features and functions have been developed.

The unnatural biaryl amino acids provide special structural features. One of the most prominent properties of this type of molecule is that axial chirality may be installed in the molecule by modifying the substituents at the proper positions (typically the *ortho*-positions of the biaryl axis) (Fig. 6.1). The axially chiral biaryl amino acid provides a unique chiral environment that differs significantly from that provided by natural and related unnatural amino acids with central chirality. The other structural feature of this type of amino acid is its rigidity along and flexibility

T. Furuta (✉)
Department of Pharmaceutical Chemistry, Kyoto Pharmaceutical University, Kyoto, Japan
e-mail: furuta@mb.kyoto-phu.ac.jp

S. Shirakawa (ed.), *Designed Molecular Space in Material Science and Catalysis*,
https://doi.org/10.1007/978-981-13-1256-4_6

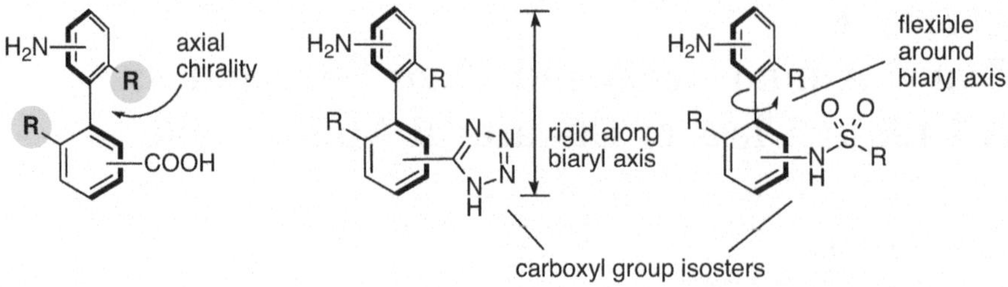

Fig. 6.1 Structural properties of the biaryl amino acid and its surrogates

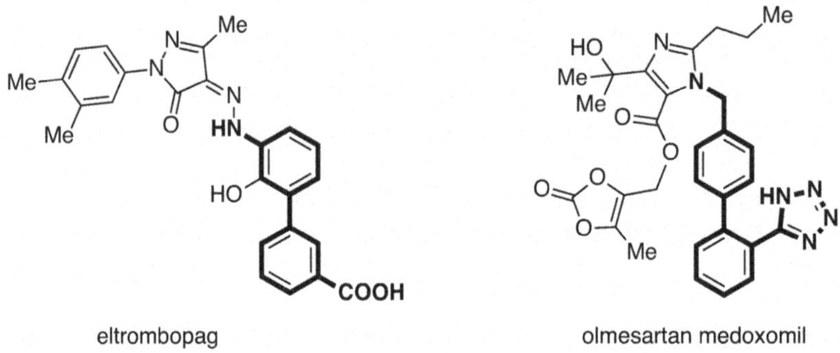

Fig. 6.2 Pharmaceuticals bearing a biaryl amino acid and its surrogate as core structures

around the biaryl axis (Fig. 6.1). The specific rigidity, flexibility, and axial chirality properties permit the positioning of the amino and carboxyl functional groups within defined spatial arrangements using a biaryl spacer with a positive or negative twist.

Carboxyl group isosteres, including tetrazole and sulfonamide groups, have been employed as acidic moieties in the pharmaceutical industry as well as in the design of catalysts [3]. Biaryl amino acid surrogates, in which a carboxyl group is replaced with tetrazole or sulfonamide groups, are discussed in this review.

The aforementioned structural properties render biaryl amino acids promising building blocks for pharmaceuticals, chiral catalysts, host molecules for host–guest chemistry, and artificial peptides for peptidomimetics. Indeed, biaryl amino acids and their surrogates bearing a tetrazole moiety have been employed as the core structures of pharmaceuticals (Fig. 6.2).

Typical examples of biaryl amino acids and their surrogates, including axially chiral compounds, are depicted in Fig. 6.3. α-Amino acids (*S*)-**1** and (S)-**2** bearing an axially chiral biaryl moiety were developed by Závada [4] and Mazaleyrat [5], respectively. Recently, the ester derivative of **2** was used as a chiral organocatalyst for catalytic asymmetric synthesis, as described in Sect. 6.2 [6, 7]. The amino acid (*R*)-**3** is a rare example of an axially chiral β-amino acid [8]. The biaryl-based axially chiral γ-amino acids (*S*)-**4a** [9, 10] and (*S*)-**4b** [11], bearing an aliphatic secondary amine in the seven-membered ring, were developed as organocatalysts for catalytic

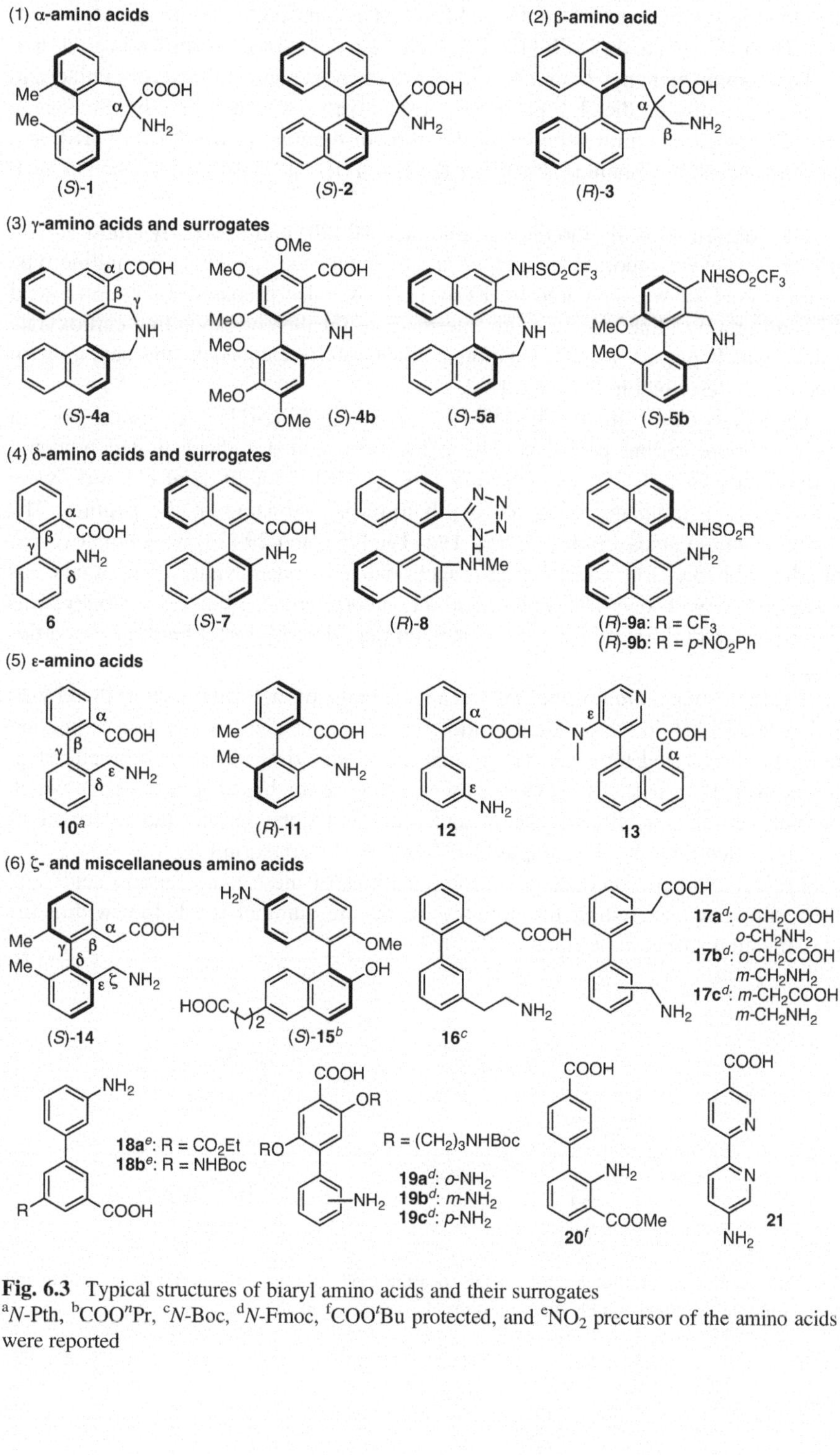

Fig. 6.3 Typical structures of biaryl amino acids and their surrogates
[a]*N*-Pth, [b]COO[n]Pr, [c]*N*-Boc, [d]*N*-Fmoc, [f]COO[t]Bu protected, and [e]NO_2 precursor of the amino acids were reported

asymmetric synthesis by Maruoka and Kano. Their sulfonamide surrogates, including (*S*)-**5a** [12, 13] and (*S*)-**5b** [14, 15], were also developed, as shown in Sect. 6.4.

The achiral biphenyl δ-amino acid **6**, which possesses aniline-type amine and carboxyl groups at the 2,2′-positions, respectively, was reported by Santagada [16]. We prepared chiral versions of this type of δ-amino acid, (*S*)-**7** [17, 18], and its tetrazole and sulfonamide surrogates (*R*)-**8** and (*R*)-**9a**,**b** [19], as described in Sect. 6.5.

The conformationally mobile ε-amino acid **10** [20] and its axially chiral variant (*R*)-**11** [21] were reported by Feigel and Závada, respectively. The aniline-type ε-amino acid **12** was prepared by Kelly [22]. We also prepared the ε-amino acid **13** bearing a pyridine moiety as an intermediate for the corresponding carboxylate anion, which played a role as a potent nucleophilic catalyst in the acylation of alcohol, as described in Sect. 6.6 [23].

The axially chiral amino acids (*S*)-**14** [21] and (*S*)-**15** [24] bear an amino group at the ζ or more distant positions. The 2,2′-, 2,3′-, and 3,3′-disubstituted biphenyl amino acids **16** [22, 25] and **17a**–**17c** [16, 24, 26], without axial chirality, were designed as a template to induce a β-hairpin-like structure in the peptide. The biphenyl amino acids **18a**,**b** [27, 28], **19a**–**19c** [29], and **20** [30] were employed as building blocks for constructing functionalized macrocyclic compounds, as discussed in Sect. 6.7. The amino acid **21**, incorporating a bipyridyl moiety, was reported as a building block for an artificial peptide with metal binding properties [31, 32].

The term "biaryl amino acid" covers a wide range of molecules exemplified by **22** [33] and **23** [34], in which the aryl moieties are connected via a carbon atom, and both the amino and the carboxyl groups are modified at the same aromatic ring, respectively (Fig. 6.4). This review, however, focuses on biaryl derivatives, in which each aryl group is modified by acidic and amino functional groups; these derivatives are connected directly via a single C–C bond, as illustrated in Fig. 6.1.

Typical synthetic procedures and applications of these biaryl amino acids and their surrogates, including the author's work, are summarized below within the classification structure outlined in Fig. 6.3.

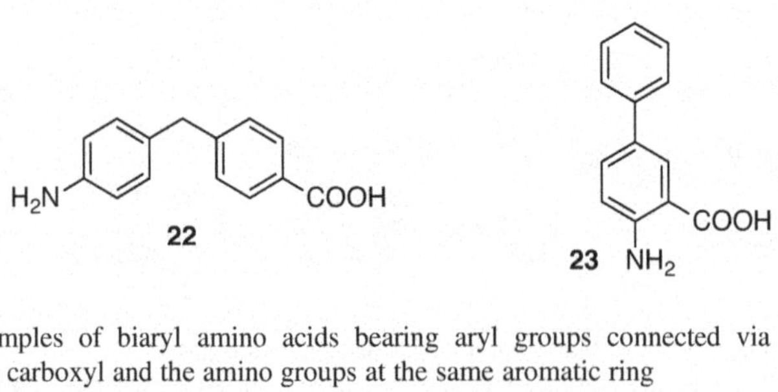

Fig. 6.4 Examples of biaryl amino acids bearing aryl groups connected via an atom, and possessing the carboxyl and the amino groups at the same aromatic ring

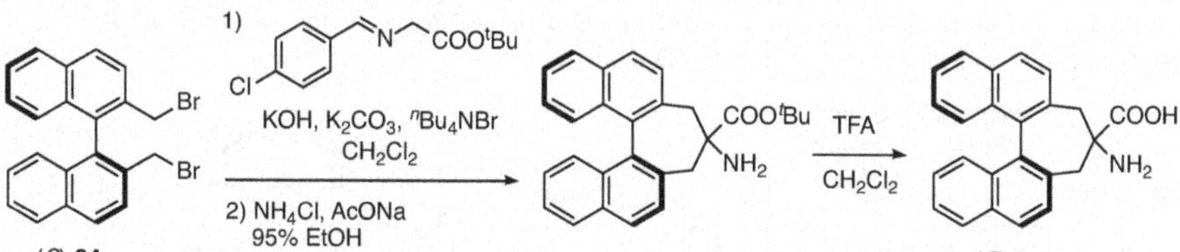

Scheme 6.1 Preparation of the axially chiral α-amino acid (*S*)-**2**

6.2 Synthesis and Application of the α-Amino Acids

Optically active axially chiral α-amino acids were prepared through the double alkylation of a glycine equivalent and the corresponding biaryl dibromide under phase transfer conditions, as is typical of the synthesis of (*S*)-**2** from (*S*)-**24** (Scheme 6.1) [4–6].

One application of these α-amino acids is as building blocks for artificial peptides. The synthesis and conformations of peptides that include these amino acids were investigated [35–37]. Recently, (*R*)-**25**, a 3,3′-disubstituted ethyl ester derivative of **2**, was applied as an organocatalyst for the asymmetric α-fluorination of an α-branched aldehyde by Shibatomi and co-workers (Table 6.1) [6]. Fluorination of the racemic 2-phenylpropanal derivatives proceeded with *N*-fluorobenzenesulfonimide (NFSI) in the presence of 10 mol% (*R*)-**25** and 10 mol% 3,5-dinitrobenzoic acid as a co-catalyst to yield the 2-fluoro-2-phenylpropanal derivatives **26** in a high conversion yield and a high enantiomeric excess. The enantiomeric excess and chemical yield were determined with the corresponding alcohol **27**.

Although the α,α-dialkyl substrates gave a low stereoselectivity or chemical yield, as observed in **27 g** and **27 h**, a variety of the substrates gave the products in excellent yields and stereoselectivities.

An enantioselective decarboxylative chlorination of the racemic **28** with NCS was also developed using cat. (*R*)-**25** to give the *C*-chlorinated derivatives **29** in a high enantioselectivity and excellent chemical yield, as shown in Scheme 6.2 [7].

6.3 Synthesis of the β-Amino Acid

The synthesis of the axially chiral β-amino acid (*R*)-**3** and its *N*-Boc derivative was achieved via double alkylation of ethyl cyanoacetate with the optically active (*R*)-**24** and the subsequent selective reduction of a cyano group in the presence of $NaBH_4$ and cobalt(II) chloride to the aminoester derivative (Scheme 6.3) [8]. Protection with a Boc group and hydrolysis of the ester group gave the *N*-Boc β-amino acid (*R*)-**30**. Removal of the Boc group gave (*R*)-**3**.

Table 6.1 α-Fluorination of the α-branched aliphatic aldehyde in the presence of the 3,3′-disubstituted α-amino acid ethyl ester (*R*)-**25**

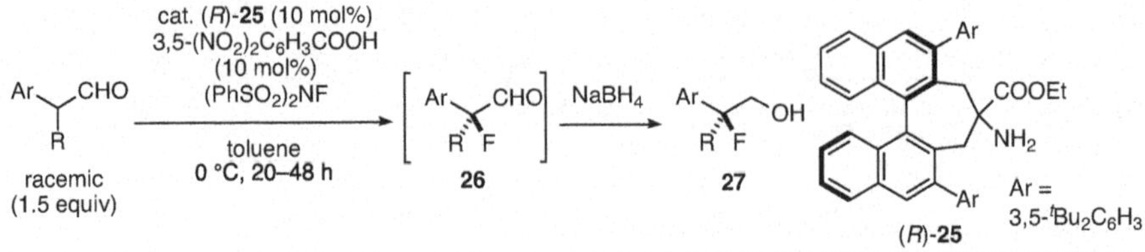

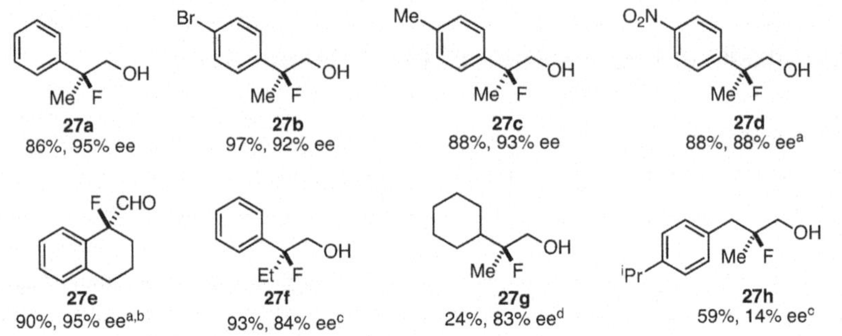

[a]The purified product contained ca. 5% of an inseparable by-product
[b]At rt for 2 h
[c]At rt for 12–24 h
[d]30 mol% catalyst

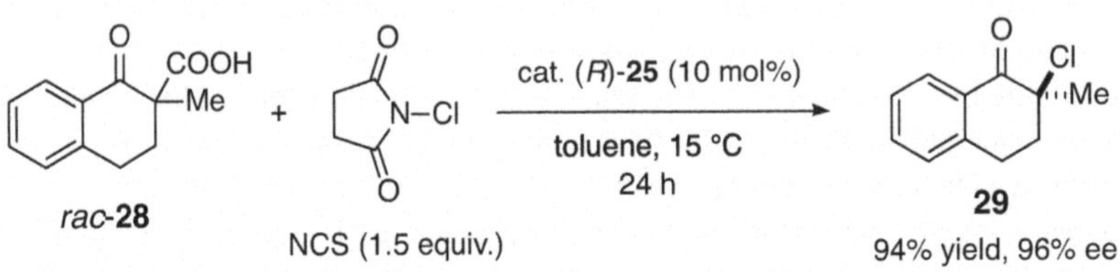

Scheme 6.2 Enantioselective decarboxylative chlorination catalyzed by (*R*)-**25**

6.4 Synthesis and Application of the γ-Amino Acids and Their Surrogates

Biaryl-based axially chiral γ-amino acids and their sulfonamide surrogates, including (*S*)-**4a** and (*S*)-**5a**, were prepared, as illustrated in Scheme 6.4 [9–11]. Bromination of the optically active (*S*)-**31** through *ortho*-magnesiation with $Mg(tmp)_2$ and subsequent trapping of the anion with bromine gave the monobromide (*S*)-**32**. Double alkylation of the allylamine with tribromide (*S*)-**33** prepared from **32** gave the seven-membered ring amine (*S*)-**34**. Carbonylation with CO in the presence of $Pd(OAc)_2$ with dppp in MeOH, and subsequent deprotection of the allyl group and

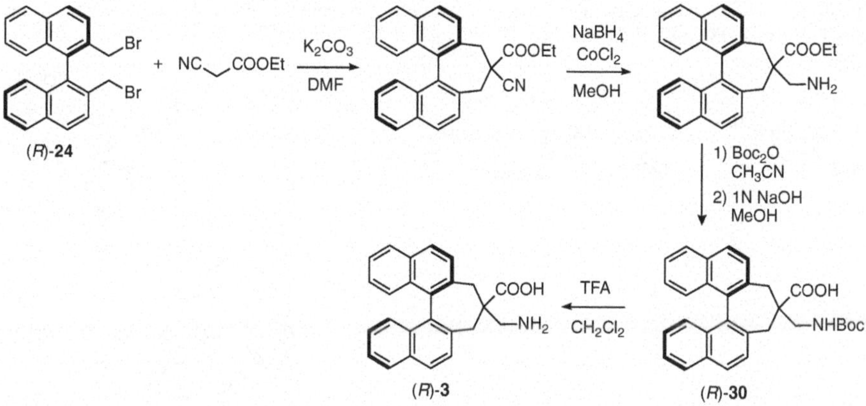

Scheme 6.3 Preparation of the axially chiral β-amino acid and *N*-Boc derivative

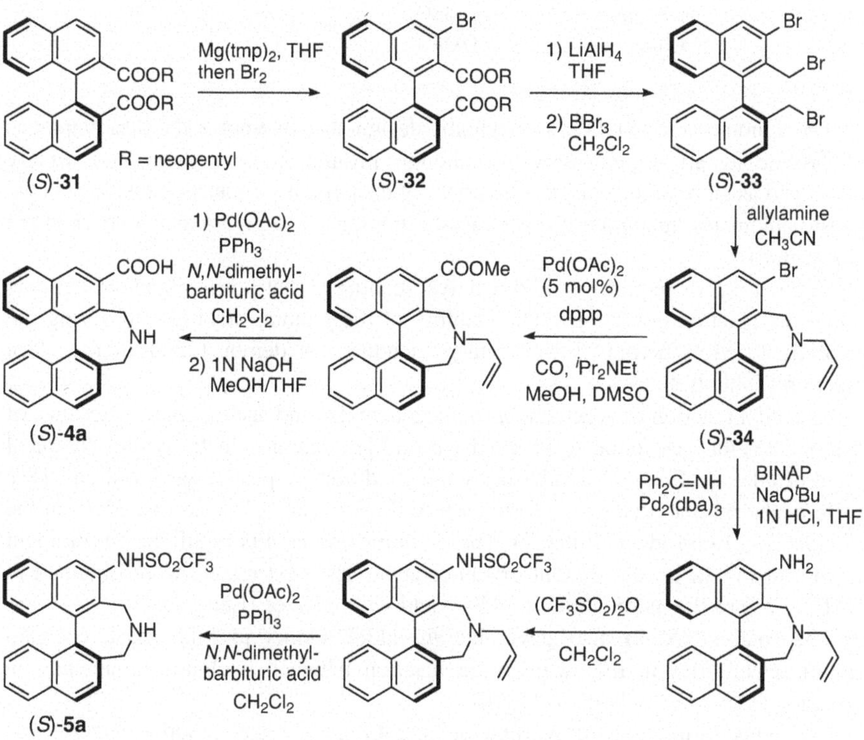

Scheme 6.4 Preparation of the axially chiral γ-amino acid and sulfonamide surrogate

hydrolysis of the ester moiety, afforded the γ-amino acid (*S*)-**4a**. Alternatively, the sulfonamide surrogate (*S*)-**5a** was synthesized through the Pd-catalyzed amination of (*S*)-**34** [12, 13].

Table 6.2 Direct asymmetric aldol reaction between the aromatic aldehyde and acetone in the presence of cat. (*S*)-**4a**

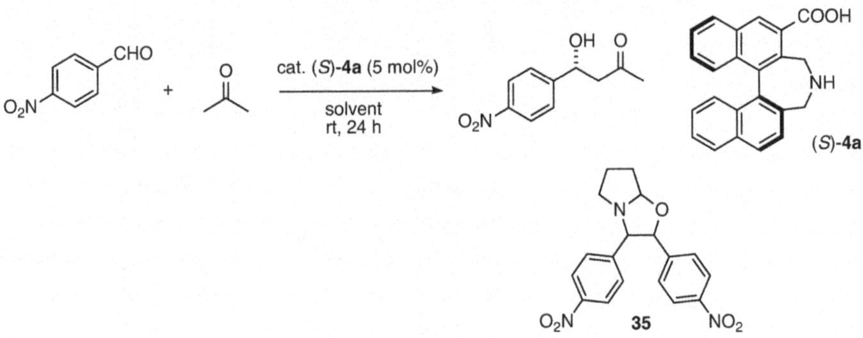

Entry	Catalyst	Solvent	Yield (%)	ee (%)
1	(*S*)-**4a**	DMSO	70	93
2	L-proline	DMSO	18	71
3	(*S*)-**4a**	DMF	82	95

The γ-amino acid (*S*)-**4a** was originally designed to overcome the disadvantages of proline as an organocatalyst. Although proline is convenient and widely employed as an organocatalyst in a variety of catalytic asymmetric transformations, substoichiometric amounts of the catalyst are required to give reasonable yields in many cases.

One reason for requiring a high catalyst loading is that proline tends to degrade under the reaction conditions. The stability of the γ-amino acid (*S*)-**4a** during the catalytic reaction permits lower catalyst loading, as demonstrated in the aldol reaction (Table 6.2).

The aldol reaction between the aromatic aldehydes and acetone in the presence of 5 mol% (*S*)-**4a** was found to proceed with a high enantioselectivity and chemical yield (Table 6.2, Entry 1). Under the same conditions, L-proline gave only an 18% yield of the product due to the formation of the bicyclic 1,3-oxazoline **35** from the catalyst as a by-product (Entry 2). These contrasting results highlight a prominent feature of (*S*)-**4a** as an efficient organocatalyst. The best results were obtained in DMF to afford the aldol adduct in 82% yield with 95% ee (Entry 3).

The catalyst (*S*)-**4a** also gave the *anti*-aldol adduct in high diastereo- and enantioselectivities in the case of the reaction between cyclohexanone and the aromatic aldehydes (Scheme 6.5) [10].

In contrast to the reaction with the cat. (*S*)-**4a**, a *syn*-selective aldol reaction was realized with the sulfonamide-type amino acid surrogate (*S*)-**5a**, as shown in Scheme 6.6 [15, 38].

One of the features of the sulfonamide surrogates is exemplified in the direct cross-aldol reaction between the α-chloro aldehyde and the aliphatic aldehyde (Scheme 6.7) [14]. The cross-aldol adduct was obtained exclusively in the presence of cat. (*S*)-**5b** in high *syn*-selectivity and enantioselectivity without forming the

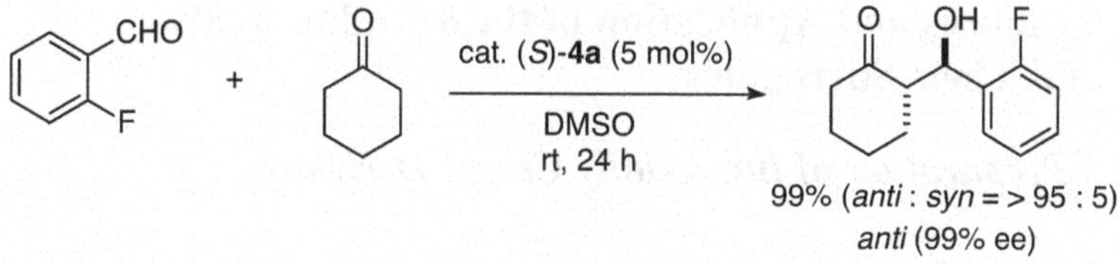

Scheme 6.5 *Anti*-selective aldol reaction catalyzed by (*S*)-**4a**

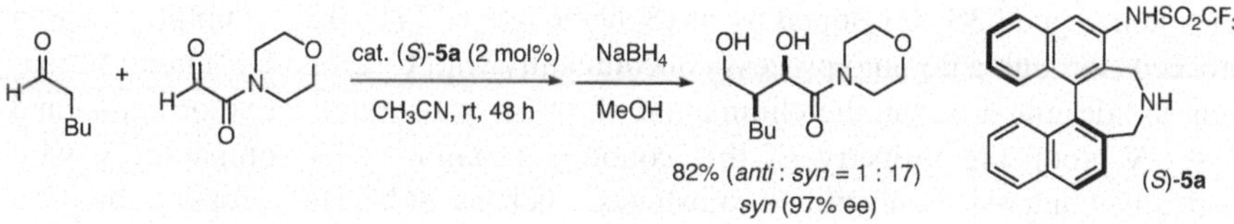

Scheme 6.6 *Syn*-selective direct cross-aldol reaction catalyzed by the amino acid surrogate (*S*)-**5a**

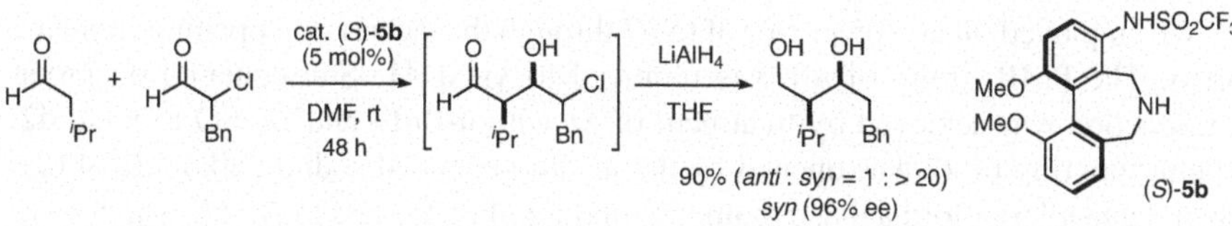

Scheme 6.7 *Syn*-selective direct cross-aldol reaction between the aliphatic aldehyde and the α-chloro aldehyde, controlled by (*S*)-**5b**

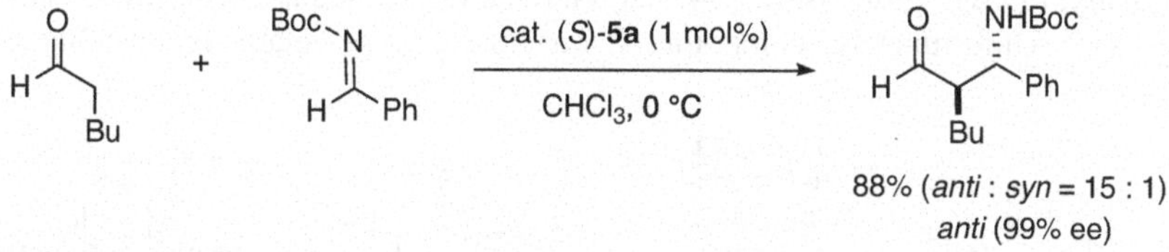

Scheme 6.8 *Anti*-selective Mannich reaction promoted by (*S*)-**5a**

homo-aldol products. Although the selective production of the cross-aldol adduct was majorly caused from the unfavorable formation of the enamine intermediate from the α-chloro aldehyde, which mainly worked as an aldol acceptor, the mild reactivity of cat. (*S*)-**5b**, which was derived from a less nucleophilic dibenzylic amine moiety, was also thought to contribute for discriminating these different aldehydes. The chloro group on the cross-aldol adduct was removed under reductive conditions.

The sulfonamide surrogate (*S*)-**5a** was applied in an *anti*-selective Mannich reaction (Scheme 6.8) [39]. This *anti*-selectivity is complementary to the proline-catalyzed reaction, which gave the *syn* diastereomer exclusively [40].

6.5 Synthesis and Application of the δ-Amino Acids and Their Surrogates

6.5.1 Preparation of the Axially Chiral Derivative

We synthesized the optically active aniline-type binaphthyl δ-amino acid (*S*)-**7**. The key reaction in the synthesis of (*S*)-**7** is a Pd-catalyzed domino coupling reaction of the *ortho*-bromo aryl amides **36** to the phenanthridinones **37** in the presence of a phosphine ligand **38**, developed by us (Scheme 6.9(1)) [41]. This coupling reaction proceeds through a domino process concomitantly with C–C and C–N bond formation and deamidation via the elimination of an isocyanate derivative or amine and CO_2. A promising property of this coupling reaction is its compatibility with sterically hindered 3-substituted substrates, such as **36b**. This property may be extended to the coupling of the naphthyl derivatives **39** to **40** (Scheme 6.9(2)) [17]. This coupling reaction proceeded quantitatively without the addition of the phosphine ligand **38**.

We employed **40** as a precursor of (*S*)-**7** through the lactam ring opening (Scheme 6.10). The PMB group of **40** was removed to yield **41**. Subsequent Boc group introduction was achieved by treatment of **41** with *n*-BuLi and Boc_2O to yield **42**. The lactam ring of **42** was opened by the in situ-generated sodium alkoxide of (*S*)-phenylethanol to afford a diastereomeric mixture of (*S*,*S*)- and (*S*,*R*)-**43**, which were separated by recrystallization to afford the optically active (*S*,*S*)-**43** as a crystalline form. The *N*-Boc amino acid (*S*)-**44** was obtained in an optically pure form via catalytic hydrogenation.

Finally, the Boc group of (*S*)-**44** was removed to give the axially chiral δ-amino acid (*S*)-**7**. Unfortunately, compound **7** was found to be relatively unstable and

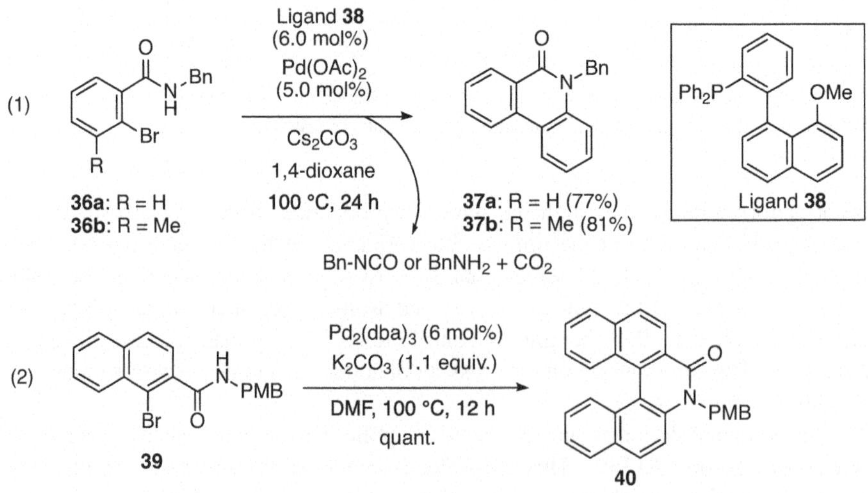

Scheme 6.9 Pd-catalyzed domino coupling reaction to the phenanthridinone derivatives

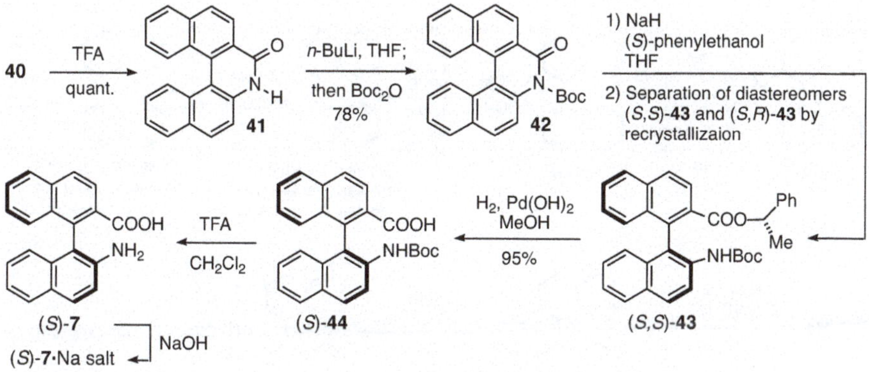

Scheme 6.10 Transformation to the axially chiral δ-amino acid (*S*)-**7**

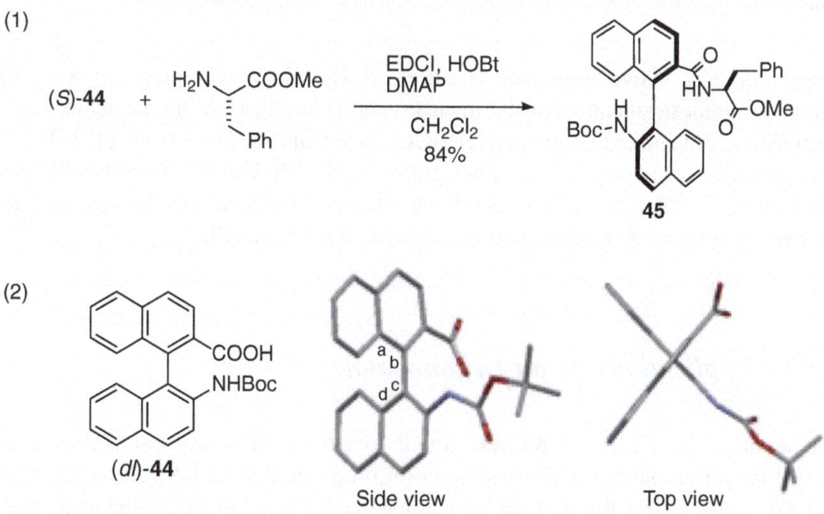

Fig. 6.5 Chemical modification of (*S*)-**44** and the crystal structure of the *racemic*-**44**

cyclized to the lactam **41** within a few days. On the other hand, gratifyingly, the sodium salt of (*S*)-**7** remained stable under ambient temperatures.

Although applications of (*S*)-**7** were difficult due to its instability, the *N*-Boc amino acid (*S*)-**44** was useful as a chiral building block for artificial peptides. For example, condensation of (*S*)-**44** with the L-phenylalanine methyl ester gave the *N*-Boc dipeptide **45** in good yield (Fig. 6.5(1)). Considering the crystal structure of (*dl*)-**44**, (*S*)-**44** could be useful for enforcing turn structures in an artificial peptide, because the moieties at C2 and C2′ are positioned orthogonally around the chiral axis (dihedral angle of a,b–c,d: 93°) (Fig. 6.5(2)).

An improved synthetic strategy without optical resolution for accessing a variety of *N*-protecting amino acid derivative was achieved through Curtius rearrangement

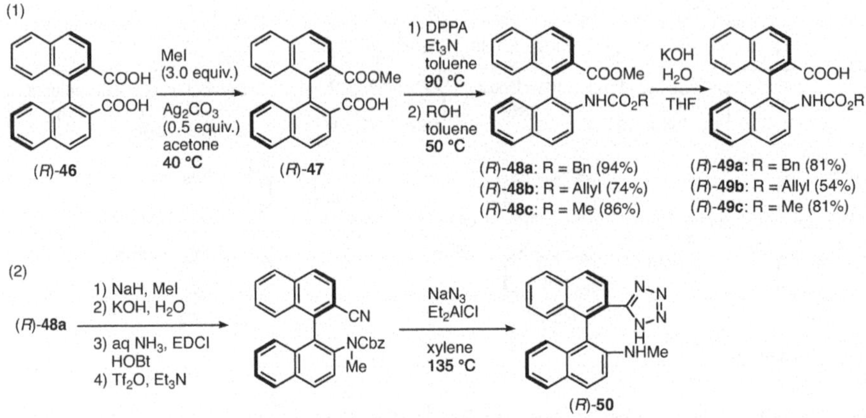

Scheme 6.11 Alternative syntheses of the axially chiral δ-amino acid derivatives

of the optically active monomethyl ester (*R*)-**47**, which was prepared from the selective monoesterification of the dicarboxylic acid (*R*)-**46** in the presence of 3.0 equivalents MeI and 0.5 equivalents Ag_2CO_3 in acetone (Scheme 6.11) [18]. The *N*-Cbz, *N*-allyl, and *N*-methoxycarbonyl amino acids (*R*)-**49a**–**49c** were readily prepared by this synthetic route in optically pure forms. The δ-amino acid surrogate (*R*)-**50**, bearing a tetrazole moiety, was also prepared from (*R*)-**48a**.

6.5.2 Applications As an Organocatalyst

The tetrazole surrogate (*R*)-**50** was found to exhibit sufficient catalytic activity toward the intramolecular aldol reaction of 1,6-hexane dial **51** [19]. In the presence of 5 mol% cat. (*R*)-**50**, the *enolexo*-intramolecular aldolization proceeded to give *ent-anti*-**52** and *syn*-**53** in DMSO (Scheme 6.12(1)). A subsequent Wittig olefination gave *ent-anti*-**54** and *syn*-**55** with a high enantioselectivity in the *syn* isomer (93% ee), although the diastereoselectivity was moderate (*ent-anti*-**54**: *syn*-**55** = 1: 1.3). During a survey of catalysts, the sulfonamide surrogate (*R*)-**9a** bearing a primary amine was identified as an excellent catalyst that afforded predominantly *anti*-**54** in a 74% yield with a high diastereo- (*anti*: *syn* = 15: 1) and enantioselectivity (97% ee) (Scheme 6.12(2)).

With the high enantio- and diastereoselectivities in hand using cat. (*R*)-**9a**, we moved on to examining the cross-aldol reaction of an unsymmetric aliphatic dial [19]. Although this reaction is valuable for the production of cyclic β-hydroxy aldehydes, control over this reaction is quite challenging due to the production of eight isomers from two regioisomers (*paths A* and *B* in Fig. 6.6), including diastereomers (*anti*/*syn*) and enantiomers of each isomer. The reaction selectively might only be achieved by controlling the diastereo- and enantioselectivities, in addition to

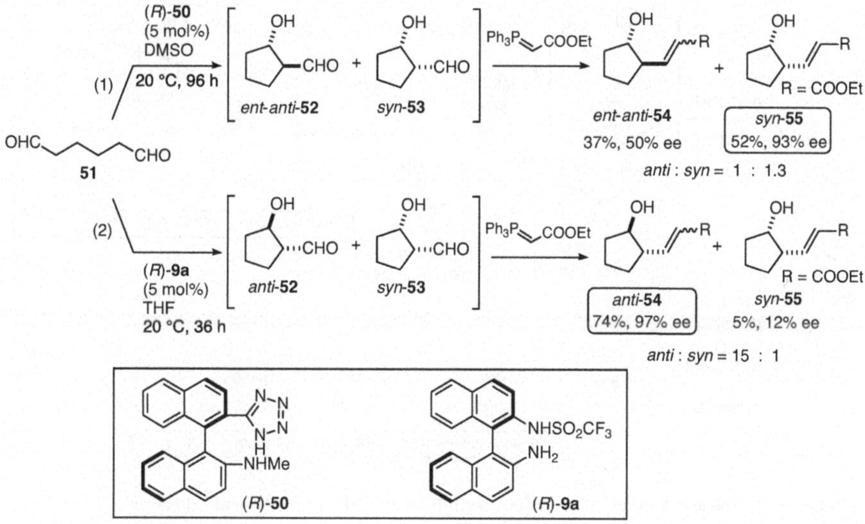

Scheme 6.12 Intramolecular aldol reaction catalyzed by the δ-amino acid surrogates

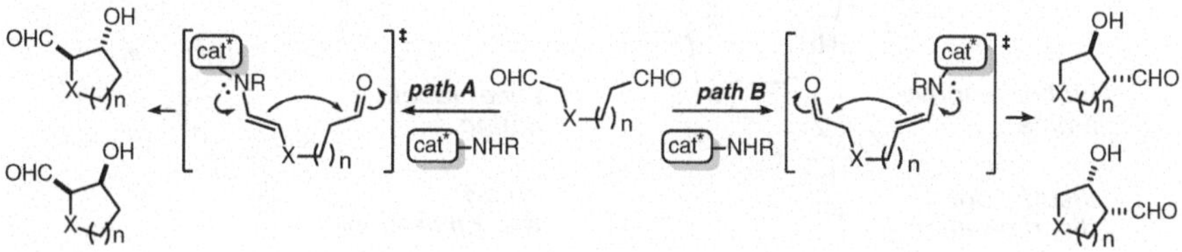

Fig. 6.6 Possible isomers from the intramolecular cross-aldol reaction of an unsymmetric aliphatic dial

controlling the regioselectivities of the products (*path A* vs. *path B*). In the amine-catalyzed reactions involving an enamine intermediate, a high regioselectivity is expected only under conditions that favor precise discrimination by the amine catalyst between two enolizable formyl groups into the enamine and the carbonyl components.

Indeed, L-proline catalyzed the reaction of the unsymmetric *N*-Ts dial **56a** to yield an undesirable mixture of products. After $NaBH_4$ reduction, the reaction mixture afforded nearly all possible regio- and stereoisomeric products, [*anti*-**57a** (9%, 6% ee), *syn*-**58** (31%, >99% ee), dehydrated **61** (2%)], and [*anti*-**59** (5%, 42% ee), *syn*-**60** (17%, 53% ee)], from the enamines of the C(6)- and C(1)-formyl groups (Scheme 6.13(2)). The regioselectivity of the reaction, (**57a** + **58** + **61**): (**59** + **60**), was 1.8: 1. This undesirable result indicated that L-proline could not discriminate between the formyl groups of **56a**.

On the other hand, the amino acid surrogate (*R*)-**9b**, bearing a *p*-Ns amide moiety as a milder acidic group compared to the Tf amide group of (*R*)-**9a**, controlled the intramolecular cross-aldol reaction well to give the 3,4-disubstituted *anti*-**57a** (59%)

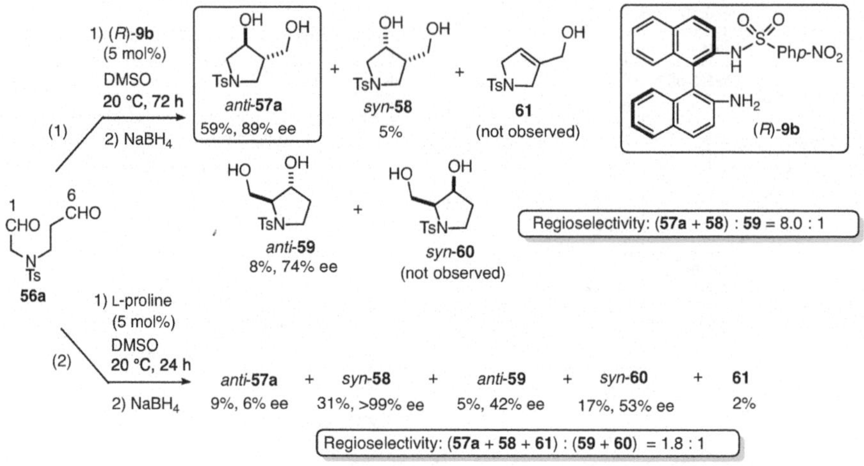

Scheme 6.13 Intramolecular cross-aldol reaction catalyzed by (*R*)-**9b** and L-proline

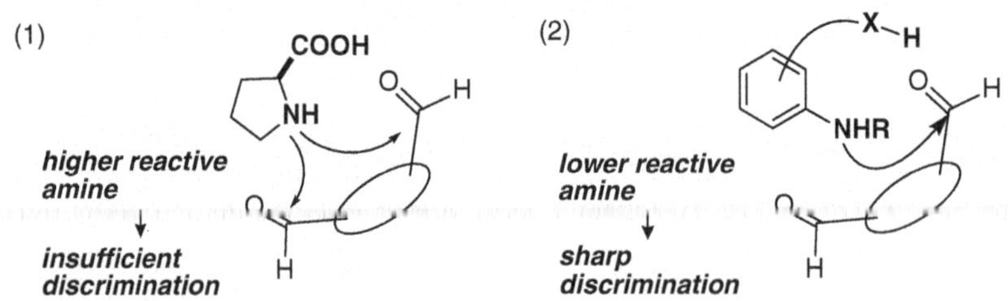

Fig. 6.7 Formyl group discrimination: Acid-base catalysts bearing an aliphatic amine vs. an aniline-type amine

as a major product in 89% ee with a high diastereoselectivity (*anti*-**57a**: *syn*-**58** = 12: 1) and the concomitant formation of the regioisomer, 2,3-disubstituted *anti*-**59** (8%) (Scheme 6.13(1)). The regioselectivity of the reaction, (**57a** + **58** + **61**): (**59** + **60**), was improved to 8.0 : 1 from the L-proline-catalyzed reaction. This selectivity indicated that cat. (*R*)-**9b** discriminated between the different formyl groups, which could not be distinguished by L-proline, converting the C(6)-formyl group into the enamine component and the C(1)-formyl group into the carbonyl component.

The excellent discrimination of the aniline-type δ-amino acid surrogate toward the formyl groups was thought to derive from the mild reactivity of the aniline-type amine.

An amino acid catalyst bearing an aliphatic amine, such as L-proline, was assumed to be too reactive to discriminate among the formyl groups bearing similar reactivities (Fig. 6.7(1)). By contrast, a catalyst bearing a mildly reactive aniline-type amine could be advantageous in discriminating between the different formyl groups (Fig. 6.7(2)).

Table 6.3 Intramolecular cross-aldol reaction of the *N*-protected dials

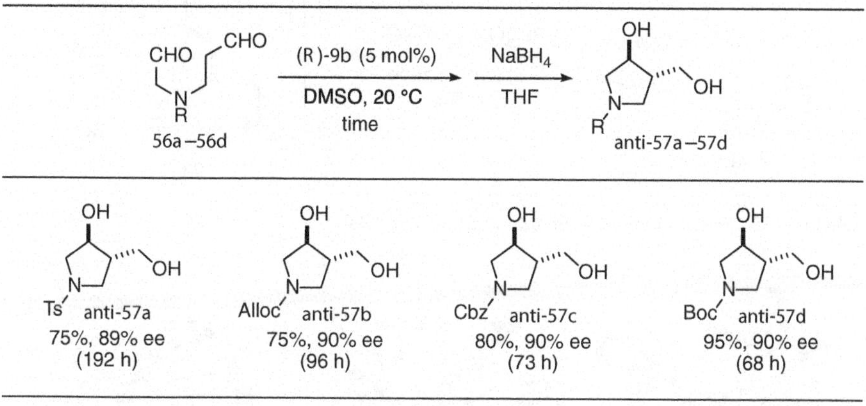

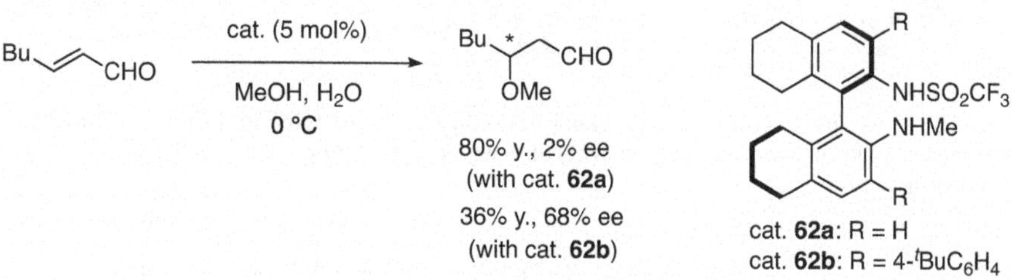

Scheme 6.14 Asymmetric oxy-Michael addition in the presence of the δ-amino acid surrogates

The yield of *anti*-**57a** increased to 75% by extending the reaction time without decreasing the enantioselectivity (Table 6.3). The *N*-Alloc-, *N*-Cbz-, and *N*-Boc dials **56b–56d** also afforded *anti*-**57b–57d** in 90% ee (Table 6.3). It should be noted that **57d** was isolated as the sole product in 95% yield, suggesting that the formyl groups of **56d** were sharply distinguished by cat. (*R*)-**9b**.

Aniline-type sulfonamide surrogates were also employed as organocatalysts for an asymmetric oxy-Michael addition via an iminium activation mechanism [42, 43]. Conjugate addition of MeOH to 2-heptenal proceeded well in the presence of 5 mol% (*R*)-**62a**, although enantioselectivity was not observed (Scheme 6.14). The enantioselectivity was improved with cat. (*R*)-**62b** having aryl substituents at the 3,3′ positions.

6.5.3 Application As a Chiral Ligand

The axially chiral δ-amino acid derivative was employed as a chiral ligand in a rhodium carboxylate catalyst. The rhodium catalyst **63** bearing *N*-methoxycarbonyl δ-amino acid (*R*)-**49c** was prepared as shown in Scheme 6.15 [44].

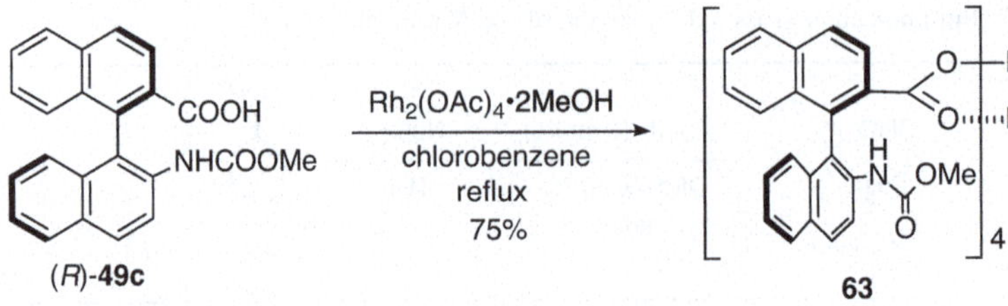

Scheme 6.15 Preparation of the rhodium carboxylate catalyst

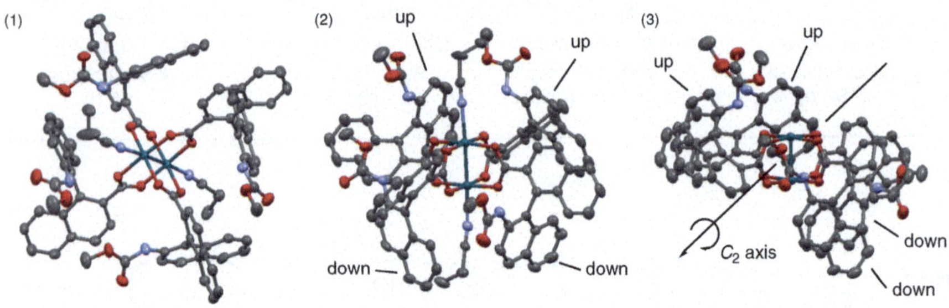

Fig. 6.8 X-ray structure of the bis(propionitrile) adduct of cat. **63** (50% probability). (1) Inclined side view. (2) Front view. (3) Side view. Hydrogen atoms and propionitrile molecules in (3) are omitted for clarity

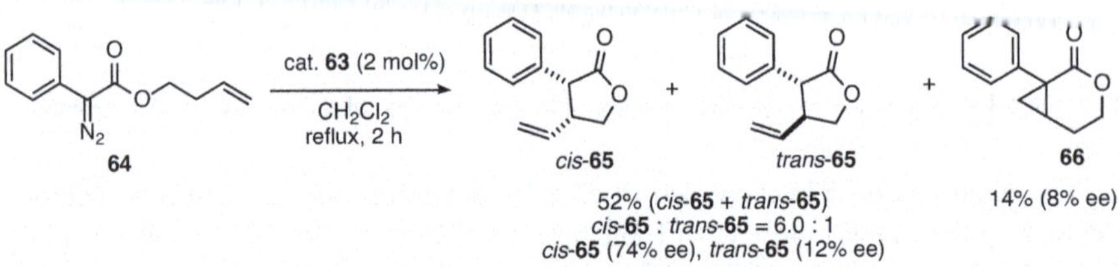

Scheme 6.16 Intramolecular asymmetric C–H insertion promoted by cat. **63**

The C_2-symmetry-like conformation of **63**, with coordination of two molecules of propionitrile at the axial sites of each octahedral rhodium, was revealed by X-ray analysis of a single crystal obtained from a propionitrile solution (Fig. 6.8).

The carbamate-substituted naphthyl rings of each bridging ligand are oriented away from the Rh carboxylate center in a down–down–up–up conformation, as shown in Fig. 6.8(2,3). The side view (Fig. 6.8(3)) suggests that a pair of carbamate-substituted naphthyl rings from adjacent bridging ligands in down–down and up–up conformations acted as a shield to control the direction of the intramolecular C–H insertion of the rhodium carbenoid intermediate.

Catalyst **63** was employed for the asymmetric intramolecular C–H insertion of diazoacetate **64** (Scheme 6.16). The reaction proceeded in the presence of 2 mol% cat. **63** to afford α,β-disubstituted γ-lactone **65** in a 52% yield with a *cis* selectivity (*cis*: *trans* = 6.0 : 1). The enantioselectivities of the *cis*- and *trans*-isomers were 74%

and 12% ee, respectively. Concomitant cyclopropanation to **66** was also observed, although five-membered ring formation via C–H insertion is the major pathway (**65**: **66** = 3.7 : 1).

6.6 Synthesis and Application of the ε-Amino Acids and Their Surrogates

The axially chiral ε-amino acid (*R*)-**11** was prepared via opening of the seven-membered lactam **68**, which was prepared through desymmetrization of the optically active dial (*R*)-**67** via an intramolecular Cannizzaro reaction (Scheme 6.17) [21].

We synthesized the ε-amino acid **13** possessing an *N*,*N*-dimethylamino pyridine (DMAP) moiety as an intermediate for the corresponding carboxylate catalyst **72a** to investigate the DMAP-catalyzed acylation (Scheme 6.18(1)) [23]. Suzuki–Miyaura cross-coupling between **69** and the aryl iodide **70** gave **71**, which was further transformed to the carboxylic acid **13**. The corresponding carboxylate ion **72a** was obtained as a tetrabutylammonium salt by treatment of an equivalent amount of tetrabutylammonium hydroxide. X-ray analysis of the ε-amino acid **13** indicated that

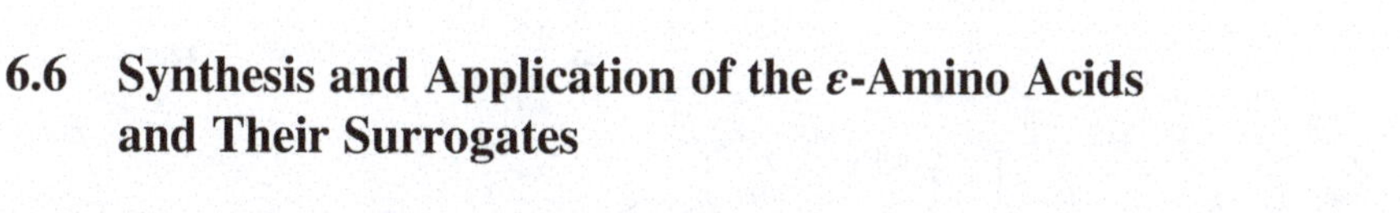

(*R*)-**67** (*R*)-**68** (*R*)-**11**

Scheme 6.17 Preparation of the axially chiral biphenyl ε-amino acid (*R*)-**11** via lactam ring opening

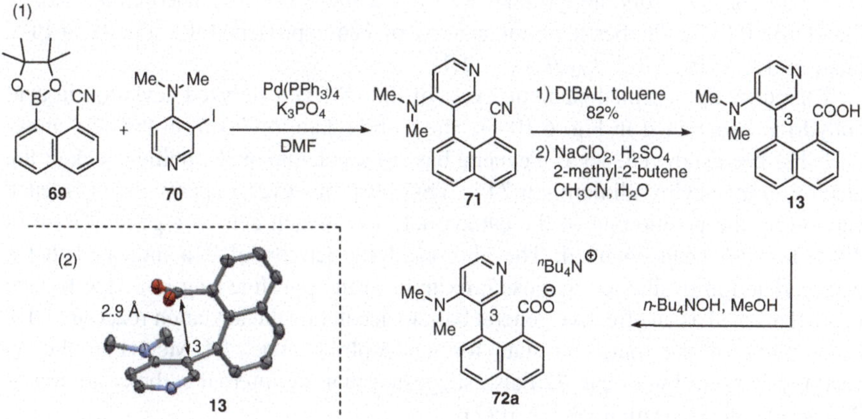

Scheme 6.18 (1) Preparation of the biaryl ε-amino acid **13** bearing a DMAP moiety and its carboxylate ion **72a**. (2) X-ray structure of **13**

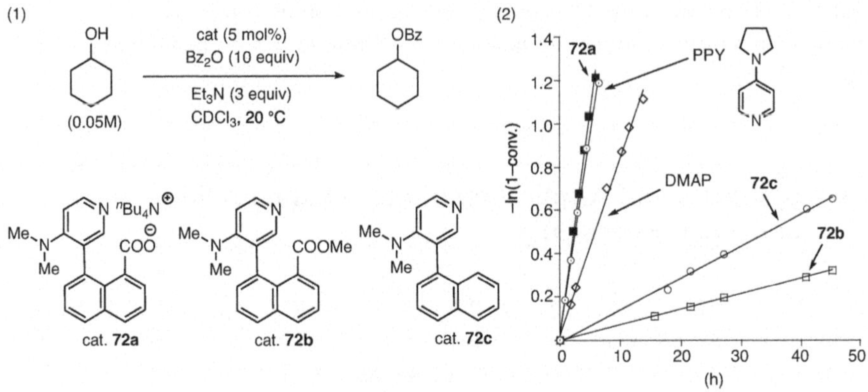

Fig. 6.9 (1) Benzoylation of cyclohexanol under pseudo-first-order conditions using 10 equivalents of Bz_2O. (2) Kinetic profiles of the cyclohexanol benzoylation, catalyzed by **72a–72c**, DMAP, or PPY

the carboxyl group was fixed in a face-to-face geometry relative to the pyridine ring with a distance of 2.9 Å between the pyridine C(3) and the carbonyl carbon (Scheme 6.18(2)).

The carboxylate catalyst **72a** was found to function as a potent nucleophilic catalyst for the acylation of alcohol by employing acid anhydride as an acylating agent [23]. The benzoylation of cyclohexanol with benzoic anhydride provides a typical example, in which cat. **72a** exhibited a higher catalytic activity than DMAP, the corresponding methyl ester catalyst **72b**, and the catalyst **72c**, without a carboxyl group (Fig. 6.9).

The relative rate constants of the benzoylation in the presence of **72a** ($k_{72a} = 2.0 \times 10^{-1}\ h^{-1}$) vs. **72b** ($k_{72b} = 7.2 \times 10^{-3}\ h^{-1}$) and **72a** vs. **72c** ($k_{72c} = 1.5 \times 10^{-2}\ h^{-1}$) were 28 ($k_{72a}/k_{72b}$) and 14 ($k_{72a}/k_{72c}$), respectively (Fig. 6.9(2)). 4-Pyrrolidinopyridine (PPY) is a more potent nucleophilic catalyst than DMAP [45]. The benzoylation activity of **72a** approached the activity of PPY ($k_{PPY} = 1.8 \times 10^{-1}\ h^{-1}$, $k_{72a}/k_{PPY} = 1.1$).

The currently accepted catalytic cycle of the DMAP-catalyzed acylation in acid anhydride is depicted in Fig. 6.10(1). The carboxylate ion **73** from the acid anhydride has been shown to act as a general base to accelerate nucleophilic attack of the alcohol to the acylpyridinium ion **74** in **TS-I** [46]; however, experimental evidence supporting the positioning of the carboxylate ion, which acts as a general base in **TS-I**, has not been obtained. The high catalytic activity of **72a** indicated that a carboxylate ion positioned in close proximity to the pyridine ring in a face-to-face geometry acted as an effective general base to accelerate the acylation reaction. DFT calculations of the transition state for nucleophilic attack of MeOH to the *N*-acetylpyridinium ion of cat. **72a** also suggested that the internal carboxylate ion of **72a** accelerated acylation (Fig. 6.10(2)).

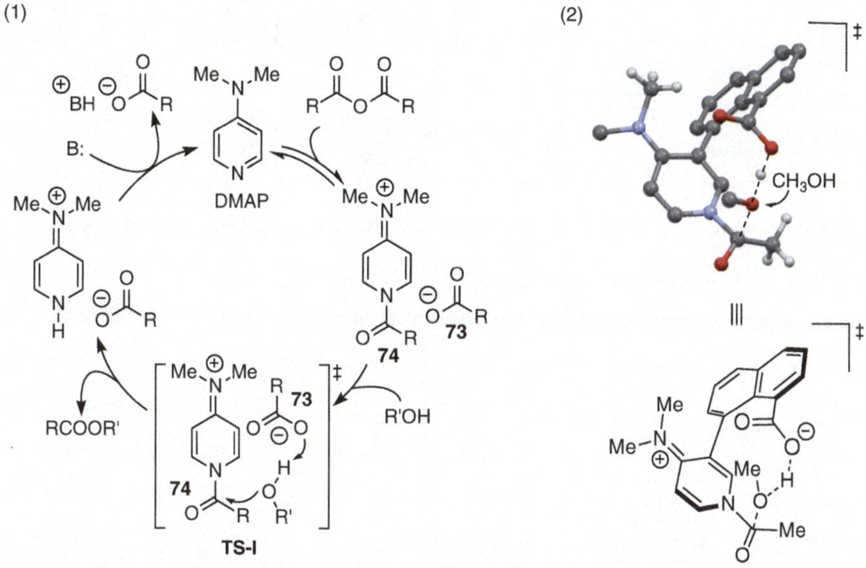

Fig. 6.10 (1) Catalytic cycle of the DMAP-catalyzed acylation in the presence of acid anhydride. (2) The transition state for nucleophilic attack of MeOH to the *N*-acetylpyridinium ion of cat. **72a**

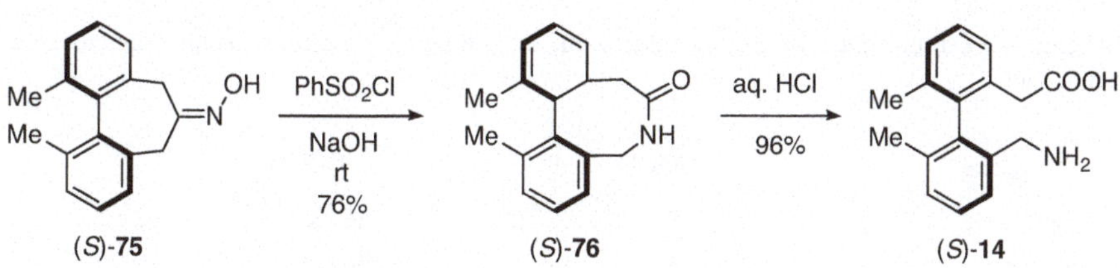

Scheme 6.19 Preparation of the axially chiral ζ-amino acid (*S*)-**14**

6.7 Synthesis and Application of the ζ- and Miscellaneous Amino Acids

The axially chiral ζ-amino acid (*S*)-**14** was prepared through the hydrolysis of the eight-membered lactam (*S*)-**76**, which was prepared from the desymmetrization of the oxime (*S*)-**75** via a Beckmann rearrangement (Scheme 6.19) [21].

Kelly and co-workers prepared a 2,3′-substituted *N*-Boc derivative of the ι-amino acid **16** (Fig. 6.3) as a template for facilitating the formation of a β-hairpin-like structure in a peptide. The peptide **77** prepared with the amino acid **16** was thought to form a β-sheet structure, as shown in Fig. 6.11, based on spectroscopic and NOE studies [25].

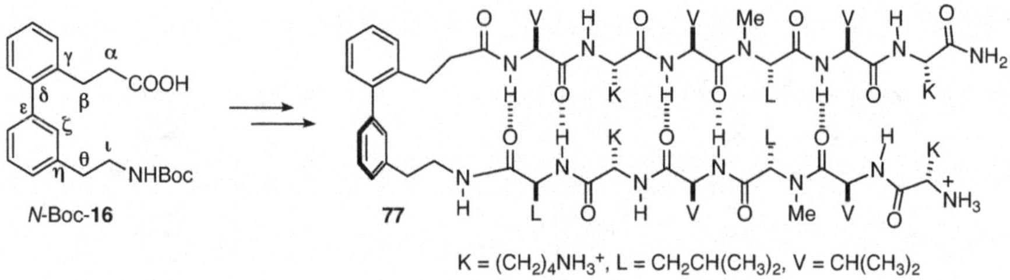

Fig. 6.11 β-Sheet structure of a peptide containing the amino acid **16**

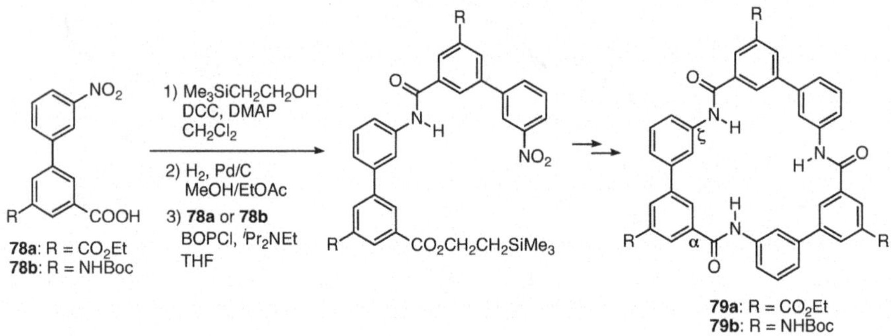

Scheme 6.20 Preparation of macrocycles with anion-binding properties, using ζ-amino acid equivalents

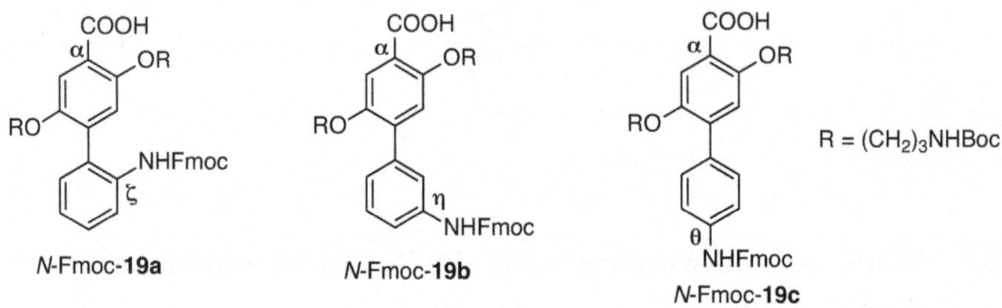

Fig. 6.12 Biphenyl Fmoc amino acids as building blocks for macrocycles

Hamilton and co-workers employed the biphenyl-type ζ-amino acids **18a** and **18b** (Fig. 6.3) as scaffolds for the macrocycles **79a** and **79b**, which display anion-binding properties through hydrogen-bonding (Scheme 6.20) [27, 28]. The biphenylcarboxylic amino acid precursors **78a** and **78b**, bearing a nitro group, were employed as building blocks for these macrocycles. Both macrocycles bound to anions, such as I^- and p-TsO^-, at their central cavity.

Nowick and co-workers designed *N*-Fmoc derivatives of the biphenyl ζ-, η-, and θ-amino acids **19a–19c** as building blocks for constructing nanometer-scale water-soluble macrocycles (Fig. 6.12) [29].

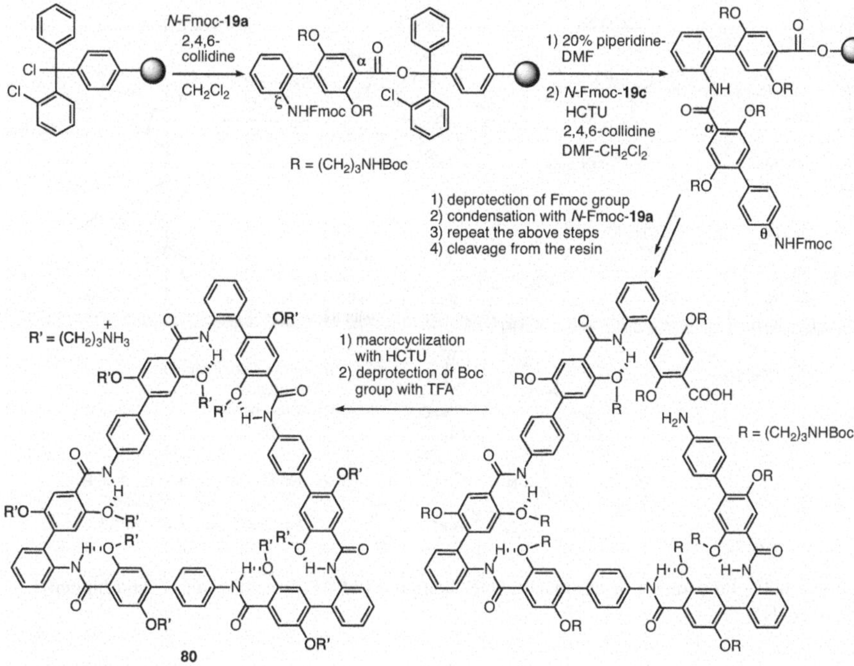

Scheme 6.21 Solid phase synthesis of macrocycles through the alternate condensation of ζ- and θ-amino acid derivatives

As a typical example, the solid phase synthesis of the triangular cyclohexamer **80** through alternate condensation of *N*-Fmoc **19a** and *N*-Fmoc **19c** is depicted in Scheme 6.21. A variety of macrocycles around 4 nm in width, up to the cyclododecamer, were also synthesized using a similar strategy.

Shionoya and co-workers used the ζ-amino acid **20** (Fig. 6.3) as a molecular scaffold to support the alternating arrangement of three proteinogenic α-amino acids in a cyclic structure (Scheme 6.22) [30]. The *C*-terminal-protected ζ-amino acid **81** was employed as a key building block to prepare the macrocycle. Repeated condensation with the natural amino acid and the ζ-amino acid gave the triangular macrocycle **83**, which was converted to the tricarboxylic acid **84** as a soluble macrocycle in basic water.

6.8 Conclusion

This chapter summarized the synthesis and applications of biaryl amino acids and their surrogates. Applications of these amino acids in fine organic synthesis, as organocatalysts or chiral ligands, highlighted the utility of the amino acid derivatives with axial chirality, as shown in Sects. 6.2, 6.3, 6.4, 6.5, and 6.6. Because these

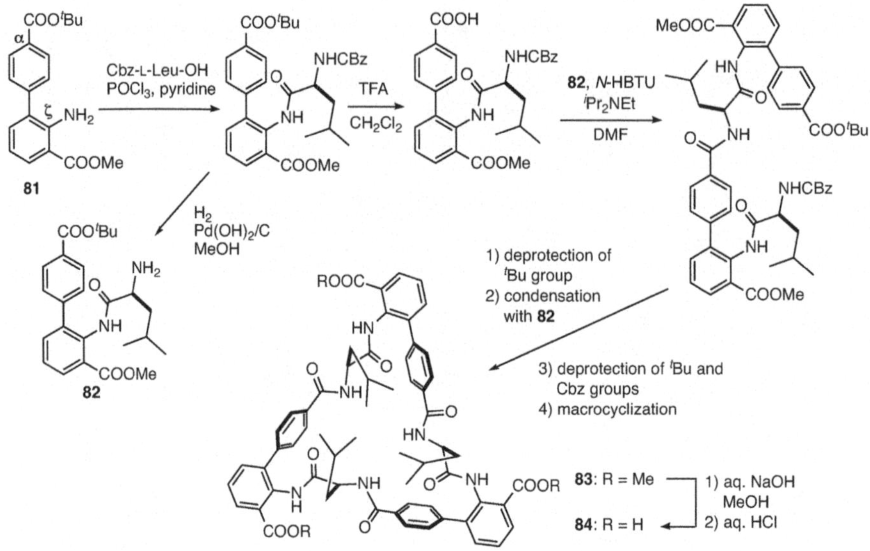

Scheme 6.22 Preparation of the macrocycles composed of the proteinogenic α- and biphenyl ζ-amino acids

structures are rigid along and flexible around the biaryl axis, biaryl amino acids may be employed as scaffolds to induce the formation of ordered conformations or cyclic structures in macromolecules, as depicted in Sect. 6.7. As exemplified in these previous studies, chemists can imagine a variety of unique structures with special functions based on the biaryl amino acids and their derivatives. The development of functional molecules based on biaryl amino acids and their surrogates show tremendous promise.

References

1. M.A.T. Blaskovich, J. Med. Chem. **59**, 10807–10836 (2016)
2. J. Vagner, H. Qu, V.J. Hruby, Curr. Opin. Chem. Biol. **12**, 292–296 (2008)
3. C. Ballatore, D.M. Huryn, A.B. Smith III, Chem. Med. Chem. **8**, 385–395 (2013)
4. L. Ridvan, N. Abdallah, R. Holakovsky, M. Tichy, J. Závada, Tetrahedron Asymmetry **7**, 231–236 (1996)
5. J.-P. Mazaleyrat, A. Gaucher, M. Wakselman, L. Tchertanov, J. Guilhem, Tetrahedron Lett. **37**, 2971–2974 (1996)
6. K. Shibatomi, K. Kitahara, T. Okimi, Y. Abe, S. Iwasa, Chem. Sci. **7**, 1388–1392 (2016)
7. K. Shibatomi, K. Kitahara, N. Sasaki, Y. Kawasaki, I. Fujisawa, S. Iwasa, Nat. Commun. **8**, 15600 (2017)
8. A. Gaucher, F. Bintein, M. Wakselman, J.P. Mazaleyrat, Tetrahedron Lett. **39**, 575–578 (1998)
9. T. Kano, J. Takai, O. Tokuda, K. Maruoka, Angew. Chem. Int. Ed. **44**, 3055–3057 (2005)
10. T. Kano, O. Tokuda, J. Takai, K. Maruoka, Chem. Asian. J. **1–2**, 210–215 (2006)
11. T. Kano, O. Tokuda, K. Maruoka, Tetrahedron Lett. **47**, 7423–7426 (2006)

12. T. Kano, Y. Yamaguchi, O. Tokuda, K. Maruoka, J. Am. Chem. Soc. **127**, 16408–16409 (2005)
13. T. Kano, Y. Yamaguchi, O. Tokuda, K. Maruoka, Chem. Eur. J. **15**, 6678–6687 (2009)
14. T. Kano, H. Sugimoto, K. Maruoka, J. Am. Chem. Soc. **133**, 18130–18133 (2011)
15. T. Kano, K. Maruoka, Chem. Sci. **4**, 907–915 (2013)
16. E. Perissuttii, F. Frecentese, A. Lavecchia, F. Fiorino, B. Severino, F.A. Angelis, V. Santagada, G. Caliendo, Tetrahedron **63**, 12779–12785 (2007)
17. T. Furuta, J. Yamamoto, Y. Kitamura, A. Hashimoto, H. Masu, I. Azumaya, T. Kan, T. Kawabata, J. Org. Chem. **75**, 7010–7013 (2010)
18. T. Furuta, M. Nikaido, J. Yamamoto, T. Kuribayashi, T. Kawabata, Synthesis **45**, 1312–1318 (2013)
19. T. Baba, J. Yamamoto, K. Hayashi, M. Sato, M. Yamanaka, T. Kawabata, T. Furuta, Chem. Sci. **7**, 3791–3797 (2016)
20. V. Brandmeier, W.H.B. Sauer, M. Feigel, Helv. Chem. Acta. **77**, 70–85 (1994)
21. M. Tichy, J. Holanová, J. Závada, Tetrahedron Asymmetry **9**, 3497–3504 (1998)
22. C. Nesloney, J.W. Kelly, J. Org. Chem. **61**, 3127–3137 (2016)
23. R. Nishino, T. Furuta, K. Kan, M. Sato, M. Yamanaka, T. Sasamori, N. Tokitoh, T. Kawabata, Angew. Chem. Int. Ed. **52**, 6445–6449 (2013)
24. M. Thoß, R.W. Seidel, M. Feigel, Tetrahedron **66**, 8503–8511 (2010)
25. C. Nesloney, J.W. Kelly, J. Am. Chem. Soc. **118**, 5836–5845 (1996)
26. N. Srinivas, K. Moehle, K. Abou-Hadeed, D. Obrecht, J.A. Robinson, Org. Biomol. Chem. **5**, 3100–3105 (2007)
27. K. Choi, A.D. Hamilton, J. Am. Chem. Soc. **123**, 2456–2457 (2001)
28. K. Choi, A.D. Hamilton, J. Am. Chem. Soc. **125**, 10241–10249 (2003)
29. C.M. Gothard, J.S. Nowick, J. Org. Chem. **75**, 1822–1830 (2010)
30. S. Tashiro, M. Chiba, M. Shionoya, Chem. Asian. J. **12**, 1087–1094 (2017)
31. A. Torrado, B. Imperiali, J. Org. Chem. **61**, 8940–8948 (1996)
32. G.R. Newkome, J. Gross, A.K. Patri, J. Org. Chem. **62**, 3013–3014 (1997)
33. S.-H. Kang, C.M. Gothard, S. Maitra, Atia-tul-Wahab, J.S. Nowick, J. Am. Chem. Soc. **129**, 1486–1482 (2007)
34. A.-L. Gérard, V. Lisowski, S. Rault, Tetrahedron **61**, 6082–6087 (2005)
35. L. Ridvan, M. Budesínsky, M. Tichy, P. Malon, J. Závada, J. Podlaha, I. Císarová, Tetrahedron **55**, 12331–12348 (1999)
36. F. Formaggio, M. Crisma, C. Toniolo, L. Tchertano, J. Guilhem, J.-P. Mazaleyrat, A. Gaucher, M. Wakselman, Tetrahedron **56**, 8721–8734 (2000)
37. F. Formaggio, C. Peggion, M. Crisma, C. Toniolo, L. Tchertanov, J. Gulhem, J.-P. Mazaleyrat, Y. Goubard, M. Wakselman, Helv. Chem. Acta. **84**, 481–501 (2001)
38. T. Kano, A. Noishiki, R. Sakamoto, K. Maruoka, Chem. Commun. **47**, 10626–10628 (2011)
39. T. Kano, Y. Yamaguchi, K. Maruoka, Angew. Chem. Int. Ed. **48**, 1838–3057 (2009)
40. J. W. Yang, M. Stadler, B. List, Angew. Chem. Int. Ed. **46**, 609–611 (2007)
41. T. Furuta, Y. Kitamura, A. Hashimoto, S. Fujii, K. Tanaka, T. Kan, Org. Lett. **9**, 183–186 (2007)
42. T. Kano, Y. Tanaka, K. Maruoka, Tetrahedron Lett. **47**, 3039–3041 (2006)
43. T. Kano, Y. Tanaka, K. Maruoka, Tetrahedron **63**, 8658–8664 (2007)
44. W. Lu, X. Pei, T. Murai, T. Sasamori, N. Tokitoh, T. Kawabata, T. Furuta, Synlett **28**, 679–683 (2017)
45. G. Höfle, W. Steglich, Synthesis **72**, 619–621 (1972)
46. S. Xu, I. Held, B. Kempf, H. Mayr, W. Steglich, H. Zipse, Chem. Eur. J. **11**, 4751–4757 (2005)

Chapter 7
Interplay of Diamides and Rare Earth Metals: Specific Molecular Spaces and Catalytic Activity

Naoya Kumagai and Masakatsu Shibasaki

Abstract A catalytic system comprising functionalized small diamides and rare earth metals (REs) exerts intriguing catalytic properties that are dictated by dynamic construction of flexible molecular spaces. The dynamic interaction of diamides and REs is characterized by broad applicability to distinct reaction systems as well as notable switchable catalysis. Structural modification of the diamide allows for enhanced intermolecular interactions to afford a self-assembled solid-phase catalyst with a specific molecular space that engages in heterogeneous asymmetric catalysis with a continuous-flow platform.

Keywords Diamide · Rare earth metals (REs) · Asymmetric catalysis · Self-assembly · Photoisomerization

7.1 Introduction

Given that the majority of organic compounds acquire inherent chirality, stereoselectivity is an indispensable feature when forging carbon-carbon or carbon-heteroatom bonds in organic synthesis. The concept of molecular space is critical to achieve stereoselective synthesis, wherein the bond-forming events reflect the stereochemical information of the specific molecular space. Asymmetric catalysis is perhaps the most successful means for transferring the stereochemical information of the promoter, i.e., the catalyst, to the reactants, producing stereochemically enriched compounds in a leveraged manner [1, 2]. Conventional asymmetric metal-based catalysts are generally harnessed with chiral ligands having a rigid framework to decorate catalytically competent metallic species, leading to the development of a number of synthetically useful protocols with a practical level of stereoselectivity. While several organocatalysts [3–11] and small peptide catalysts [12–20] with conformational flexibility are available as synthetically useful

N. Kumagai (✉) · M. Shibasaki
Institute of Microbial Chemistry (BIKAKEN), Tokyo, Shinagawa-ku, Tokyo, Japan
e-mail: nkumagai@bikaken.or.jp

S. Shirakawa (ed.), *Designed Molecular Space in Material Science and Catalysis*,
https://doi.org/10.1007/978-981-13-1256-4_7

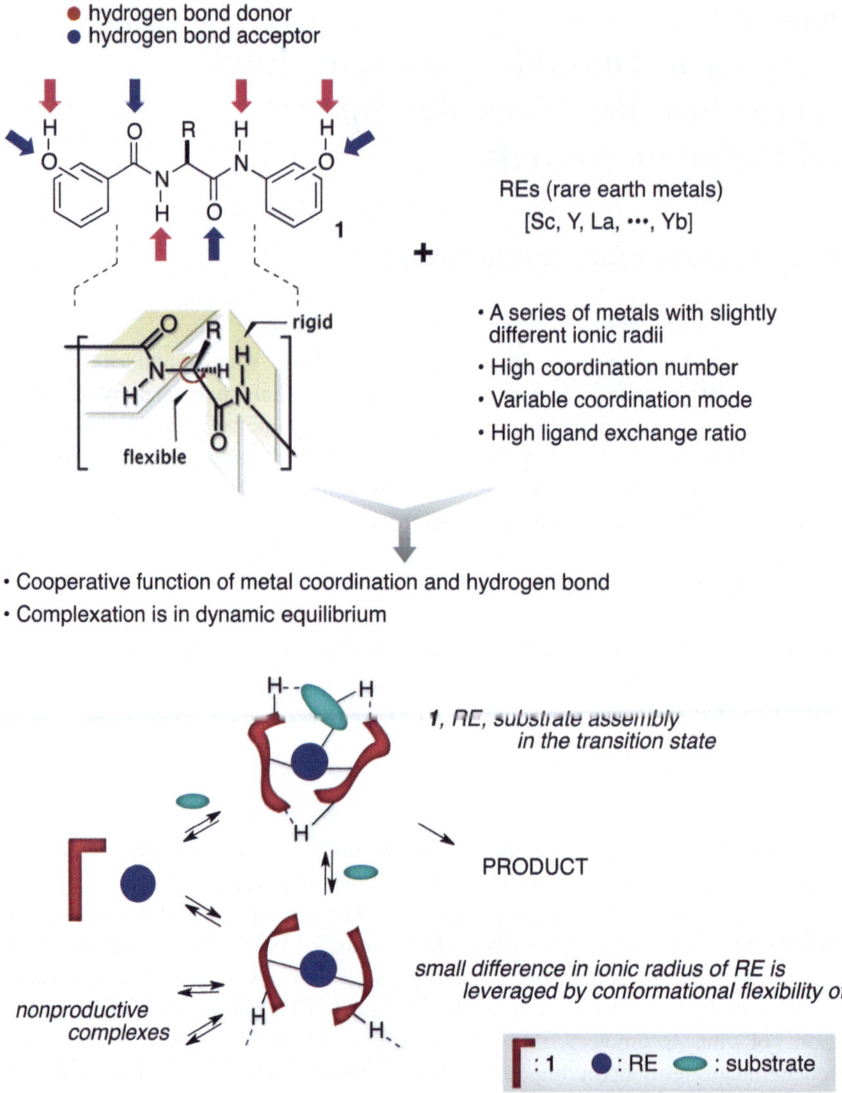

Fig. 7.1 Catalytic system comprising conformationally flexible diamide ligand **1** and rare earth metals (REs)

catalysts, this is not the case for metal-based catalysts in which structural flexibility plays a key role in catalysis [21–25]. This chapter spotlights a peculiar catalytic system comprising structurally flexible diamide ligands **1** with two phenol units and rare earth metals (REs) that exerts specific physical and chemical properties (Fig. 7.1) [26, 27]. Diamide **1** is readily prepared from the corresponding α-amino acids without chromatographic purification. **1** possesses two secondary amides as well as two phenolic hydroxyl groups, displaying multifaceted coordination properties by hydrogen bonding interactions and metal coordination. As for the conformation of the entire framework, **1** has two planar amide moieties that are interconnected by a flexible α-carbon and exhibits reasonable rigidity and flexibility. On the other hand, REs, a series of 17 metals from Sc, Y, La, Ce, to Lu, typically exhibiting trivalency and Lewis acidity with slightly different ionic radii, are characterized by a diverse coordination mode and an unparalleled high coordination number (6–12) [28–30]. The combination of conformationally flexible diamide ligand **1** and REs leads to a dynamic catalytic system, in which loose hydrogen bonding interactions, including reactants, and metal coordination with REs orchestrate to form a transition-state assembly in a chiral molecular space. The availability of REs with small variations in their ionic radii allows an optimal RE/**1** catalytic system to be found for specific reactions without requiring tedious structural modifications of **1**.

7.2 Properties of Bis(hydroxyphenyl)diamides

7.2.1 Heterochiral Aggregation

Diamide **1** possesses a unique structure and is rich in hydrogen bonding donor and acceptor sites, resulting in solvent-dependent association/dissociation properties [31]. (*S*)-**1a**, derived from natural L-Val, instantaneously forms insoluble aggregates with antipodal (*R*)-**1a** in halogenated solvents, e.g., chloroform (Fig. 7.2). In contrast, in ethereal and alcoholic solvents, a homogeneous racemic solution develops, implying that hydrogen bonding interactions are involved in this heterochiral aggregation. X-ray crystallographic analysis of the heterochiral aggregates formed in $CHCl_3$ confirmed the presence of tight hydrogen bonding interactions in an alternating array of (*S*)- and (*R*)-**1a** (Fig. 7.3). Indeed, self-discrimination efficiently proceeds and a sample of **1a** with 4% ee can be enriched to 91.2% ee in solution phase. Of note is the high fidelity of this heterochiral aggregation, in which stringent recognition of molecular architectures is demonstrated in an ensemble of structurally similar molecules (Fig. 7.4). In the presence of the same amount of six different dummy molecules with an *S*-configuration, ester analogs (*S*)-**2**,**3**, diamides (*S*)-**4**,**5** lacking one phenolic hydroxyl group, and analogs (*S*)-**6**,**7** embedded with a methylene group, only (*R*)-**1a** and (*S*)-**1a** form heterochiral aggregates. This molecular recognition suggests that these two secondary amides and two phenols play a central role as a privileged framework for aggregation through a tight hydrogen bonding network.

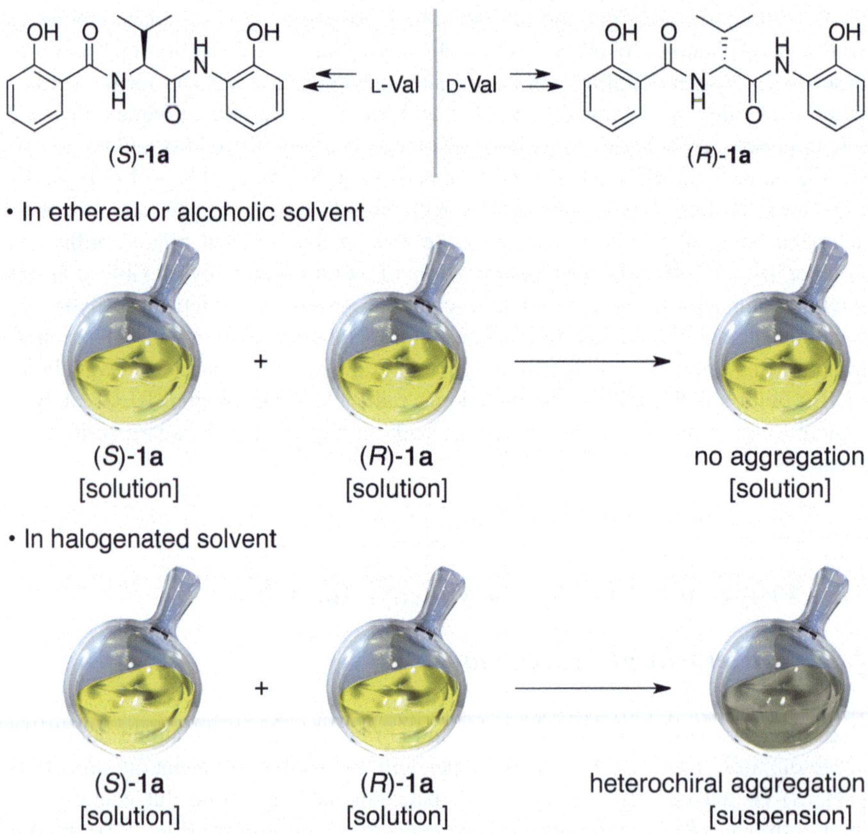

Fig. 7.2 Heterochiral aggregation of bis(hydroxyphenyl)diamide **1a**

7.2.2 Photoresponsive Aggregation/Dissociation

This molecular recognition process inspired the development of a photoresponsive aggregation/dissociation system, enabling spatiotemporal manipulation of the molecular behavior by an external stimulus. Designed diamide (*S*)-**1b** has a photoresponsive azobenzene unit, while retaining the privileged framework for aggregation (Fig. 7.5) [32]. When the azobenzene unit is in the *E*-form, (*S*)-(*E*)-**1b** readily forms heterochiral aggregates with antipodal (*R*)-(*E*)-**1b** in $CHCl_3$, as expected. X-ray crystallographic analysis of the aggregate reveals a tight hydrogen bonding network as well as favorable π-contact due to the presence of an electron-deficient 3,5-bis(trifluoromethyl)phenyl unit at the azo functional group. In sharp contrast, the mixture of corresponding *Z* isomers of (*S*)-**1b** and (*R*)-**1b** in the same solvent had high solubility, as evidenced by the disappearance of solid (*S*)-(*E*)-**1b**/(*R*)-(*E*)-**1b** aggregates after UV (365 nm) irradiation. This is presumably due to the

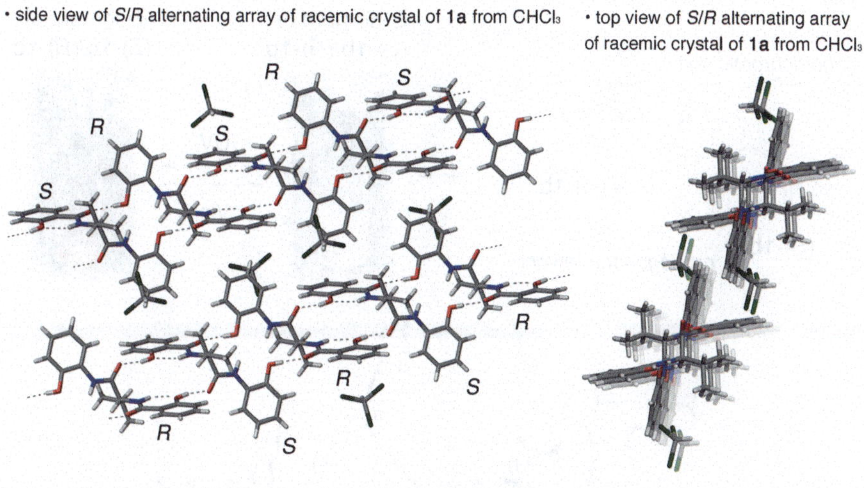

Fig. 7.3 Intermolecular hydrogen bonding interactions of **1a** found in X-ray crystallographic analysis. Color code for crystal structures: hydrogen, white; carbon, gray; nitrogen, blue; oxygen, red; chlorine, dark green

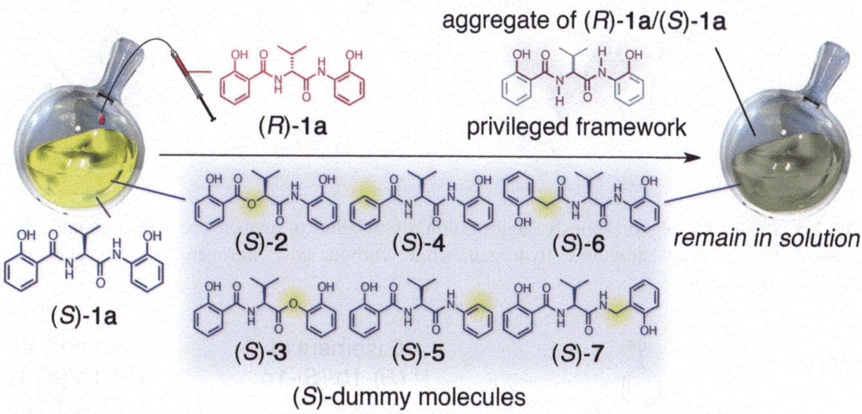

Fig. 7.4 High-fidelity heterochiral recognition of diamide (*S*)-**1a**/(*R*)-**1a**

geometric isomerization of azobenzene unit, which significantly impedes the requisite hydrogen bonding network to maintain tight aggregation. Visible light irradiation (>422 nm) of the developed solution induces re-isomerization to *E* isomers, resulting in re-aggregation of the diamide molecules, which can be repeatedly manifested in both directions. (*R*)-(*E*)-**1b** forms "cross" aggregates with (*S*)-(*E*)-**1c**, having a 3,5-dimethylphenyl unit instead of the 3,5-bis(trifluoromethyl)phenyl unit, with reversible features, further supporting the notion that the heterochiral aggregation is primarily caused by the privileged hydrogen bonding array (Fig. 7.6).

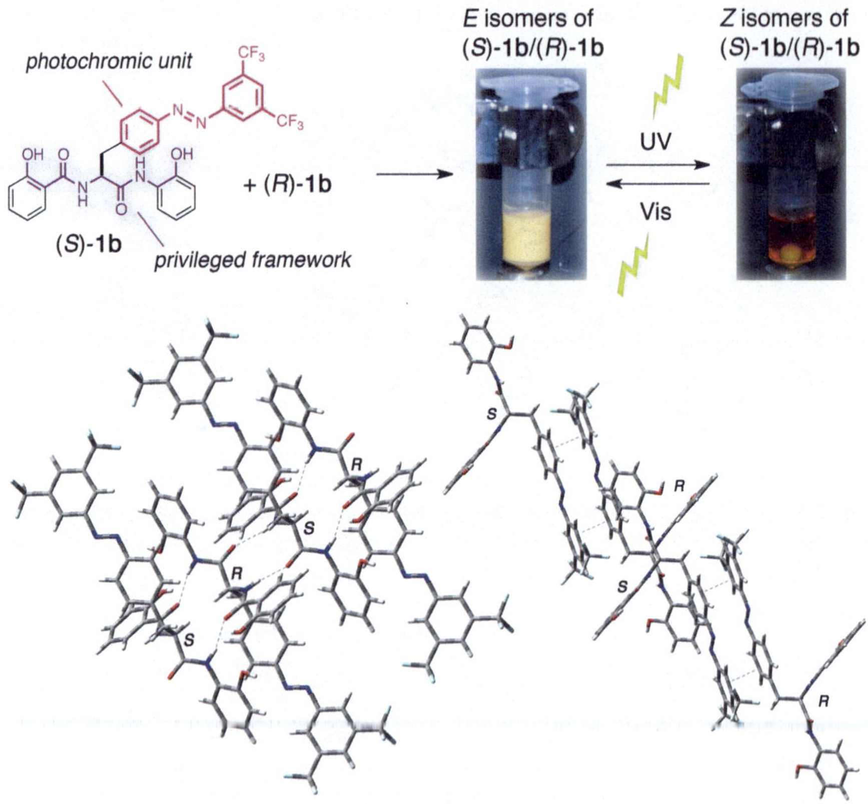

Fig. 7.5 Reversible photoresponsive aggregation/dissociation of azobenzene-armed diamide **1b**. Color code for crystal structures: hydrogen, white; carbon, gray; nitrogen, blue; oxygen, red; chlorine, dark green

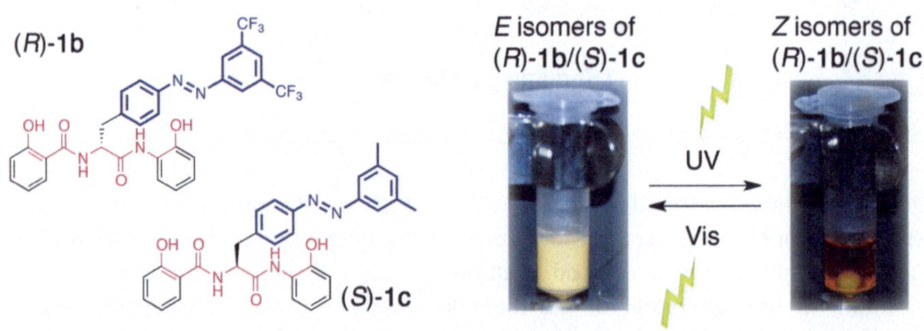

Fig. 7.6 Cross heterochiral aggregation/dissociation of (*R*)-**1b** and (*S*)-**1c**

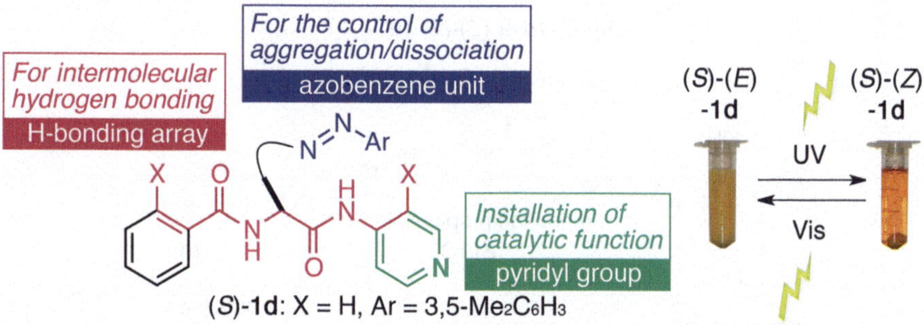

Fig. 7.7 Designed diamide (*S*)-**1d** armed with a nucleophilic pyridyl group and its aggregation/dissociation dynamics

7.2.3 Photoresponsive Switchable Catalysis

Diamide **1b** discussed above shows photoresponsive activity by incorporating the azobenzene unit. The diamide molecule still contains a manipulative moiety, i.e., aromatic rings of phenol units at both ends (Fig. 7.7) [33, 34]. Molecular catalysis is generally faster in a homogeneous ensemble and aggregated heterogeneous catalysts likely exhibit lower catalytic activity compared with the homogeneous counterparts having identical molecular architectures. This idea led to the production of diamide **1d** equipped with a pyridyl unit that is responsible for nucleophilic catalysis, allowing for spatiotemporal control of the catalyst activity of interest. (*S*)-**1d** without two phenolic hydroxyl groups displays inherently lower solubility than the other derivatives, and even its homochiral *E* isomer is almost insoluble in *n*-hexane/propionitrile = 4/1, producing only self-aggregated material from the solution. Upon irradiation, the aggregates become soluble by photoisomerization from *E* to *Z* isomers, leading to a completely homogeneous solution rich in *Z* isomers. This aggregation/dissociation process is repeatedly manifested without degradation. Given that the differential catalytic activity of the pyridine-armed (*S*)-**1d** is confirmed in *O*-Boc formation with Boc_2O, the feasibility for in situ modulation of the catalytic activity was evaluated (Fig. 7.8). The initially observed catalytic activity by (*S*)-(*Z*)-**1d** gradually diminishes under visible light (>422 nm) irradiation due to the aggregation of the thus-formed (*S*)-(*E*)-**1d,** as visually observed. Subsequent UV (365 nm) irradiation reverts (*S*)-(*E*)-**1d** to soluble (*S*)-(*Z*)-**1d**, which allows the reaction to proceed again. This spatiotemporal on/off catalysis using (*S*)-**1d** is also valid for Steglich rearrangement of 2-acyloxybenzofurans to 3-acyl-2-benzofuranones.

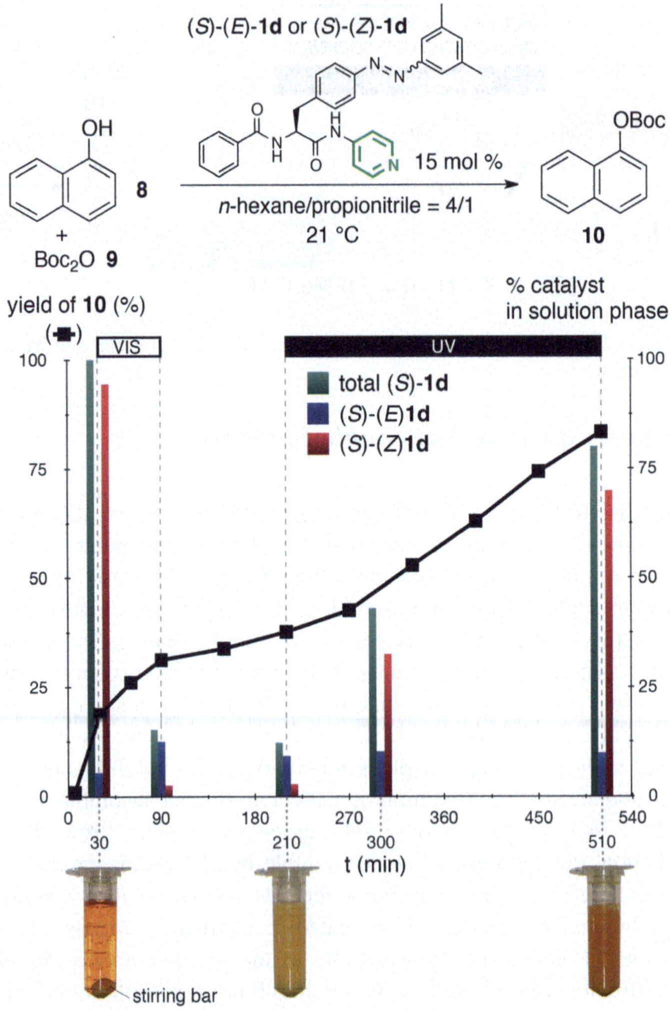

Fig. 7.8 In situ on/off catalysis exerted by photoresponsive catalyst (*S*)-**1d**

7.3 Homogeneous Complex

7.3.1 Monometallic Catalysis

As discussed in the preceding sections, diamide (*S*)-**1a** displays notable hydrogen bonding properties. The utility of (*S*)-**1a** as a fundamental unit to construct a specific molecular space is further expanded by successful marriage with REs. Five different reactions (or stereoselectivities) are realized using only the single molecular architecture of (*S*)-**1a** by changing the REs (Scheme 7.1) [35–42]. This is likely because

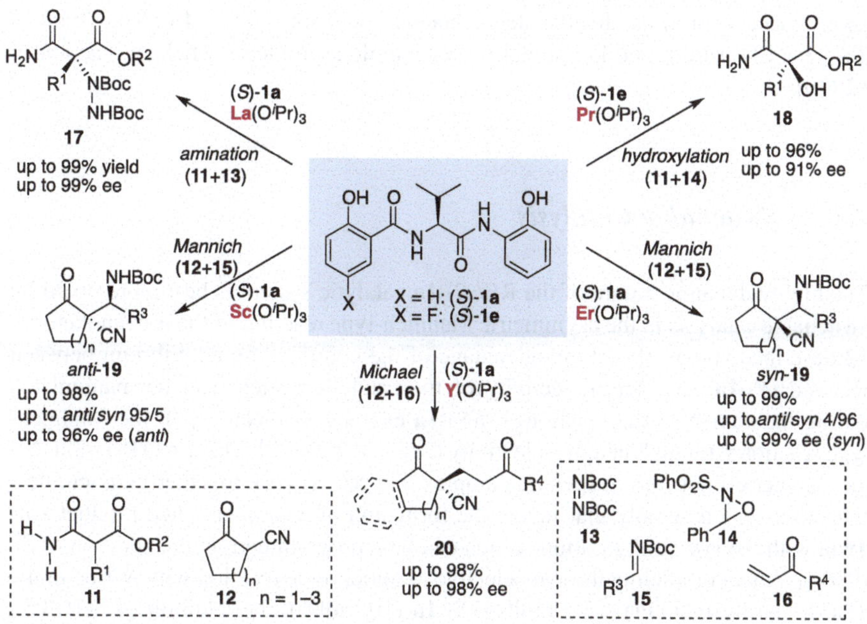

Scheme 7.1 Monometallic RE/(*S*)-**1a** catalysts for five different asymmetric reactions

(1) the coordination of (*S*)-**1a** to RE(O^{i}Pr)$_3$ is in equilibrium and the dissociated state is dominant; (2) (*S*)-**1a** and RE(O^{i}Pr)$_3$ work together to assemble a transition-state complex including substrates; and (3) at the transition state, due to a high coordination number and flexible coordination mode of RE to respond to the chemical environment, the most suitable chemical space for high stereoselectivity can be easily accessed by employing REs of different ionic radii. In this regard, this catalytic system favors the use of substrates with high coordination ability. Specifically, in the asymmetric amination promoted by the La/(*S*)-**1a** catalytic system, primary amide **11** is a privileged substrate, and the corresponding secondary or tertiary amide produces a significantly diminished reaction outcome [35, 37, 38]. This substrate specificity is also observed in the catalytic asymmetric hydroxylation promoted by Pr/(*S*)-**1e,** while minor electronic modifications of diamide are essential for achieving high stereoselectivity [42]. Strongly basic La(O^{i}Pr)$_3$ can be replaced by the combined use of neutral La(NO$_3$)$_3$ hydrate and weak amine base, suggesting that no specific metal complex is formed in the reaction mixture and all of the chemical species are in equilibrium. An Eyring plot reveals large negative activation entropy of the amination, further supporting the involvement of a highly organized transition state [38]. Of particular note is the drastic reversal of diastereoselectivity in the Mannich-type reaction of α-cyanoketones **12** just by changing REs [36, 39]. Circular dichroism (CD) spectra of Sc/(*S*)-**1a** and Er/(*S*)-**1a** give largely different patterns, implying that chiroptically different assemblies are

taking place to produce diverse stereochemical outcomes. Y/(*S*)-**1a** also exhibits a different CD pattern and is optimal for a catalytic asymmetric Michael reaction of **12** [40].

7.3.2 Switchable Catalysis

The highly dynamic nature of the RE/(*S*)-**1a** catalytic system is best represented by switchable catalysis in the asymmetric Mannich-type reaction of the α-cyanoketones **12** described above. By taking advantage of the coordination equilibrium between RE and (*S*)-**1a**, the stereoselectivity of the catalytic system can be manipulated during the course of the reaction. Er/(*S*)-**1a** exerts *syn*-selectivity in the Mannich-type reaction with high enantioselectivity (Scheme 7.2) [39]. This $Er(O^iPr)_3$ and (*S*)-**1a** together do not form a specific complex, but rather work together to assemble a transition-state assembly that favors the formation of enantioenriched product *syn*-**19aa** with the *syn*-configuration, which can be reprogrammed by the addition of $Sc(O^iPr)_3$. After completing the *syn*-selective Mannich-type reaction with *N*-Boc imine **15a** driven by $Er(O^iPr)_3$ (2.5 mol%)/(*S*)-**1a** (10 mol%), the addition of $Sc(O^iPr)_3$ (2.5 mol%) overrides the coordination of (*S*)-**1a** to $Er(O^iPr)_3$, switching the diastereoselectivity from *syn* to *anti* exerted by Sc/(*S*)-**1a** catalysis. In situ tracing by CD measurement reveals that the CD fingerprint changes from an Er/(*S*)-**1a** pattern to an Sc/(*S*)-**1a** pattern. Subsequent addition of different imine **15b** allows

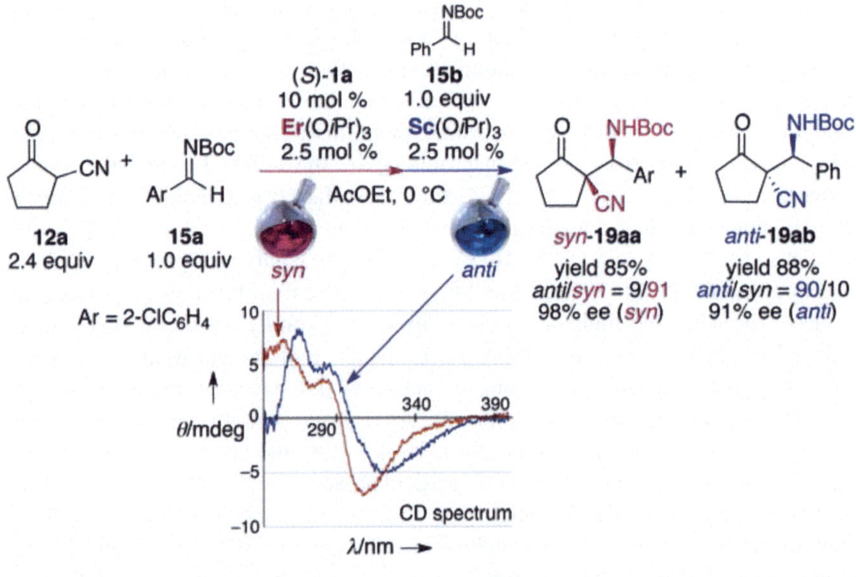

Scheme 7.2 In situ dynamic switching of the diastereoselectivity in catalytic asymmetric Mannich-type reactions

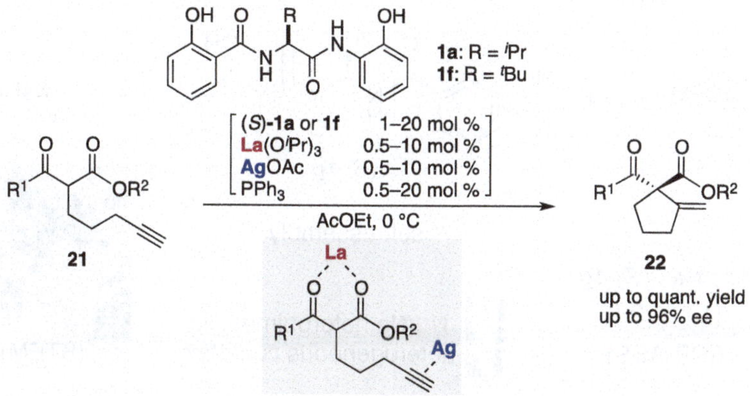

Scheme 7.3 Catalytic asymmetric Conia-ene reaction promoted by La/Ag/(*S*)-**1a** homogeneous bimetallic catalysts

the second Mannich-type reaction to proceed in an *anti*-selective manner with high enantioselectivity.

7.3.3 Bimetallic Catalysis

The use of another metal cation expands the utility of the RE/(*S*)-**1a** catalytic system. Given the hard Lewis basic and acidic nature of (*S*)-**1a** and RE, the incremental use of a soft Lewis acidic metal cation exerts orthogonal catalytic activity. This combination of catalytic functions finds ambident molecules to be suitable substrates that can be doubly activated by hard and soft Lewis acids. The La/(*S*)-**1a** catalytic system, the optimal catalyst for catalytic asymmetric amination of 1,3-dicarbonyl compounds (Scheme 7.1), finds a new dimension of utility in a catalytic asymmetric Conia-ene reaction in the presence of soft Ag(I) salt (Scheme 7.3) [43]. AgOAc performs best with PPh_3 to activate a soft Lewis basic terminal alkyne functionality. This alkyne activation is integrated with enolate formation as well as the stereochemical control exerted by RE/(*S*)-**1a**, thereby affording the desired cyclized product **22**. Minor structural modifications of the amino acid residue part of the diamide are beneficial for some substrates.

7.4 Heterogeneous Complex

7.4.1 RE/Na Heterobimetallic Complex and Catalysis

Although the preceding discussion of the bimetallic catalytic system is intriguing in terms of its dual activation mode, the catalysis itself is a conventional organic

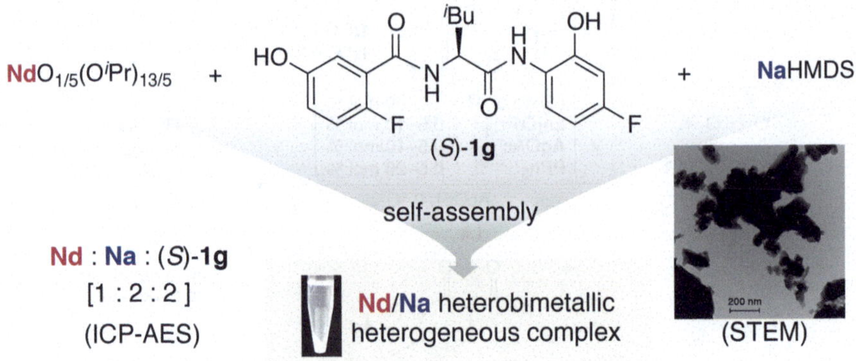

Fig. 7.9 Preparation and characterization of the insoluble heterobimetallic complex containing diamide (*S*)-**1g**, Nd, and Na cations

transformation in homogeneous media. This section examines the heterogeneous complex comprising RE, alkali metal, and diamide **1**. While heterogeneous catalysis has some practical advantages, e.g., repetitive use and application in a continuous-flow platform, most of the asymmetric catalysts are applicable only in homogeneous media. A common strategy for developing heterogeneous asymmetric catalysts is to chemically ligate soluble catalysts of interest onto chemically inert solid materials, which frequently diminishes the catalytic activity and stereoselectivity. By taking advantage of the multiple hydrogen bonding sites of diamide **1** and the high coordination number of RE, under certain conditions their mixture affords an insoluble complex. Designed amide (*S*)-**1g**, in which one hydroxyl group is translocated from the *ortho*- to *meta*-position and two fluorine atoms are installed on the aromatic rings, forms a highly insoluble complex in the presence of Nd and Na cations (Fig. 7.9) [44, 45]. The ESI-MS analysis indicates that the thus-formed solid material contains both metal cations, which is consistent with the elemental analysis by ICP-AES as well as EDS mapping in STEM imaging. The molecular space composed of chiral diamide (*S*)-**1g**, Nd, and Na cations serves as an effective heterogeneous catalyst for *anti*-selective asymmetric nitroaldol reactions of aldehydes **23** with broad substrate generality, in which the catalyst suspended in THF functions as a solid-phase catalyst (Scheme 7.4). The nitroaldol products **25** are direct precursors of *anti*-vicinal amino alcohols via facile reduction of the nitro group, which represents a privileged structural motif in the field of medicinal chemistry, as exemplified by the enantioselective synthesis of the anti-influenza drug Relenza® [46]. $NdO_{1/5}(O^iPr)_{13/5}$ and NaHMDS can be replaced with much less expensive alternatives, such as $NdCl_3$ hydrate, and NaO^tBu, with almost identical catalytic performance [47, 48]. The combination of REs and alkali metals is important for inducing self-assembly to make infinite coordination networks, and another combination, Pr/Na/(*S*)-**1g**, could be utilized as a self-assembled solid-phase catalyst in an asymmetric nitroaldol reaction using trifluoromethyl ketones **26** as electrophiles (Schemes 7.5) [49].

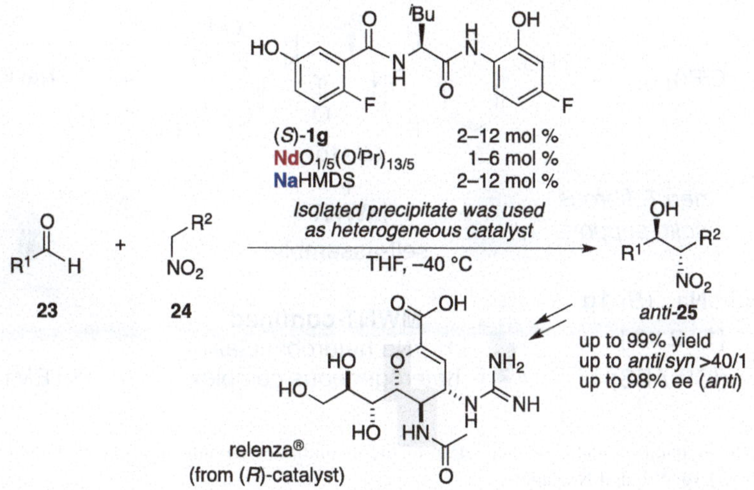

Scheme 7.4 Catalytic asymmetric nitroaldol reaction of aldehydes **23** promoted by Nd/Na/(*S*)-**1g** heterogeneous heterobimetallic catalysts

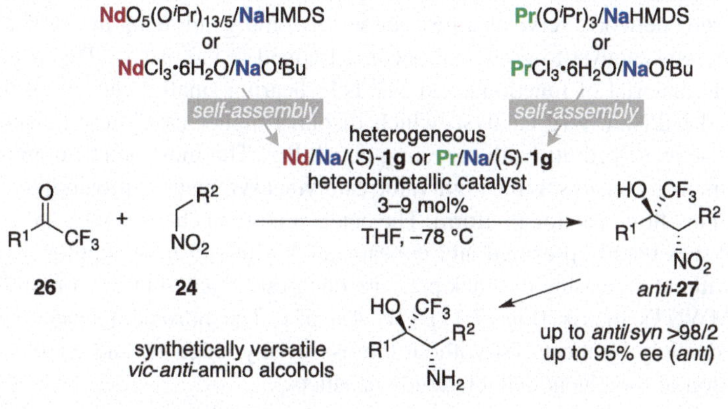

Scheme 7.5 Catalytic asymmetric nitroaldol reaction of trifluoromethyl ketones **26** promoted by Nd/Na/(*S*)-**1g** or Pr/Na/(*S*)-**1g** heterogeneous heterobimetallic catalysts

7.4.2 Composite Material with Multi-walled Carbon Nanotube

The Nd/Na/(*S*)-**1g** heterogeneous catalyst is noteworthy for its simple preparation protocol and particular utility in medicinal chemistry. Although the protocol makes use of the self-assembly process and simple mixing can afford the requisite heterogeneous catalyst in a highly reproducible manner, recycling of the catalyst is operationally difficult because the particle size of the catalysts is too small to filter efficiently. By taking advantage of the self-assembly process, strategic use of multi-

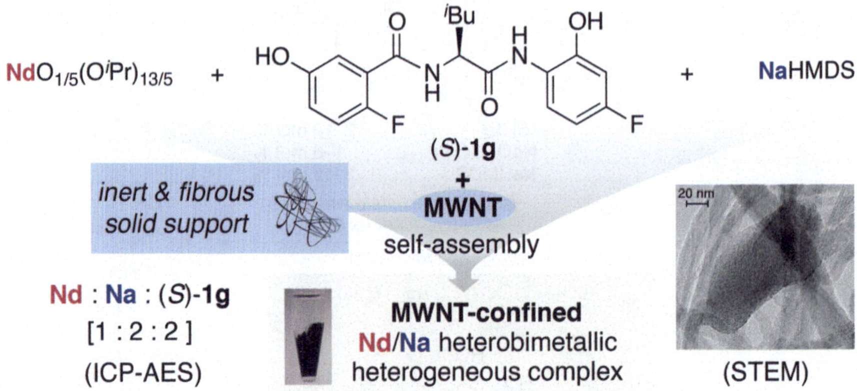

Fig. 7.10 Preparation and characterization of the insoluble heterobimetallic complex containing diamide (*S*)-**1g**, Nd, and Na cations

walled carbon nanotubes (MWNTs) is beneficial for producing more durable and recyclable catalysts (Fig. 7.10) [50, 51]. MWNTs with a higher aspect ratio, featuring a fibrous network, serve as a nest for self-assembly, allowing the Nd/Na/(*S*)-**1g** heterogeneous catalyst to grow and become trapped in the matrix. This results in a composite material of functionalized MWNTs bearing smaller clusters of the self-assembled Nd/Na/(*S*)-**1g** catalyst, which displays higher catalytic efficiency with identical stereoselectivity with recycling capability. The most practical application of the composite catalyst is the high-value catalytic asymmetric nitroaldol reaction in a continuous-flow reaction platform. The catalyst column charged with the composite catalyst is readily prepared and concatenated with substrate supplies in a THF stream without increasing the back pressure due to sufficient numbers of void spaces in the MWNTs matrix (Fig. 7.11) [47, 48, 52]. The nitroaldol reaction can be promoted solely by a catalyst without the use of any reagents, and thereby simple evaporation of the eluent delivers crude products.

This self-assembly-driven confinement strategy for the preparation of heterogeneous catalysts is valid for monometallic catalysts by increasing the amide functional groups. The newly designed diamide (*S*,*S*)-**1h**, a linked derivative of (*S*)-**1a**, exhibits self-assembling properties with $Er(O^iPr)_3$ (Fig. 7.12) [53]. In contrast to the Er/(*S*)-**1a** mixture, Er/(*S*,*S*)-**1h** readily assembles to form insoluble solid materials composed of the catalytically active complex. The use of MWNTs renders the self-assembly process in the fibrous MWNTs matrix, resulting in a composite solid catalyst favorable for continuous-flow reactions. The *syn*-selective Mannich-type reaction of α-cyanoketones **12**, efficiently promoted by Er/(*S*)-**1a** catalytic system, proceeds smoothly with the Er/(*S*,*S*)-**1h** catalyst with a comparable level of stereoselectivity due to its similar ligand architecture. The linker length of (*S*,*S*)-**1h** has a significant effect on the assembly efficiency as well as catalytic activity and

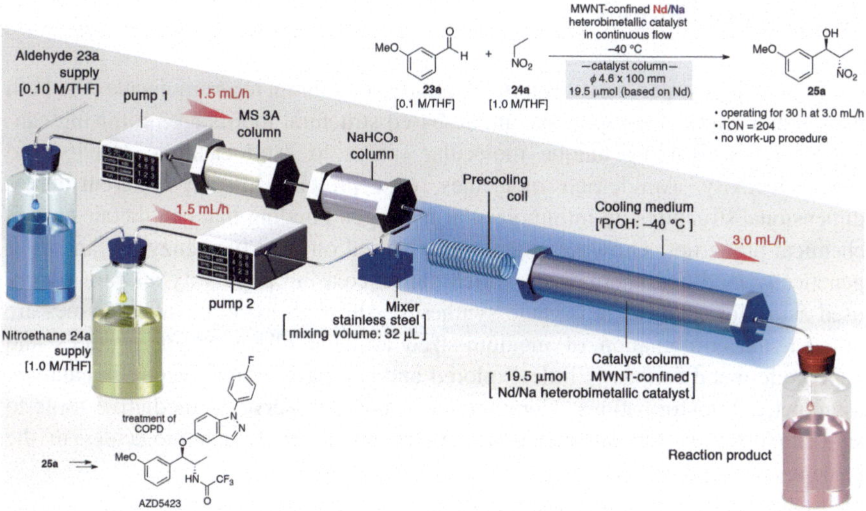

Fig. 7.11 Utility of the MWNT-confined composite catalyst for a catalytic asymmetric nitroaldol reaction in a continuous-flow platform and its application for the synthesis of a drug candidate for COPD treatment

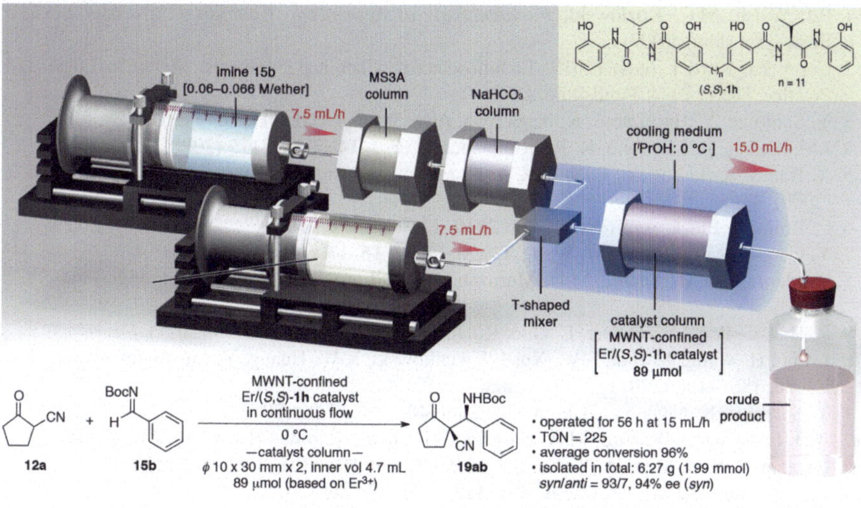

Fig. 7.12 Utility of MWNT-confined composite catalyst for catalytic asymmetric Mannich-type reaction of α-cyanoketone **12a** in a continuous-flow platform

stereoselectivity. The successful demonstration of flow reaction with **12a** in a multi-gram scale highlights the synthetic utility of the MWNT confinement strategy in stereoselective organic synthesis.

7.5 Future Outlook

Given the highly pervasive nature of amide-containing macromolecules in both natural and artificial compounds, amide-based structural motifs are promising candidates for scaffolding unique molecular spaces to exert catalytic activity and stereoselectivity. Amide-rich molecules, e.g., proteins, display inherent three-dimensional structures by intramolecular hydrogen bonding that can dictate stereochemical properties. Growing attention is focused on wild-type enzymes and their genetically modified alternatives, which gain increasing popularity and are widely used as useful catalysts in organic syntheses. Although many metalloenzymes are utilized, the combination of medium-sized artificial oligo- or polypeptides and xenobiotic metal cations is little explored and can pave a new avenue in catalytic asymmetric transformations. Computational aids will serve as predictive tools to simulate the three-dimensional structures of novel amide-rich molecules in the presence of metal cations.

References

1. P.J. Walsh, M.C. Kozlowski, *Fundamentals of Asymmetric Catalysis* (University Science Books, Sausalito, 2009)
2. Corey EJ, Kürti L (eds) (2010) Enantioselective Chemical Synthesis: Methods, Logic and Practice. Direct Book Publishing, Dallas
3. Y. Sohtome, Y. Hashimoto, K. Nagasawa, Adv. Synth. Catal. **347**, 1643–1648 (2005)
4. J.M. Andres, R. Manzano, R. Pedrosa, Chem. Eur. J. **14**, 5116–5119 (2008)
5. X. Han, J. Kwiatkowski, F. Xue, K.W. Huang, Y. Lu, Angew. Chem. Int. Ed. **48**, 7604–7607 (2009)
6. C. Bolm, A. Flock, A. Krebs, Synlett **2010**, 1219–1222 (2010)
7. Y. Gao, Q. Ren, L. Wang, J. Wang, Chem. Eur. J. **16**, 13068–13071 (2010)
8. R. Manzano, J.M. Andrés, M.D. Muruzábal, R. Pedrosa, Adv. Synth. Catal. **352**, 3364–3372 (2010)
9. B. Han, C. Peng, W. Huang, L. Guo, R. Hu, Synlett **2011**, 2981–2984 (2011)
10. J. Luo, H. Wang, X. Han, L.W. Xu, J. Kwiatkowski, K.W. Huang, Y. Lu, Angew. Chem. Int. Ed. **50**, 1861–1864 (2011)
11. Y. Sohtome, K. Nagasawa, Chem. Commun. **48**, 7777–7789 (2012)
12. A.H. Hoveyda, A.W. Hird, M.A. Kacprzynski, Chem. Commun. **16**, 1779–1785 (2004)
13. E.C. Davie, S.M. Mennen, Y. Xu, S.J. Miller, Chem. Rev. **107**, 5759–5812 (2007)
14. C.R. Shugrue, S.J. Miller, Chem. Rev. **117**, 11894–11951 (2017)
15. F.D. Toste, M.S. Sigman, S.J. Miller, Acc. Chem. Res. **50**, 609–615 (2017)
16. S.J. Miller, G.T. Copeland, N. Papaioannou, T.E. Horstmann, E.M. Ruel, J. Am. Chem. Soc. **120**, 1629–1630 (1998)
17. M.B. Fierman, D.J. O_Leary, W.E. Steinmetz, S.J. Miller, J. Am. Chem. Soc. **126**, 6967–6971 (2004)
18. Y. Zhao, J. Rodrigo, A.H. Hoveyda, M.L. Snapper, Nature **443**, 67–70 (2006)
19. A.J. Chinn, B. Kim, Y. Kwon, S.J. Miller, J. Am. Chem. Soc. **139**, 18107–18114 (2017)
20. A.J. Metrano, N.C. Abascal, B.Q. Mercado, E.K. Paulson, A.E. Hurtley, S.J. Miller, J. Am. Chem. Soc. **139**, 492–516 (2017)
21. A.W. Hird, A.H. Hoveyda, J. Am. Chem. Soc. **127**, 14988–14989 (2005)

22. L.C. Wieland, H. Deng, M.L. Snapper, A.H. Hoveyda, J. Am. Chem. Soc. **127**, 15453–15456 (2005)
23. P. Fu, M.L. Snapper, A.H. Hoveyda, J. Am. Chem. Soc. **130**, 5530–5541 (2006)
24. S.J. Malcolmson, S.J. Meek, E.S. Sattely, R.R. Schrock, A.H. Hoveyda, Nature **456**, 933–937 (2008)
25. O. Cusso, M.W. Giuliano, X. Ribas, S.J. Miller, M. Costas, Chem. Sci. **8**, 3660–3667 (2017)
26. N. Kumagai, Chem. Pharm. Bull. **59**, 1–22 (2011)
27. N. Kumagai, M. Shibasaki, Angew. Chem. Int. Ed. **52**, 223–234 (2013)
28. H.C. Aspinall, Chem. Rev. **102**, 1807–1850 (2002)
29. H. Tsukube, S. Shinoda, Chem. Rev. **102**, 2389–2403 (2002)
30. S.A. Cotton, *Lanthanide and Actinide Chemistry* (Wiley, Chichester/Hoboken, 2006)
31. A. Matsuzawa, A. Nojiri, N. Kumagai, M. Shibasaki, Chem. Eur. J. **16**, 5036–5042 (2010)
32. A. Nojiri, N. Kumagai, M. Shibasaki, Angew. Chem. Int. Ed. **51**, 2137–2141 (2012)
33. N. Kumagai, M. Shibasaki, Cat. Sci. Technol. **3**, 41–57 (2013)
34. A. Nojiri, N. Kumagai, M. Shibasaki, Chem. Commun. **49**, 4628–4630 (2013)
35. T. Mashiko, K. Hara, D. Tanaka, Y. Fujiwara, N. Kumagai, M. Shibasaki, J. Am. Chem. Soc. **129**, 11342–11343 (2007)
36. N.K. Akihiro Nojiri, M. Shibasaki, J. Am. Chem. Soc. **130**, 5630–5631 (2008)
37. T. Mashiko, N. Kumagai, M. Shibasaki, Org. Lett. **10**, 2725–2728 (2008)
38. T. Mashiko, N. Kumagai, M. Shibasaki, J. Am. Chem. Soc. **131**, 14990–14999 (2009)
39. A. Nojiri, N. Kumagai, M. Shibasaki, J. Am. Chem. Soc. **131**, 3779–3784 (2009)
40. Y. Kawato, N. Takahashi, N. Kumagai, M. Shibasaki, Org. Lett. **7**, 1484–1487 (2010)
41. F. Berhal, S. Takechi, N. Kumagai, M. Shibasaki, Chem. Eur. J. **17**, 1915–1921 (2011)
42. S. Takechi, N. Kumagai, M. Shibasaki, Tetrahedron Lett. **52**, 2140–2143 (2011)
43. A. Matsuzawa, T. Mashiko, N. Kumagai, M. Shibasaki, Angew. Chem. Int. Ed. **50**, 7616–7619 (2011)
44. T. Nitabaru, N. Kumagai, M. Shibasaki, Tetrahedron Lett. **49**, 272–276 (2008)
45. T. Nitabaru, A. Nojiri, M. Kobayashi, N. Kumagai, M. Shibasaki, J. Am. Chem. Soc. **131**, 13860–13869 (2009)
46. T. Nitabaru, N. Kumagai, M. Shibasaki, Angew. Chem. Int. Ed. **51**, 1644–1647 (2012)
47. A. Nonoyama, K. Hashimoto, A. Saito, N. Kumagai, M. Shibasaki, Tetrahedron Lett. **57**, 1815–1819 (2016)
48. A. Nonoyama, N. Kumagai, M. Shibasaki, Tetrahedron **73**, 1517–1521 (2017)
49. T. Karasawa, N. Kumagai, M. Shibasaki, Org. Lett. **19**, 710–713 (2017)
50. T. Ogawa, N. Kumagai, M. Shibasaki, Angew. Chem. Int. Ed. **52**, 6196–6201 (2013)
51. D. Sureshkumar, K. Hashimoto, N. Kumagai, M. Shibasaki, J. Org. Chem. **78**, 11494–11500 (2013)
52. K. Hashimoto, N. Kumagai, M. Shibasaki, Org. Lett. **16**, 3496–3499 (2014)
53. K. Hashimoto, N. Kumagai, M. Shibasaki, Chem. Eur. J. **21**, 4262–4266 (2015)

Chapter 8
Controlling the Chiral Molecular Space Using Helical Polymers

Yuuya Nagata and Michinori Suginome

Abstract Controlling the chiral molecular space using helical polymers has attracted much attention in the context of developing advanced chiral functional materials. So far, various helical polymers have been developed in order to investigate the screw-sense induction in helical backbones. In general, helical polymers can be classified based on their helix inversion barriers into static and dynamic helical polymers; however, this classification does not consider the chemical structure of the helical polymers, and consequently, there are several examples of static and dynamic helical polymers that exhibit the same polymer backbone. Poly (quinoxaline-2,3-diyl)s (PQXs) are particularly promising candidates to control the chiral molecular space, as their static and/or dynamic behavior can be controlled via the substituents. Furthermore, PQXs with chiral side chains exhibit a solvent-dependent helix inversion that can be used to establish control over the chiral molecular space via external stimuli. In this chapter, we describe recent progress concerning PQXs that exhibit a solvent-dependent helix inversion and their applications in the development of advanced chiral functional materials.

Keywords Helical chirality · Helix inversion · Circularly polarized light · Asymmetric catalysis

8.1 Introduction

Biomacromolecules such as DNA, RNA, proteins, and polysaccharides fulfill various essential functions that can be attributed to their precisely defined three-dimensional structures. In particular, their helical structures play important roles in, e.g., molecular recognition and chemical transformations, as well as storage, retrieval, and duplication of genetic information that is based on controlling the chiral molecular space. As these biomacromolecules consist of homochiral sugars

Y. Nagata (✉) · M. Suginome
Department of Synthetic Chemistry and Biological Chemistry, Graduate School of Engineering, Kyoto University, Kyoto, Japan

S. Shirakawa (ed.), *Designed Molecular Space in Material Science and Catalysis*,
https://doi.org/10.1007/978-981-13-1256-4_8

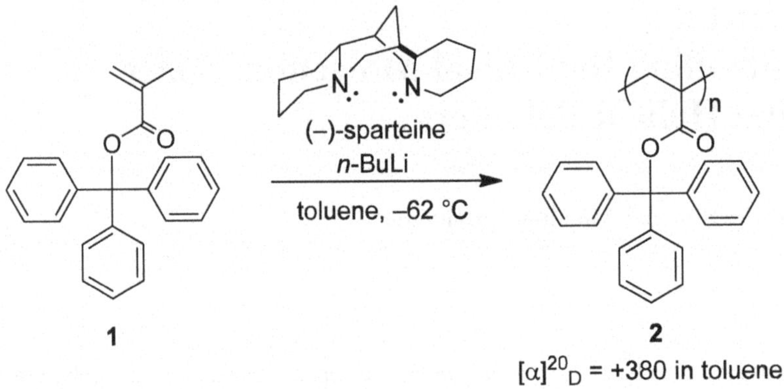

Fig. 8.1 Synthesis of a static helical polymer: asymmetric polymerization of triphenylmethyl methacrylate (**1**)

and amino acids, their right- and left-handed helical structures are bound by a diastereomeric relationship, which means that a thermodynamically stable single-handed helical conformation is induced preferentially.

Recently, much attention has been attributed to controlling the helical chirality of synthetic polymers. Given their precisely controlled helical structures, these synthetic helical polymers have become increasingly attractive for applications in chiral stationary phases, asymmetric catalysts, and chiroptical materials. In general, helical polymers are classified into two subclasses: (i) static helical polymers with high helix-inversion barriers and (ii) dynamic helical polymers with low helix-inversion barriers. In other words, static helical polymers do not show a helix inversion at room temperature, while dynamic helical polymers may exhibit a helix inversion even at room temperature [1, 2].

In 1979, Okamoto and co-workers reported that the anionic polymerization of triphenylmethyl methacrylate (**1**) in the presence of (−)-sparteine afforded the single-handed helical polymer **2** (Fig. 8.1) [3]. Interestingly, the optical rotation of **2** became undetectable upon removal of the triphenylmethyl groups, which suggested that the helical structure of **2** was kinetically stabilized. Since then, **2** has been regarded as a representative example for static helical polymers.

In 1988, Green and co-workers reported that poly(isocyanate) **4**, whose side chains exhibited chirality only by virtue of hydrogen/deuterium substitution, showed a high specific rotation compared to its monomer **3** (Fig. 8.2) [4]. Poly(isocyanate) **4** is a representative example of a dynamic helical polymer, and its helix sense is a result of the accumulation of very small energy differences per repeating unit between *P*- and *M*-helices. This cooperative effect on the screw-sense induction is a characteristic feature of dynamic helical polymers.

It should be noted that this classification into static or dynamic helical polymers is based on the energy of the helix inversion barriers and not on the chemical structure of the helical polymers. Consequently, the aforementioned classification requires careful attention, as boundaries may sometimes be ambiguous. For example,

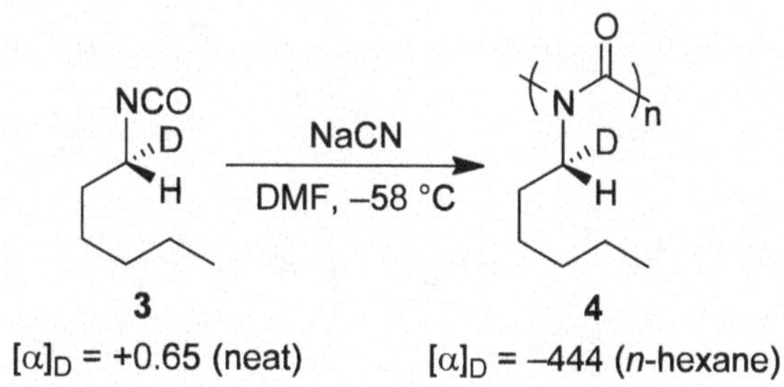

Fig. 8.2 Synthesis of a dynamic helical polymer: synthesis of the single-handed poly(isocyanate) **4**, whose side chains exhibit chirality due to the hydrogen/deuterium substitution pattern

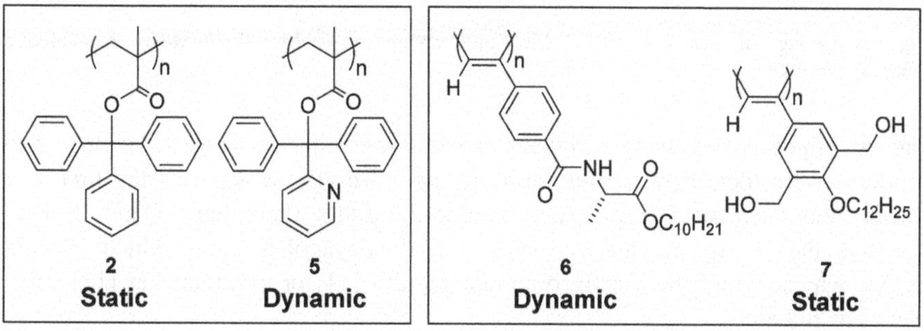

Fig. 8.3 Selected examples of static and dynamic helical polymers with identical backbones

triphenylmethyl methacrylate **2** can be classified as a static helical polymer, while polymer **5**, which bears slightly smaller side chains, exhibits a helix inversion at room temperature and is thus a dynamic helical polymer (Fig. 8.3) [5]. On the other hand, poly(acetylene)s such as polymer **6**, developed by Yashima and co-workers, are generally known as typical dynamic helical polymers [6]. Aoki and co-workers have reported that polymer **7** behaves as a static helical polymer owing to the strong hydrogen bonds between the hydroxyl groups on the side chains [7]. These reports suggest that the static or dynamic behavior of helical polymers that exhibit identical backbones may depend on the nature of the side chains.

8.2 Poly(quinoxaline-2,3-diyl)s

8.2.1 Aromatization Polymerization of 1,2-Diisocyanobenzenes

Isocyanides are isoelectronic to carbon monoxide and often exhibit similar reactivity in various transition-metal-catalyzed reactions [8]. However, carbon monoxide hardly engages in multiple insertions into the carbon-metal bond, while isocyanides readily participate in such multiple insertions to afford poly(isocyanide)s [9]. In 1999, Ito and co-workers reported that 1,2-diisocyano-3,4,5,6-tetramethylbenzene

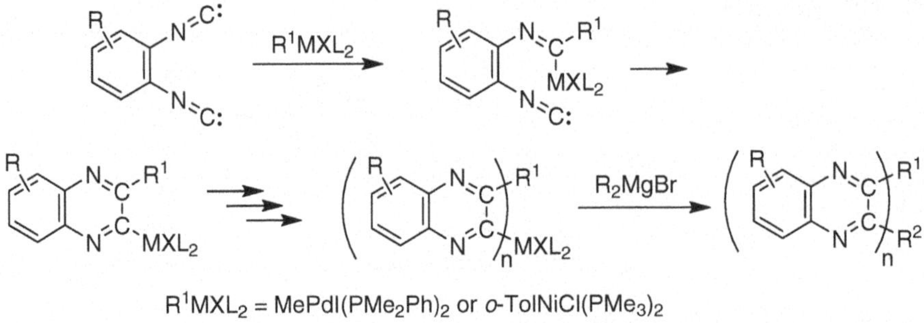

Fig. 8.4 Living polymerization of 1,2-diisocyanobenzenes using $MePdI(PPhMe_2)_2$ or *o*-TolNiCl $(PMe_3)_2$ as initiators

can be oligomerized using Grignard reagents as initiators (Fig. 8.4) [10]. The products were identified as oligo(quinoxaline-2,3-diyl)s, which was the first step toward the chemistry of poly(quinoxaline-2,3-diyl)s (hereafter PQXs). Subsequently, the living polymerization of 1,2-diisocyanobenzenes, which affords PQXs, was accomplished using organopalladium [11] or organonickel [12] complexes as initiators.

8.2.2 Single-Handed Poly(quinoxaline-2,3-diyl)s As Static Helical Polymers

The helicity of the main chain of PQXs arises from the steric repulsion between the two substituents at the 5- and 8-positions of the quinoxaline rings. This feature was unequivocally determined by the synthesis and isolation of the *P*- and *M*-helices of a PQX (Fig. 8.5) [13]. For that purpose, 3,6-di-*p*-tolyl-1,2-diisocyanobenzene was initially oligomerized using a palladium complex with chiral ligands, and oligomer **8** was isolated by gel permeation chromatography (GPC). Here, *P*- and *M*-helical **8** are bound by a diastereomeric relationship due to the same chiral ligands on the terminal palladium moieties. The diastereomer mixture was successfully separated using high-pressure liquid chromatography (HPLC), and the circular dichroism (CD) spectra of the thus obtained *P*- and *M*-helical **8** were mirror images of each other. Using *P*- and *M*-helical **8** as initiators, 1,2-diisocyano-3,6-dimethyl-4,5-bis (propoxymethyl)benzene was polymerized to give (*P*)- and (*M*)-**10**. Even after removal of the chiral terminal groups by treatment with a Grignard reagent, the CD spectra of (*P*)- and (*M*)-**10** were perfect mirror images of each other. As (*P*)- and (*M*)-**10** do not show helix inversion at room temperature owing to the bulky tolyl side chains, these polymers are classified as static helical polymers.

Given that it is very laborious to separate the *P*- and *M*-helical oligomers using a combination of GPC and HPLC, an improved method to prepare single-handed helical PQXs had been desired. In order to synthesize single-handed helical PQXs,

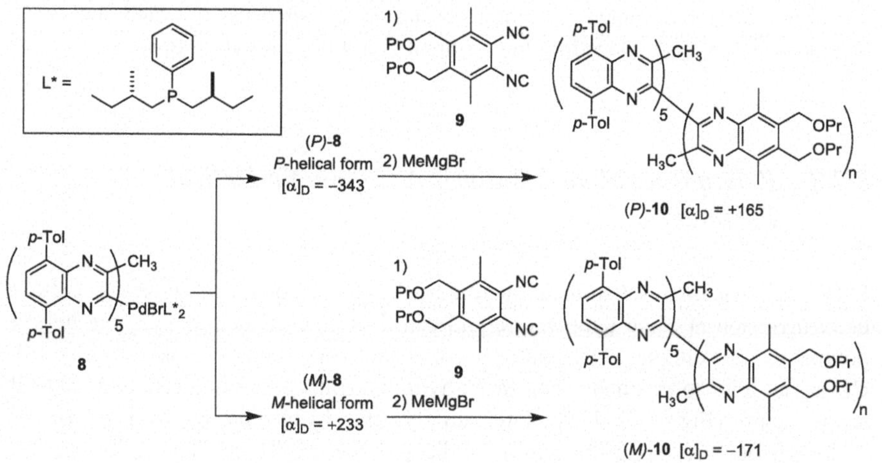

Fig. 8.5 Living polymerization of 1,2-diisocyanobenzenes using chiral organopalladium complexes as initiators

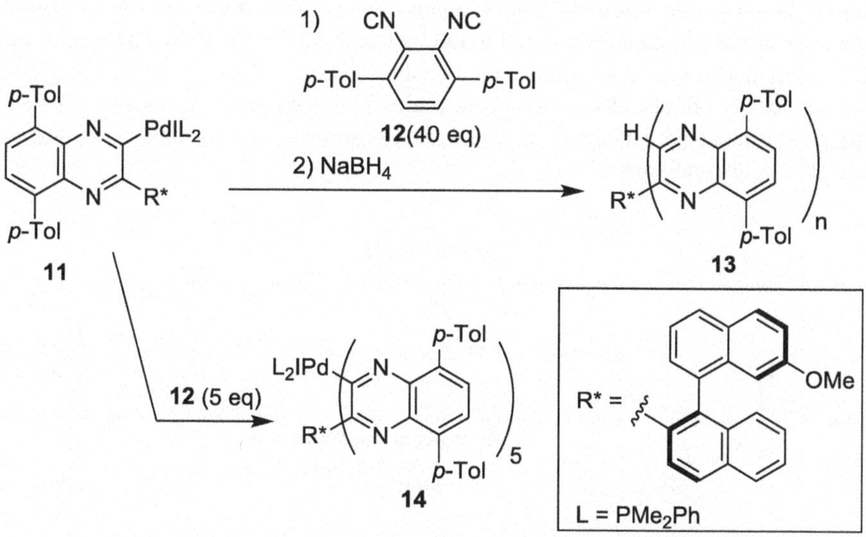

Fig. 8.6 Synthesis of single-handed helical PQXs with a chiral binaphthyl derivative as a terminal group

palladium complexes with chiral binaphthyl substituents were developed as chiral initiators (Fig. 8.6) [14]. In this case, the chiral substituent is not in the proximity of the polymerization growth terminal during the polymerization. Nevertheless, based on the CD intensity, polymer **13** showed ca. 90% screw-sense excess. Furthermore, pentamer **14** was successfully isolated, and its molecular structure, which was unambiguously determined by single-crystal X-ray analysis, revealed that a *P*-helical structure was induced by the terminal chiral binaphthyl groups.

8.2.3 Poly(quinoxaline-2,3-diyl)s As Dynamic Helical Polymers

PQXs that bear small substituents such as methyl groups on the 5,8-positions exhibit helix inversion at room temperature and can thus be classified as dynamic helical polymers. For example, PQXs with a boronyl end group such as **15** did not show any CD signal prior to treatment with (*R*,*R*)-hydrobenzoin, which induced an *M*-helical screw sense (Fig. 8.7) [15]. Interestingly, the hydrolytic removal of (*R*,*R*)-hydrobenzoin caused racemization of the polymer backbone, and a subsequent reaction with (*S*,*S*)-hydrobenzoin induced a *P*-helical conformation.

Screw-sense induction in PQXs was also achieved via the introduction of chiral side chains (Fig. 8.8) [16]. A series of random copolymers (**21–25**) of achiral **9** and chiral **20**, which contain (*R*)-2-butoxymethyl groups as side chains, was prepared using an achiral organonickel initiator and various ratios of **9** and **20**. CD spectra of these polymers revealed a nonlinear relationship between the screw-sense excess and the number of chiral units per polymer chain. This nonlinear relationship showed good agreement with Green's theory, which describes the sergeants-and-soldiers effect in helical polymers.

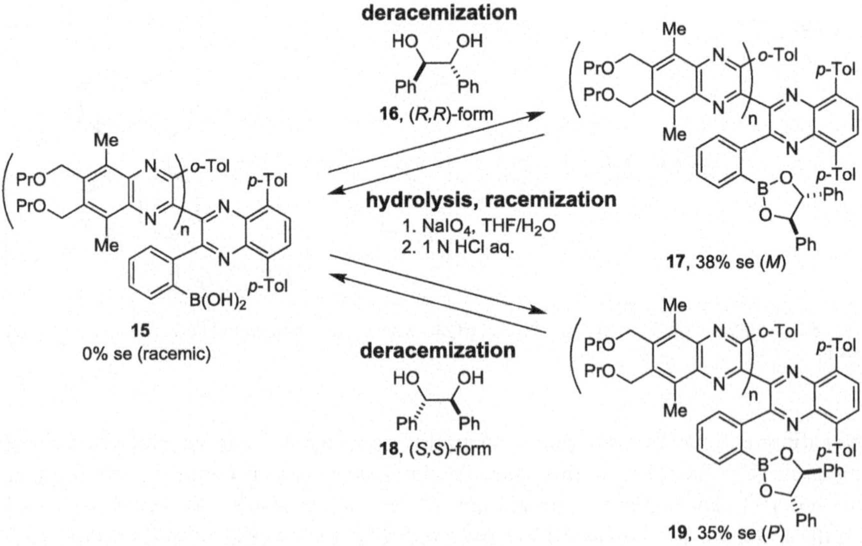

Fig. 8.7 Screw-sense induction in PQX with a boronyl end group (**15**) by introduction of (*R*,*R*)- or (*S*,*S*)-hydrobenzoin

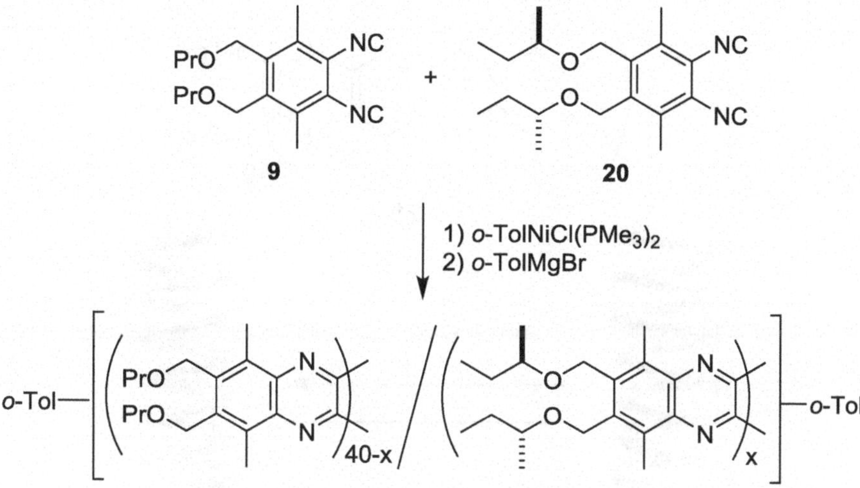

21, x = 0, racemic (0% se)
22, x = 4, *P*-helix (24% se)
23, x = 12, *P*-helix (89% se)
24, x = 20, *P*-helix (>99% se)
25, x = 40, *P*-helix (>99% se)

Fig. 8.8 PQXs with achiral and chiral side chains and their screw-sense excess (se) in $CHCl_3$

8.3 Solvent–Dependent Helix Inversion of Poly (quinoxaline-2,3-diyl)s

A polymer that bears (*R*)-2-butoxymethyl groups as side chains (**25**) shows a *P*-helical conformation in $CHCl_3$. Intriguingly, when **25** is dissolved in 1,1,2-trichloroethane, an opposite CD signal is observed, which is indicative of an inversion of the helical backbone to an *M*-helical conformation (Fig. 8.9) [16]. This solvent-dependent helix inversion is completely reversible, i.e., after removing 1,1,2-trichloroethane under reduced pressure and adding $CHCl_3$, the helical backbone of **25** inverts to a *P*-helical conformation again.

PQX copolymers that bear various chiral units and common achiral units were synthesized to investigate the efficiency of the screw-sense induction and its dependence on the solvent ($CHCl_3$ or 1,1,2-trichloroethane; Fig. 8.10) [17]. Polymer **26** exhibited a helix inversion, indicating that oxygen atoms in the side chain were not essential for the helix inversion. Polymers **27** and **28** were prepared from the inexpensive chiral source (*S*)-2-methylbutanol, which is easily available as a by-product of alcoholic fermentation. Although polymer **28** with (*S*)-2-methylbutoxymethyl groups did not show any helix inversion, polymer **27** with (*S*)-2-methylbutoxy groups exhibited a helix inversion. Polymer **29**, which contains (*S*)-2-butoxy groups, did not show any helix inversion. Here, chirality-switchable

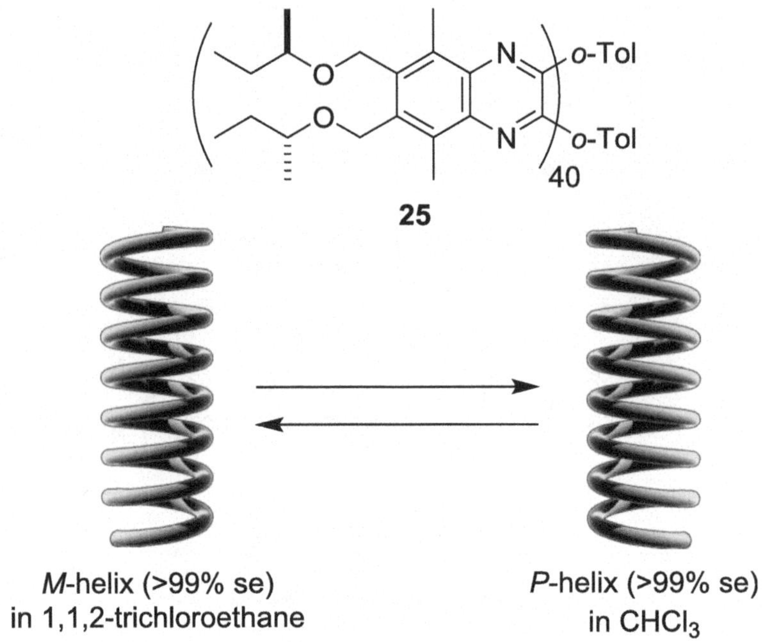

Fig. 8.9 Solvent-dependent helix inversion of PQX **25**

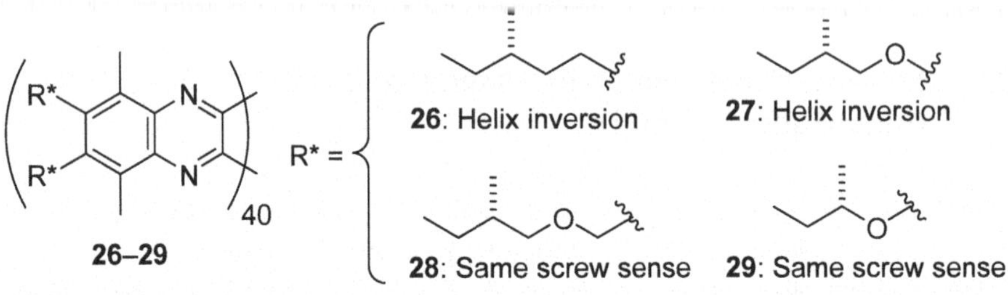

Fig. 8.10 PQXs **26–29** with various chiral side chains and their dependence of the screw sense on the solvent ($CHCl_3$ or 1,1,2-trichloroethane)

polymers **25–27** have in common that their chiral center resides at the third position in the chiral side chains on the quinoxaline ring. It has accordingly been suggested that the position of the chiral center in the side chain may be essential for the solvent-dependent helix inversion.

PQXs bearing (*R*)-pentyloxymethyl, (*R*)-hexyloxymethyl, or (*R*)-octyloxymethyl side chains were also synthesized to investigate the efficiency of the screw-sense induction. Among these, PQXs with (*R*)-octyloxymethyl side chains exhibited the most efficient screw-sense induction, suggesting that longer side chains efficiently stabilize the single-handed helical structure. Polymer **30** showed a *P*-helical

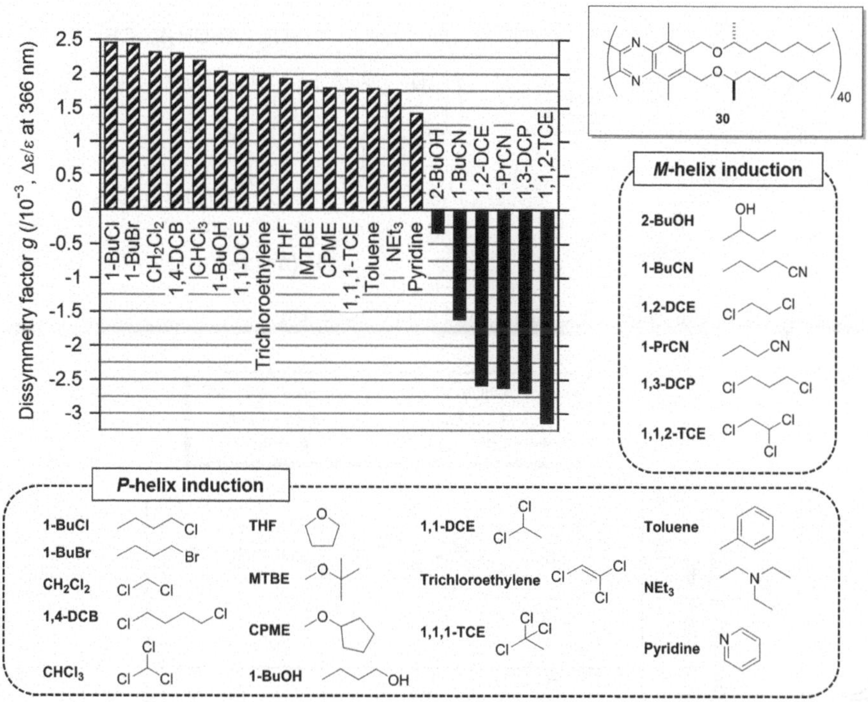

Fig. 8.11 Solvent-dependent values for the dissymmetry factor, *g*, of polymer **30** at 366.0 nm; DCE, dichloroethane; DCP, dichloropropane; DCB, dichlorobutane; TCE, trichloroethane; MTBE, methyl *tert*-butyl ether; and CPME, cyclopentyl methyl ether

conformation in a variety of common organic solvents, while some nitrile solvents (1-BuCN and 1-PrCN) and halogenated solvents (1,1,2-trichloroethane, 1,3-dichloropropane, and 1,2-dichloroethane) induced *M*-helical structures (Fig. 8.11).

As most of the naturally occurring chiral compounds are available in only one of the two enantiomeric forms, it is very important from a practical perspective to be able to induce both helical senses (*P*- and *M*-helices) via the introduced enantiopure chiral side chains. For example, PQXs with chiral side chains derived from naturally occurring L-lactic acid were prepared in order to investigate the screw-sense induction in various ether solvents (Fig. 8.12) [18].

Polymer **31** adopted *M*-helical structures in both $CHCl_3$ and 1,1,2-trichloroethane, and a solvent-dependent helical inversion could not be observed (Fig. 8.13). However, in ether solvents, the *g* values crucially depended on the length of alkyl chains and showed a clear odd-even effect in the region of $n = 1$–5, i.e., from methyl to pentyl esters. For higher alkyl groups ($n \geq 6$), such an odd-even effect was not observed anymore. In addition, the use of *tert*-butyl methyl ether generally induced a *P*-helical structure. This solvent-dependent helical induction was most clearly observed for **31**, which carries a pentyl ester ($n = 5$) in the side chain: the use

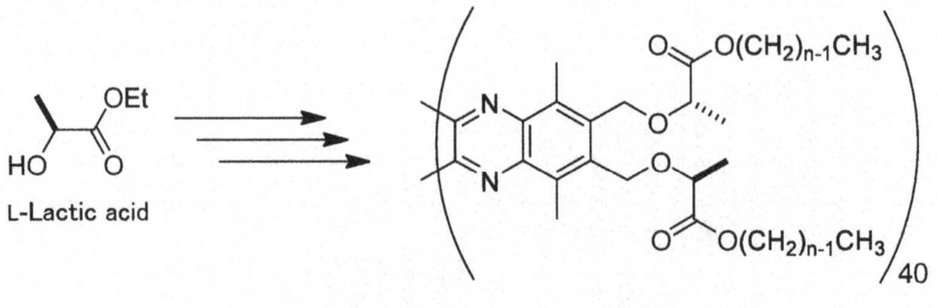

Fig. 8.12 PQXs with chiral side chains derived from L-lactic acid

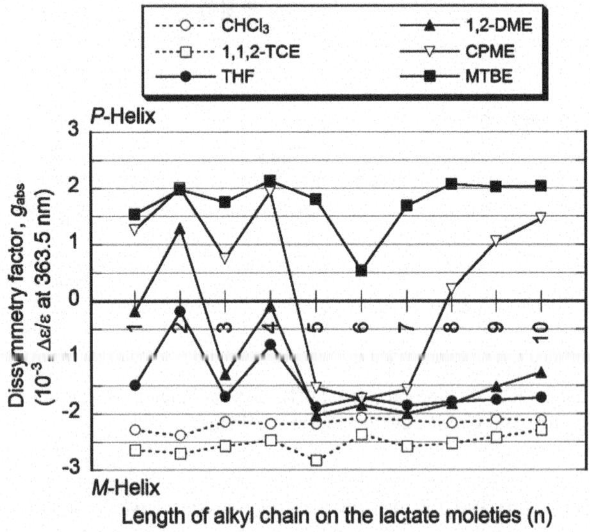

Fig. 8.13 Solvent-dependent correlation between the length of the alkyl side chains, *n*, and the *g* values of **31**; 1,1,2-TCE, 1,1,2-trichloroethane; THF, tetrahydrofuran; 1,2-DME, 1,2-dimethoxyethane; CPME, cyclopentyl methyl ether; and MTBE, *tert*-butyl methyl ether

of either *tert*-butyl methyl ether or 1,2-dimethoxyethane selectively induced *P*- or *M*-helical structures, respectively.

Saturated hydrocarbons are generally considered as highly hydrophobic solvents, which allows excluding most solvent-solute interactions. This is also reflected in the family name of the hydrocarbons, which are commonly referred to as "paraffins," originating from the Latin "parum affinis," which means "little affinity" [19]. Due to the minimal solvent-solute interactions, a change between saturated hydrocarbon solvents does usually not result in appreciable differences regarding the outcome of chemical events or conformational changes. Nevertheless, a PQX bearing (*S*)-3-octyloxymethyl side chains (**32**) exhibited a solvent-dependent helix inversion in

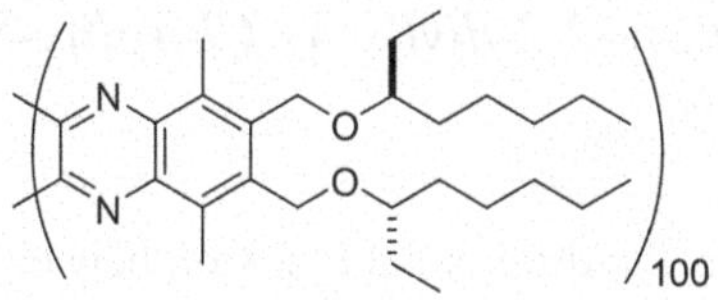

32

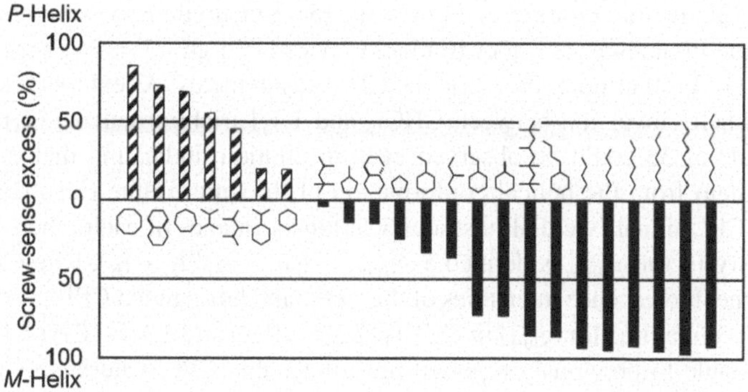

Fig. 8.14 Screw-sense excess (%) of polymer **32** in various saturated hydrocarbons

saturated hydrocarbons (Fig. 8.14) [20]. For example, a series of linear *n*-alkanes induced an *M*-helical conformation, while for branched or cyclic alkanes, trends with respect to the handedness of the helical chirality of **32** were more complicated. Whereas methyl-, ethyl-, propyl-, and butyl-substituted cyclohexanes induced an *M*-helical conformation, unsubstituted cyclohexane, as well as isopropylcyclohexane, induced a *P*-helical conformation. On the basis of these results, it seems feasible to assume that the handedness of the induced screw sense depends on the molecular shape of the alkane solvents.

8.4 Poly(quinoxaline-2,3-diyl)s As Chirality-Switchable Chiroptical Materials

Single-handed PQXs can show significant chiral molecular functions. For instance, a single-handed PQX bearing chiral side chains and diarylphosphino pendants can be used as highly effective chiral ligands in asymmetric catalysis [21–26]. Recently, their helically chiral scaffold was applied to the generation of a chiral organocatalyst [27]. Yet, the single-handed helical scaffold of PQXs is not only attractive for applications in asymmetric catalysis but also in chiroptical materials. In this section, PQXs exhibiting circularly polarized luminescence (CPL) and selective reflection are described.

8.4.1 Poly(quinoxaline-2,3-diyl)s As Chirality-Switchable CPL Materials

CPL [28, 29] is defined as the differential emission between right- and left-handed circularly polarized light. CPL has attracted considerable interest due to potential applications in chemical sensors [30], biological probes [31], and three-dimensional displays [32]. In this context, a PQX with (*S*)-2-methylbutoxy side chains (**33**) showed blue luminescence under irradiation with UV light (λ = 365 nm) in dilute solution [33]. Furthermore, based on the CD measurements, **33** exhibited a solvent-dependent helix inversion between $CHCl_3$ and 1,1,1-trichlroroethane (1,1,1-TCE). CPL signals of **33** could be observed even in dilution, indicating that the signals observed stem from the helical conformation of the quinoxaline rings (Fig. 8.15), and these CPL signals were almost identical mirror images of each other. The CPL dissymmetry factor, g_{lum}, is defined as $g_{lum} = 2(I_L - I_R)/(I_L + I_R)$, where I_L and I_R represent the fluorescence intensities of the right- and left-handed CPL, respectively. The observed maxima for $|g_{lum}|$ in $CHCl_3$ (2.3×10^{-4}) and 1,1,1-TCE (4.1×10^{-4}) are comparable to previously reported non-switchable CPL materials in solutions, which are typically $|g_{lum}| = 10^{-5}$–10^{-2} [34, 35].

Polymer **33** exhibited blue CPL, whose handedness could be switched by the solvent-dependent helix inversion of the polymer main chain. However, the main chain of **33** exhibited only blue emission with a low photoluminescent quantum yield (<0.7%). However, the introduction of simple π-groups to the quinoxaline rings enhanced the photoluminescent quantum yield and thus accomplished more

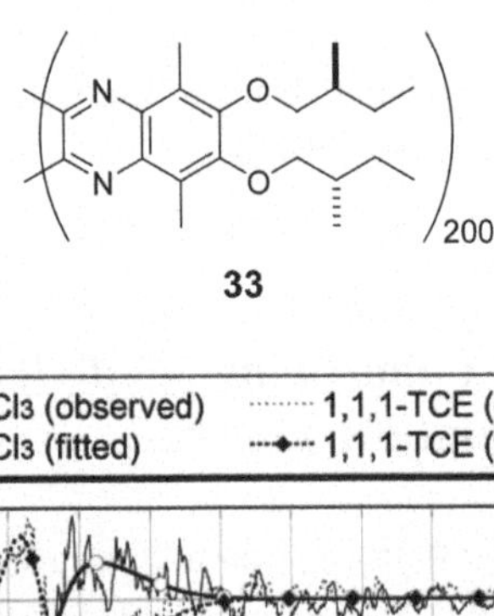

33

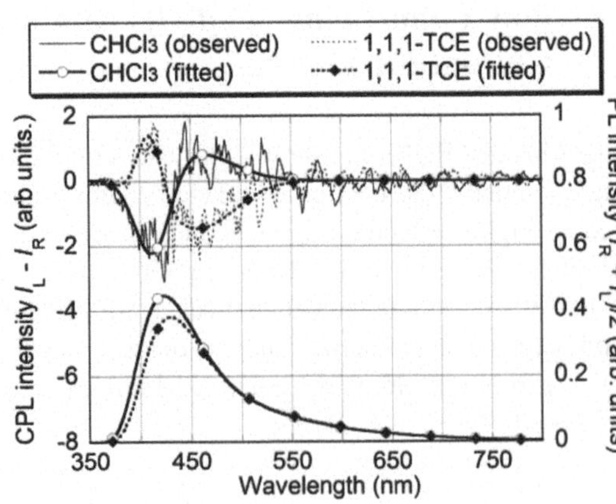

Fig. 8.15 CPL and photoluminescence (PL) spectra of **33** in dilute (9.1×10^{-5} M) $CHCl_3$ and 1,1,1-TCE solutions (λ_{ex} = 300.0 nm)

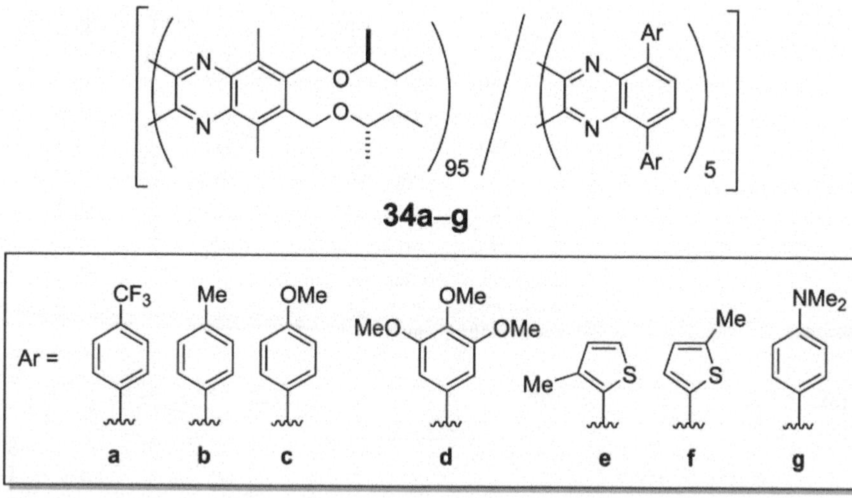

Fig. 8.16 Structures of co-PQXs **34a–g**, which bear luminophore units

efficient CPL using the PQX scaffold (Fig. 8.16) [36]. Random co-100mers **34a–g**, which bear common chiral units containing (*S*)-2-butoxymethyl side chains along with various 5,8-diarylqunoxalaine units, were prepared to investigate their optical properties. These polymers exhibited virtually identical UV-vis absorption spectra, although **34f** and **34g** showed broad, weak peaks around 400–500 nm, which were assigned to donor-accepter interactions in the 5-methylthienyl- or 4-dimethylaminophenyl-substituted quinoxaline units. On the other hand, PL spectra of these polymers covered the entire visible light region. The mean emission wavelength was bathochromically shifted with decreasing electron-donating ability of the aryl groups at the 5,8-positions of the quinoxaline ring. While **34a** and **34b** with 4-trifluoromethylphenyl and *p*-tolyl groups exhibited blue emission, **34c** and **34d**, which contain electron-rich 4-methoxyphenyl and 3,4,5-trimethoxyphenyl groups, showed green emission. **34e**, **34f**, and **34g**, which carry more electron-rich 3- or 5-methylthienyl, and 4-dimethylaminophenyl groups, exhibited orange to red emission. This trend may be rationalized based on the mechanism of the fluorescence emission involving the donor-accepter interaction between the aryl substituents and the electron-deficient quinoxaline ring. The absolute photoluminescent quantum yield, Φ_{PL}, was improved up to 49.7% by random incorporation of five 5,8-diarylquinoxaline units in a 100mer (**34f**, Table 8.1). It should be noted that most part of the excitation light (350 or 351 nm, which was set to the absorption maxima) was initially absorbed by the chiral units with no π-substituents at 5- and 8-positions. Subsequently, the photoexcited energy was efficiently transferred to the 5,8-diarylquinoxaline units to exhibit photoluminescence.

CD spectra of **34a–g** showed that these polymers adopted pure *M*- or *P*-helical conformations in $CHCl_3$ or 1,1,2-TCE, respectively (**34a**, **34b**, and **34g** were dissolved in a mixed solvent, 1,1,2-TCE/THF = 8/2, due to their low solubility).

Table 8.1 Optical properties of **34a–g** in $CHCl_3$

		λ_{abs} [a] (nm)	λ_{PL} [b] (nm)	Φ_{PL} [c] (%)
1	**34a**	351, 290	(646), 426	0.5
2	**34b**	351, 291	473	3.1
3	**34c**	350, 291	512	33.0
4	**34d**	351, 291	537	12.7
5	**34e**	350, 290	569	40.7
6	**34f**	(459), 351, 291	577	49.7
7	**34 g**	(455), 350, 291	600	24.6

[a]Maximum absorption wavelength
[b]Maximum PL wavelength
[c]Absolute PL quantum yield

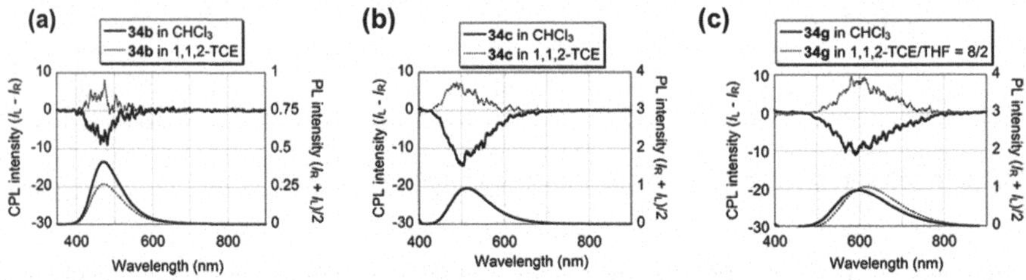

Fig. 8.17 CPL and PL spectra of (**a**) **34b,** (**b**) **34c,** and (**c**) **34g** in $CHCl_3$ or 1,1,2-TCE. For **34g**, a mixed solvent (1,1,2-TCE/THF = 8/2, v/v) was used instead of 1,1,2-TCE due to the low solubility of **34g** in the latter

It should be noted that the incorporation of 5,8-diarylquinoxaline units into the polymer backbone did not affect the screw-sense induction and the solvent-dependent helix inversion. CPL spectra of **34a–g** were recorded in dilute solutions of $CHCl_3$ and 1,1,2-TCE. **34a** afforded weak CPL signals, which mainly originated from chiral units in the main chain [33]. It should be noted that **34b–g** clearly showed CPL peaks, suggesting that the helical backbones are involved in the emission process. All polymers exhibited positive CPL signals in $CHCl_3$, due to the influence of their *M*-helical main chain. The g_{CPL} values at maximum PL wavelength in $CHCl_3$ (-3.2 to -11.5×10^{-4}) were comparable to those previously reported for non-switchable organic CPL materials, which typically exhibit $|g_{CPL}|$ values of 10^{-5} -10^{-3} [34, 35], except for particularly efficient helical molecules ($|g_{CPL}| = 10^{-2}$–10^{-1}) [37, 38]. In 1,1,2-TCE or a mixed solvent (1,1,2-TCE/THF = 8/2, v/v), these polymers showed positive peaks ($g_{CPL} = +1.2$ to $+10.5 \times 10^{-4}$) except for **1f**, which showed no CPL emission; the reason for this exception is not clear at this moment. As representative examples, the CPL and PL spectra of **34b** (blue emission), **34c** (green emission), and **34 g** (red-orange emission) are shown in Fig. 8.17, indicating the solvent-dependent chirality switch of CPL handedness. Here, the chiral side

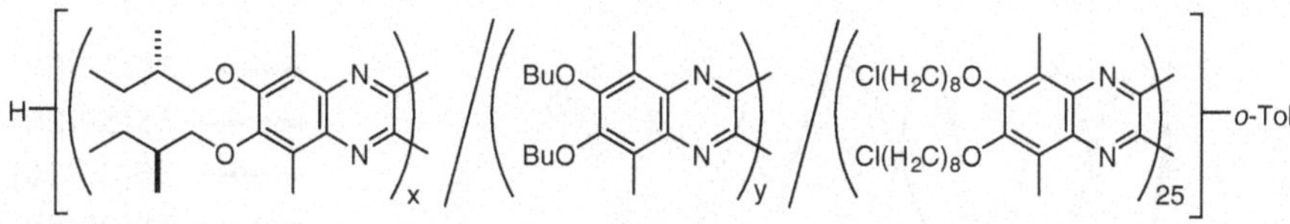

35(x/y)

Fig. 8.18 Structures of PQX-based ternary co-100mers **35(x/y)**

chains of **34** induce the main-chain helical chirality, which provides chiral environments for the luminescent diaryl-substituted quinoxalines to exhibit the CPL signals.

8.4.2 Poly(quinoxaline-2,3-diyl)s As Chirality-Switchable Cholesteric Materials

Physical interaction of light with a chiral nematic (cholesteric) nanostructure, whose pitch falls within a wavelength range of visible light, shows coloration by selective reflection of CPL. The wavelength of the reflected light is thereby determined by the product of the average refractive index and the pitch of the cholesteric structure [39]. Such structural coloration is widely found in nature, as exemplified by the colored skin tissue of jeweled beetles [40] and certain plant types [41].

Thin films of PQXs consisting of three components containing (*S*)-2-methylbuthyloxy, *n*-butoxy, and 8-chlorooctyloxy groups **35(x/y)** exhibited selective reflection of right-handed CPL in the visible region after annealing in $CHCl_3$ vapor at room temperature [42]. Firstly, **35(x/y)** containing various ratios of chiral (x) and achiral units (y), along with a fixed number of chlorooctyl units (25), were prepared to investigate the role of chiral units (Fig. 8.18). The color of the thin films after $CHCl_3$ vapor annealing varied from blue to green and red according to the ratio of chiral monomer units (Fig. 8.19). The change in color was clearly correlated to the screw-sense excess of the helical macromolecules and, in turn, to the degree of the screw sense of the cholesteric-like superstructures.

Polymer **35(x/y)** also showed a solvent-dependent helix inversion between $CHCl_3$ and 1,2-dichloroethane (1,2-DCE). The $CHCl_3$-treated film reflected right-handed CPL, whereas the 1,2-DCE-treated film reflected left-handed CPL. Given that it is possible to tune the macromolecular screw-sense purity by changing the ratio of the two solvents [16], we tried to prepare a single film for which the color and handedness of reflected light are fully tunable by the solvent effect alone. For polymer **35(30/45)**, which showed reflection of right-handed blue CPL upon exposure to 100% $CHCl_3$ vapor (Fig. 8.20, a), increasing the ratio of 1,2-DCE to 33% and

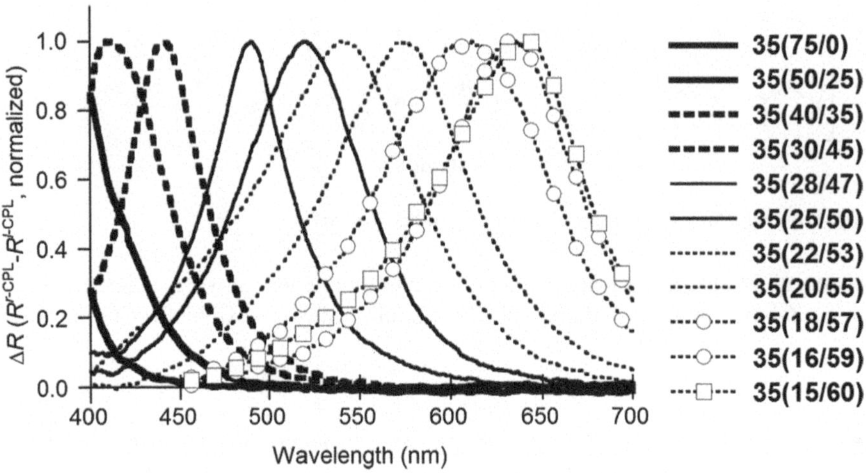

Fig. 8.19 Difference reflection spectra of **35(x/y)** annealed in $CHCl_3$

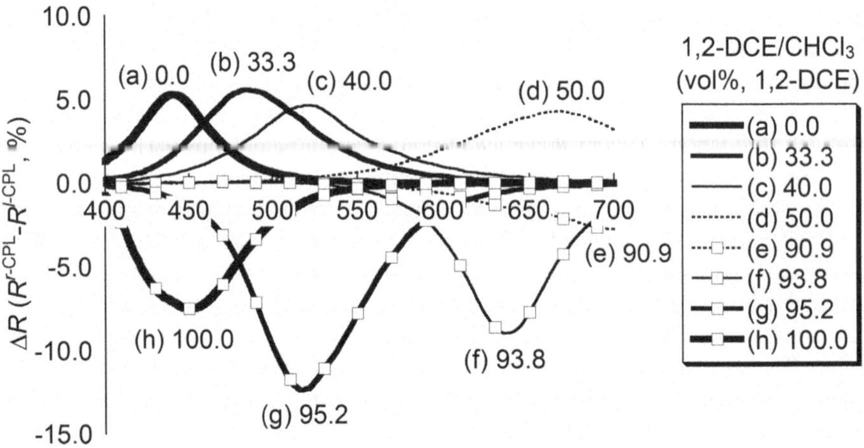

Fig. 8.20 Difference reflection spectra of **35(x/y)** annealed in $CHCl_3$

50% resulted in a bathochromic shift of the reflected wavelength, i.e., green (b) and red (d) color, respectively. Further increases in the ratio of 1,2-DCE caused the reflection of visible light to disappear. When the ratio of 1,2-DCE was further increased to 90% (e), a right-handed helical structure with left-handed helical superstructure becomes predominant, which results in the reflection of left-handed red CPL. A hypsochromic shift of the reflection of left-handed CPL was observed upon increasing the ratio of 1,2-DCE (f–h), which resulted in the observation of green (g) and blue (h) color. Consequently, both the wavelength and the chirality of the reflected light could be fully tuned by changing the ratio of the mixed solvents for the vapor annealing process.

8.5 Conclusion

PQXs can provide unique and highly effective chiral molecular space for asymmetric reactions as well as the generation of CPL. These characteristic features of PQXs are required in precise syntheses via, e.g., living polymerizations, the almost perfect screw-sense induction, and the solvent-dependent switch of the helical backbone. At this stage, it seems difficult to achieve these features using any other synthetic helical polymers. Nevertheless, in order to develop new functional materials based on the control over the chiral molecular space, challenges for the design and synthesis of new helical polymers remain to be overcome. In the near future, the concept of controlling the chiral molecular space using not only PQXs but also various other helical polymers will be extended and applied to a variety of chiral functional materials.

References

1. E. Yashima, K. Maeda, H. Iida, Y. Furusho, K. Nagai, Chem. Rev. **109**, 6102–6211 (2009)
2. E. Yashima, N. Ousaka, D. Taura, K. Shimomura, T. Ikai, K. Maeda, Chem. Rev. **116**, 13752–13990 (2016)
3. Y. Okamoto, K. Suzuki, K. Ohta, K. Hatada, H. Yuki, J. Am. Chem. Soc. **101**, 4763–4765 (1979)
4. M.M. Green, C. Andreola, B. Munoz, M.P. Reidy, K. Zero, J. Am. Chem. Soc. **110**, 4063–4065 (1988)
5. Y. Okamoto, H. Mohri, T. Nakano, K. Hatada, J. Am. Chem. Soc. **111**, 5952–5954 (1989)
6. K. Okoshi, S. Sakurai, S. Ohsawa, J. Kuniaki, E. Yashima, Angew. Chem. Int. Ed. **45**, 8173–8176 (2006)
7. T. Aoki, T. Kaneko, N. Maruyama, A. Sumi, M. Takahashi, T. Sato, M. Teraguchi, J. Am. Chem. Soc. **125**, 6346–6347 (2003)
8. V.P. Boyarskiy, N.A. Bokach, K.V. Luzyanin, V.Y. Kukushkin, Chem. Rev. **115**, 2698–2779 (2015)
9. E. Schwartz, M. Koepf, H.J. Kitto, R.J.M. Nolte, A.E. Rowan, Polym. Chem. **2**, 33–47 (2011)
10. Y Ito, E Ihara, M Hirai, H Ohsaki, A Ohnishi, M Murakami, J. Chem. Soc. Chem. Commun. 403–405 (1990)
11. Y. Ito, E. Ihara, M. Murakami, M. Shiro, J. Am. Chem. Soc. **112**, 6446–6447 (1990)
12. Y. Ito, E. Ihara, M. Murakami, Polym. J. **24**, 297–299 (1992)
13. Y. Ito, E. Ihara, M. Murakami, Angew. Chem. Int. Ed. **31**, 1509–1510 (1992)
14. Y. Ito, T. Miyake, S. Hatano, R. Shima, T. Ohara, M. Suginome, J. Am. Chem. Soc. **120**, 11880–11893 (1998)
15. Y. Nagata, S. Ohashi, M. Suginome, J. Polym. Sci. Part A: Polym. Chem. **50**, 1564–1571 (2012)
16. T. Yamada, Y. Nagata, M. Suginome, Chem. Commun. **46**, 4914–4916 (2010)
17. Y. Nagata, T. Yamada, T. Adachi, Y. Akai, T. Yamamoto, M. Suginome, J. Am. Chem. Soc. **135**, 10104–10113 (2013)
18. Y. Nagata, T. Kuroda, K. Takagi, M. Suginome, Chem. Sci. **5**, 4953–4956 (2014)
19. A.E. Shilov, G.B. Shul'pin, Chem. Rev. **97**, 2879–2932 (1997)
20. Y. Nagata, T. Nishikawa, M. Suginome, J. Am. Chem. Soc. **136**, 15901–15904 (2014)
21. T. Yamamoto, M. Suginome, Angew. Chem. Int. Ed. **48**, 539–542 (2009)

22. T. Yamamoto, T. Yamada, Y. Nagata, M. Suginome, J. Am. Chem. Soc. **132**, 7899–7901 (2010)
23. T. Yamamoto, Y. Akai, Y. Nagata, M. Suginome, Angew. Chem. Int. Ed. **50**, 8844–8847 (2011)
24. Y. Akai, T. Yamamoto, Y. Nagata, T. Ohmura, M. Suginome, J. Am. Chem. Soc. **134**, 11092–11095 (2012)
25. M. Suginome, T. Yamamoto, Y. Nagata, T. Yamada, Y. Akai, Pure Appl. Chem. **84**, 1759–1769 (2012)
26. T. Yamamoto, Y. Akai, M. Suginome, Angew. Chem. Int. Ed. **53**, 12785–12788 (2014)
27. T. Yamamoto, R. Murakami, M. Suginome, J. Am. Chem. Soc. **139**, 2557–2560 (2017)
28. J.P. Riehl, F.S. Richardson, Chem. Rev. **86**, 1–16 (1986)
29. J.P. Riehl, F. Muller, *Comprehensive Chiroptical Spectroscopy*, vol 1 (Wiley, New York, 2012)
30. H. Maeda, Y. Bando, K. Shimomura, I. Yamada, M. Naito, K. Nobusawa, H. Tsumatori, T. Kawai, J. Am. Chem. Soc. **133**, 9266–9269 (2011)
31. G. Muller, Dalton Trans. **44**, 9692–9707 (2009)
32. M. Schadt, Annu. Rev. Mater. Sci. **27**, 305–379 (1997)
33. Y. Nagata, T. Nishikawa, M. Suginome, Chem. Commun. **50**, 9951–9953 (2014)
34. E.M. Sánchez-Carnerero, F. Moreno, B.L. Maroto, A.R. Agarrabeitia, M.J. Ortiz, B.G. Vo, G. Muller, M. Sdl, J. Am. Chem. Soc. **136**, 3346–3349 (2014)
35. J.F. Kögel, S. Kusaka, R. Sakamoto, T. Iwashima, M. Tsuchiya, R. Toyoda, R. Matsuoka, T. Tsukamoto, J. Yuasa, Y. Kitagawa, T. Kawai, H. Nishihara, Angew. Chem. Int. Ed. **55**, 1377–1381 (2016)
36. T. Nishikawa, Y. Nagata, M. Suginome, ACS Macro Lett. **6**, 431–435 (2017)
37. K. Nakamura, S. Furumi, M. Takeuchi, T. Shibuya, K. Tanaka, J. Am. Chem. Soc. **136**, 5555–5558 (2014)
38. Y. Morisaki, M. Gon, T. Sasamori, N. Tokitoh, Y. Chujo, J. Am. Chem. Soc. **136**, 3350–3353 (2014)
39. L.M. Blinov, *Structure and Properties of Liquid Crystals* (Springer Science and Business Media, Heidelberg, 2011)
40. V. Sharma, M. Crne, J.O. Park, M. Srinivasarao, Science **325**, 449–451 (2009)
41. S. Vignolini, P.J. Rudall, A.V. Rowland, A. Reed, E. Moyroud, R.B. Faden, J.J. Baumberg, B.J. Glover, U. Steiner, Proc. Natl. Acad. Sci. U. S. A. **109**, 15712–15715 (2012)
42. Y. Nagata, K. Takagi, M. Suginome, J. Am. Chem. Soc. **136**, 9858–9861 (2014)

Chapter 9
Functional Supramolecular Materials Formed by Non-covalent Bonds

Yoshinori Takashima, Yuichiro Kobayashi, Motofumi Osaki, and Akira Harada

Abstract Molecular recognition is essential for realizing functional supramolecular materials. Non-covalent host-guest interactions are effective tools to introduce various functions and properties into materials. This review focuses on the functions such as selective molecular adhesions, self-healing, toughness, and actuation properties of the supramolecular polymeric materials. These functions have been achieved by using reversible bond formations between cyclodextrins (CDs) and guest molecules. The host-guest interactions involving CDs can be used to achieve efficient stimuli-responsive behaviors and self-healing properties. Furthermore, the supramolecular materials have been found to exhibit macroscopic rapid expanding and contracting driven by external stimuli under semidry conditions. Supramolecular actuator using host-guest complexations can be prepared by two approaches. The first is the functionalization of a supramolecular gel, which changes the cross-linking density between the polymers. The second is the utilization of a topological gel to change length of the polymer chain between cross-linked points. Both types of the actuators exhibit bending behaviors by external stimuli. This review summarizes the advancements within the past 10 years in supramolecular materials that utilize the host-guest interactions and the sliding motion of ring molecules functionalized by chemical or physical processes.

Keywords Non-covalent bond · Host-guest interaction · Polymeric material · Adhesion · Stimuli-responsive material · Functional material

9.1 Introduction

Molecular recognitions between small molecules and macromolecules play important roles to show various functions in biological systems [1–6]. Herein, we focus on dynamics of muscle motion [7–10], self-healing properties, and cell adhesion

Y. Takashima (✉) · Y. Kobayashi · M. Osaki · A. Harada
Osaka University, Toyonaka, Osaka, Japan
e-mail: takasima@chem.sci.osaka-u.ac.jp; harada@chem.sci.osaka-u.ac.jp

S. Shirakawa (ed.), *Designed Molecular Space in Material Science and Catalysis*,
https://doi.org/10.1007/978-981-13-1256-4_9

[11, 12] through molecular recognitions. Myosin, kinesin, and dynein convert energy from ATP hydrolysis to mechanical work. In muscle fibrils, myosin and actin filaments form a complex with an alternating layered structure, components of which slide over one another with molecular recognitions, leading to macroscopic contraction and expansion of the muscles [7–9]. The sliding motion of myosin and actin filaments inspired the development of artificial linear motors. The molecular recognitions of cells also inspired self-healing property, which is similar to cell adhesion via molecular recognitions on the cell surfaces. Cellular adhesion, through which cells form clumps, is essential for maintaining multicellular structures. These cell clumps eventually form organs through cell sorting. Moreover, these behaviors can be controlled through molecular recognitions. Inspired by these mechanical movement and selective assembly, development of stimuli-responsive supramolecular materials has been attempted by using molecular recognitions.

Recently, molecular recognition chemistry [13–15] and supramolecular chemistry [16, 17] have received much attentions from scientists, owing to their dynamic effects on molecular structures, catalytic activities, photochemical behaviors, and switching of those properties. Macrocyclic molecules (crown ethers [13], cyclophanes [14], cryptands [16], and cucurbiturils [18]) are typical host molecules, which are widely used to achieve functional properties in supramolecular chemistry. In this field, controls of supramolecular assemblies and polymer structures [19, 20] by external stimuli, such as pH [21–23], redox [24–28], and light [29–38], are issues of increasing interest. Stimuli-responsive supramolecular polymers are relevant not only for mimicking biological functions but also for a range of applications in materials science. Supramolecular chemists including us employed cyclodextrins (CDs) as supramolecular components. CDs are a family of macrocyclic oligosaccharides, the most common of which are composed of 6 (αCD), 7 (βCD), or 8 (γCD) α-1,4-linked D-glucopyranose units. Table 9.1 shows the physical properties of CDs [39–42]. Inclusion complexes of CDs are formed through hydrophobic interaction and *van der Waals* force between inner surface of the CD ring and hydrophobic guest molecule with a suitable size for the CD cavity. In early studies of CD inclusion complexes, mimicking enzyme catalysis was attempted [43, 44]. CDs were also used for the study of phase-boundary supramolecular catalysis and developed as artificial enzymes [45].

For the development of functional supramolecular materials controlled by host-guest interactions, we have realized sol-gel switching, self-healing, adhesion control, and contraction-expansion movements through the formation of inclusion complexes with CDs in polymeric materials [46]. Our research aims at observing macroscopic phenomena triggered by molecular recognitions without using a microscope. There are three approaches to prepare the functional supramolecular polymers. The first is use of polyrotaxane-type materials, the second is employing reversible complex formation on polymer side chains, and the third is formation of sequential supramolecular complexes (supramolecular polymers) (Fig. 9.1).

Study of incorporations between CDs and polymers were provided a few decades ago. It was found that CD-modified poly(acrylic acid) forms ester guests via complexation and that the CD polymer acts as an artificial hydrolysis enzyme to

Table 9.1 Chemical structure, approximate geometric dimensions, and physical properties of α, β, and γCD

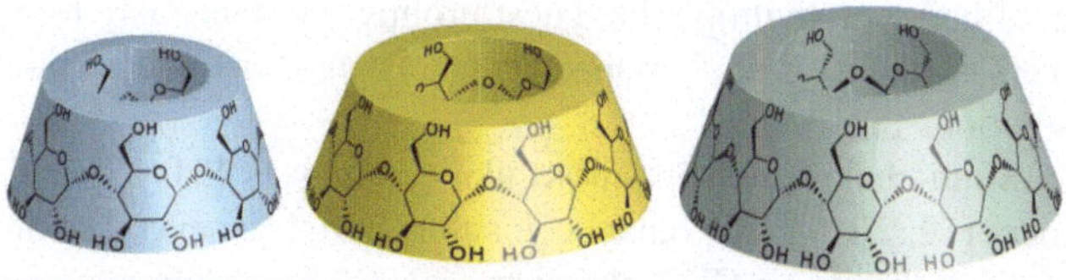

	α-CD	β-CD	γ-CD
Molecular Weight	972	1135	1297
No. of Glucose Units	6	7	8
Cavity Diameter (nm)	0.47	0.60	0.75
Height of Torus (nm)	0.79	0.79	0.79

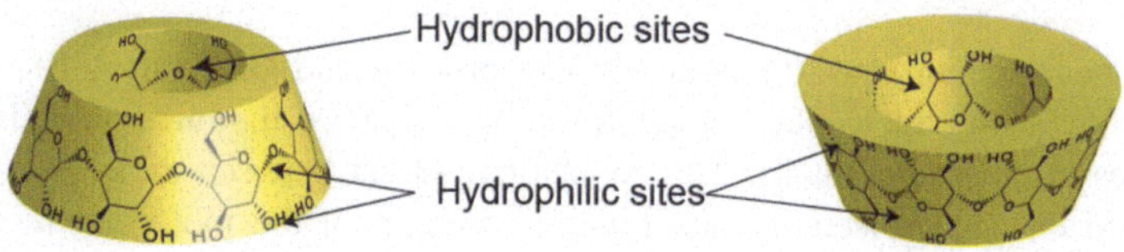

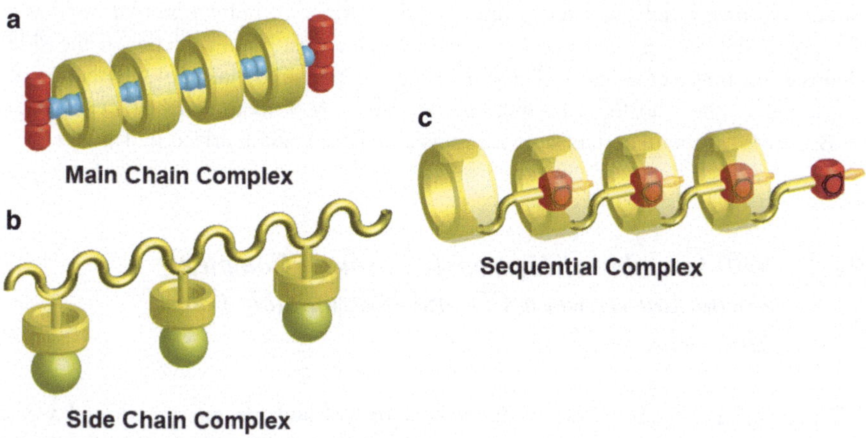

Fig. 9.1 Preparation of functional supramolecular materials by using CDs. (**a**) Polyrotaxane-type materials, (**b**) reversible complex formation on a polymer side chain, and (**c**) sequential supramolecular complexes (supramolecular polymers)

promote hydrolysis of the guest molecules [47]. Native CDs accurately recognize saturated aliphatic guest units on polymer side chains. Furthermore, it was found that the selectivity of CDs with aliphatic guest units attached to polymer chain is higher than that with the small guest molecules isolated in solutions [48]. In the study of polymer chemistry with CDs, water-soluble polymer and CDs in aqueous solutions were found to form poly-*pseudo*-rotaxanes in 1990 [49, 50]. The research concerning sequential-type supramolecular polymers with CDs revealed that CDs modified with hydrophobic guest groups spontaneously form supramolecular polymers, which were connected through host-guest interactions in aqueous solutions [51].

Main-chain-type supramolecular complexes [51–53] and supramolecular polymers [19, 20, 54–56] have been summarized in several reviews previously. On the basis of the above CD's chemistry, herein, we summarize our recent researches on functional supramolecular polymeric materials controlled by the formation of complexes between CD and guest molecules in the polymer networks.

9.2 Molecular Recognition of Polymer Side Chains

Complex formation and self-assembly like proteins and DNAs have been demonstrated by using molecular recognitions of polymers modified with host/guest moieties at their side chains. The cooperative effect between the side chains was found to enhance selectivity and binding strength of the molecular recognitions [57, 58].

On the basis of these effects, we have investigated the formation of macroscopic assemblies using molecular recognition on the surface of polymeric hydrogel. We prepared CD-host gels and guest gels, which have host/guest moieties at the side chains of the polymer network. These materials are prepared via copolymerization of acrylamide, the host/guest monomers, and *N,N′*-methylenebis(acrylamide) (MBAAm). All the materials shown in this chapter are summarized in Table 9.2.

9.2.1 Self-Assembly of Hydrogels Through Complex Formation Between CD and Hydrophobic Guest Molecules

Macroscopic adhesions between the host hydrogel and the guest hydrogel were investigated. We prepared the host hydrogels (αCD gel and βCD gel), which have αCD or βCD moieties at the side chains of polyacrylamide, respectively. The guest hydrogels were also prepared by corresponding functional monomers to obtain Ad gel (having adamantly group), *n*Bu gel (*n*-butyl groups), and *t*Bu gel (*t*-butyl group). Adhesive behaviors were tested by shaking gels floating on water in a petri dish. Ad gel showed selective adhesion to βCD gel not to other Ad gel (Fig. 9.2b), indicating that host-guest complex formation between the gels led to the adhesive behavior. When αCD gel, βCD gel, *n*Bu gel, and *t*Bu gel were shaken in water at the same

Table 9.2 Chemical structures of the host and guest gels showing adhesive behaviors

Host gels	Guest gels	Ref.
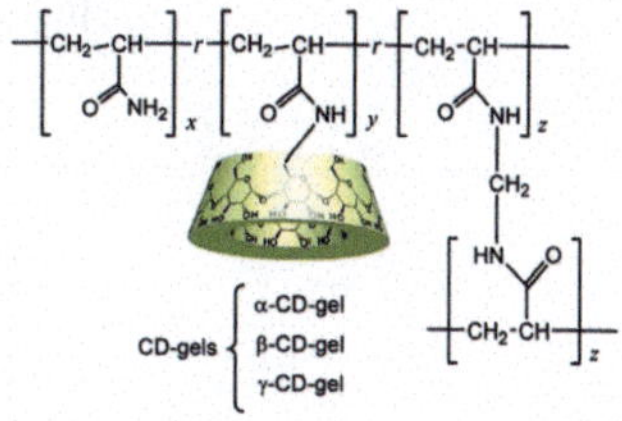	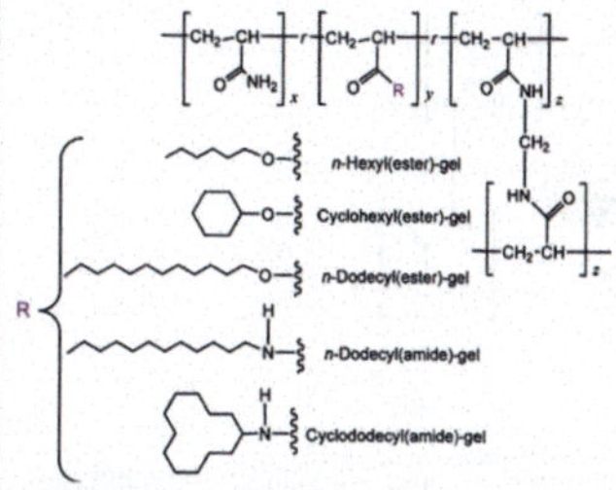	[60]
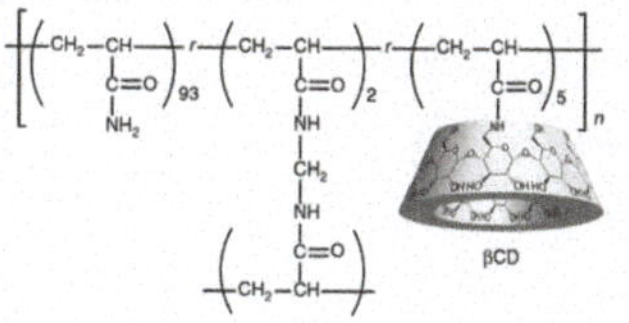 βCD gel	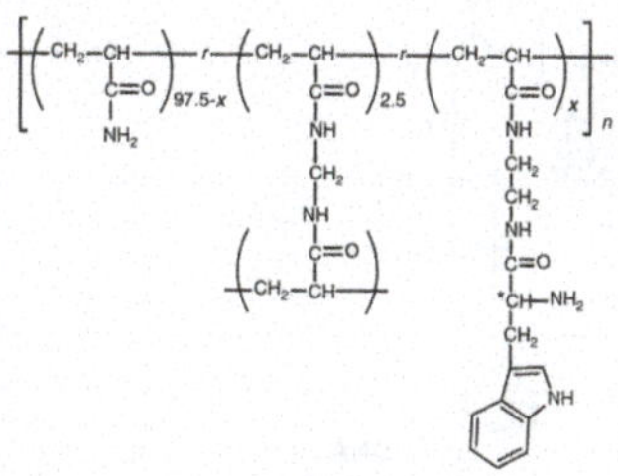 Trpgel	[61]
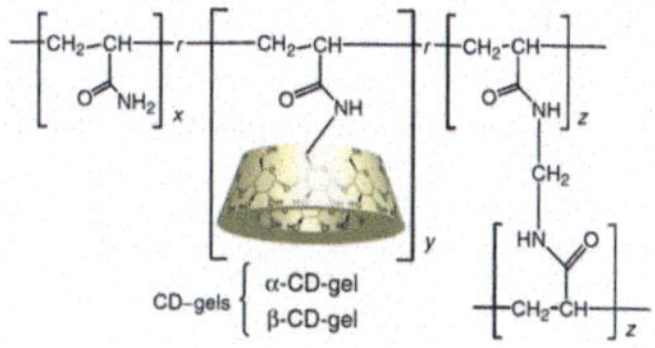	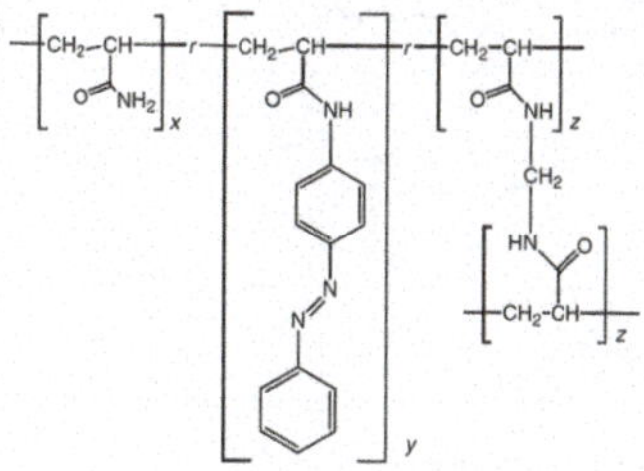 Azo-gel	[62]
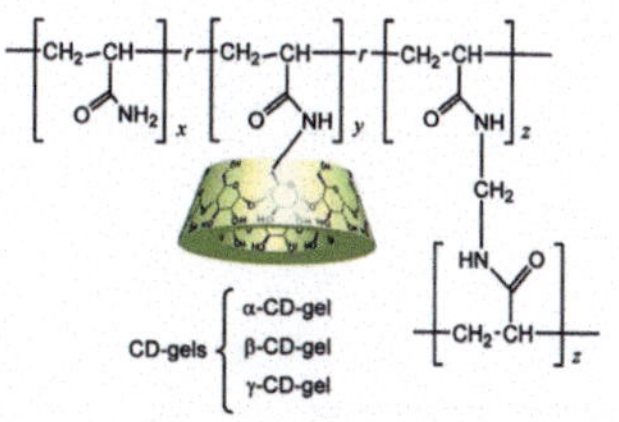	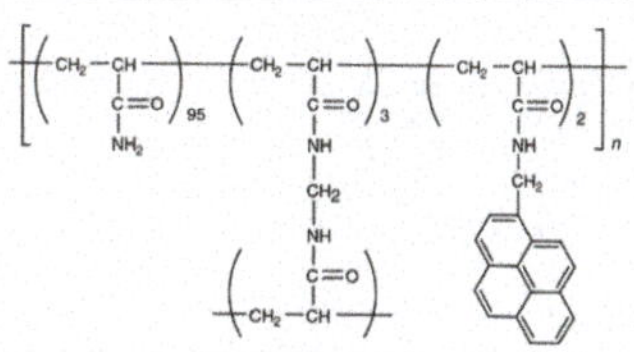 Pygel	[63]
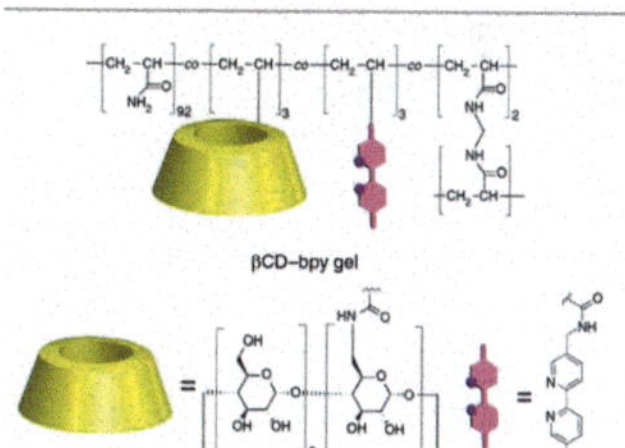	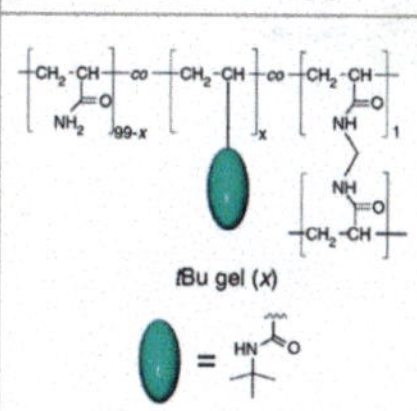	[64]

(continued)

Table 9.2 (continued)

Host gels	Guest gels	Ref.
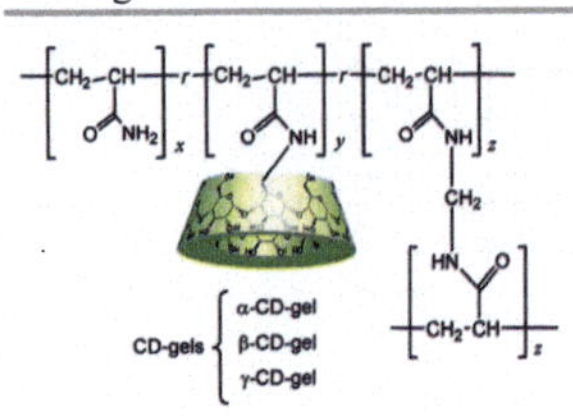	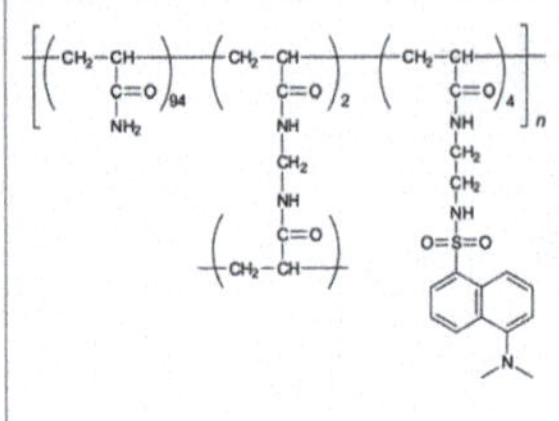 Dns-gel	[65]
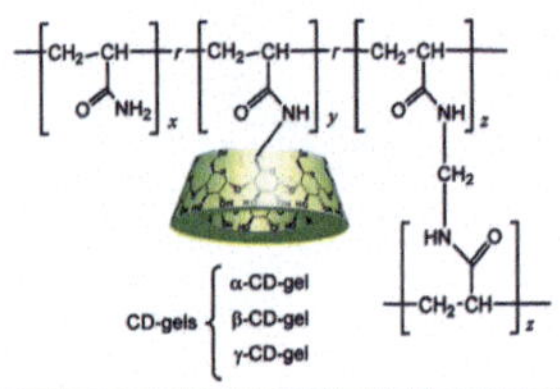	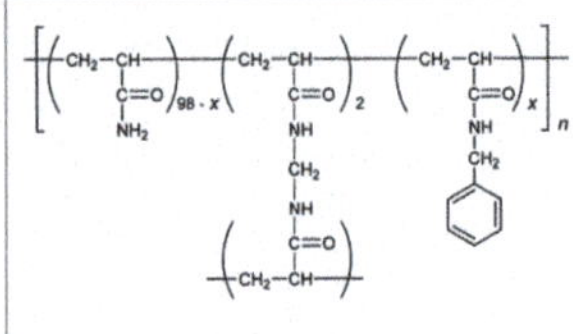 Bz-gel	[66]
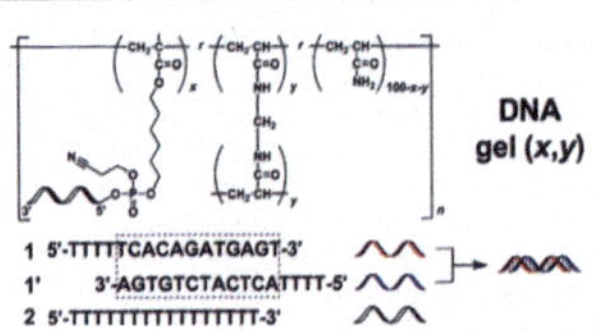	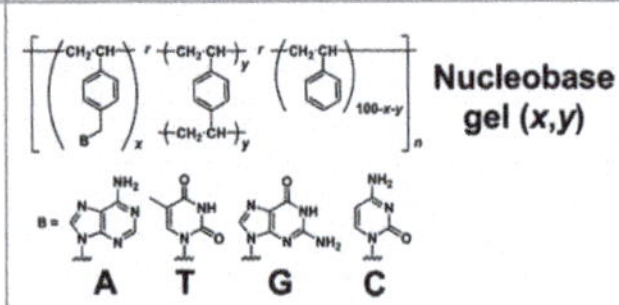	[80]
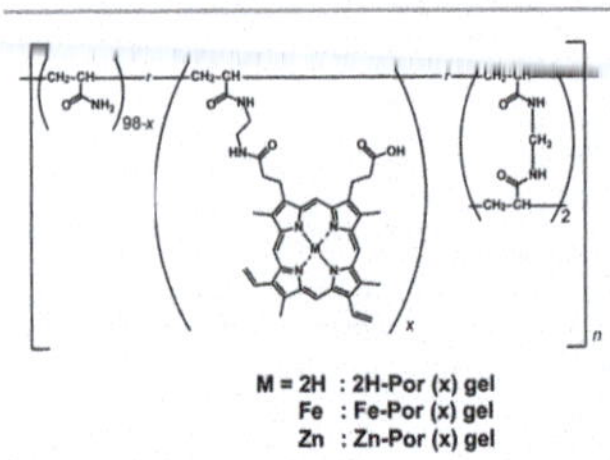 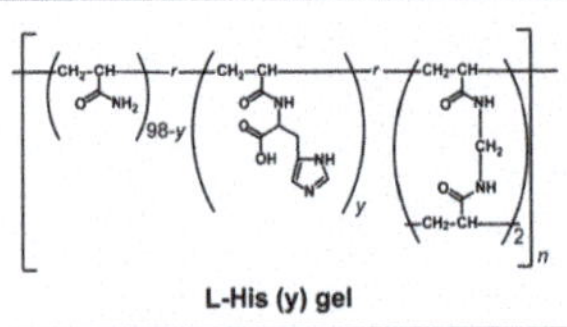	–	[81]
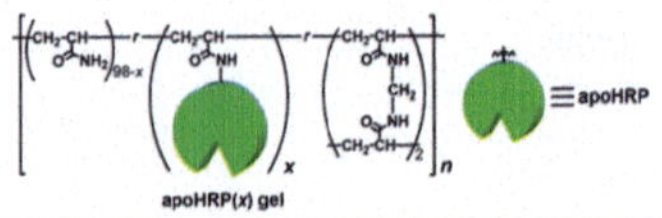	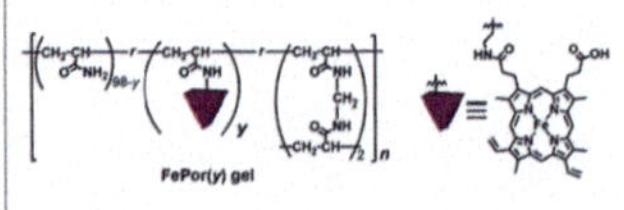	[83]
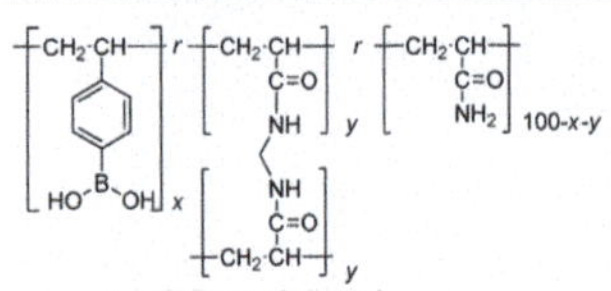 PB gel (x,y)	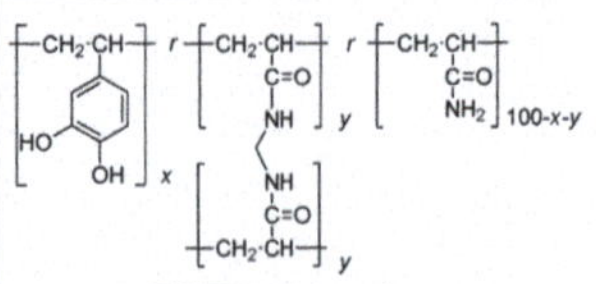 CAT gel (x,y)	[88]

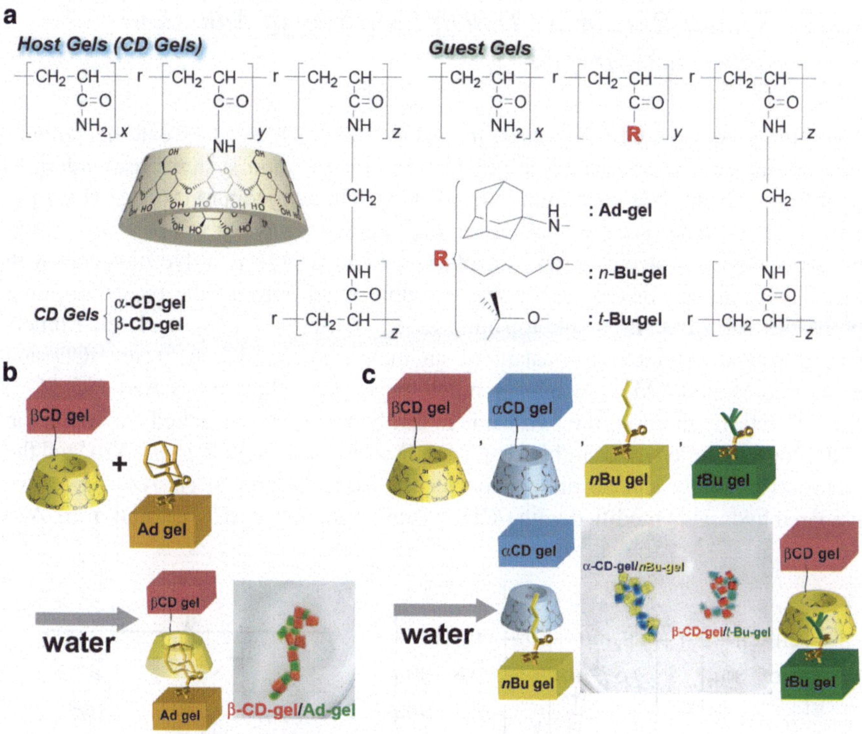

Fig. 9.2 (**a**) Chemical structures of host gels (αCD gel and βCD gel) and guest gels (adamantyl gel (Ad gel), *normal*-butyl-gel (*n*Bu gel), and *tert*-butyl-gel (*t*Bu gel)). (**b**) Macroscopic self-assembly between βCD gels and Ad gels. (**c**) Macroscopic self-assembly of βCD gel/*t*Bu gel and αCD gel/*n*Bu gel

time, αCD gel and βCD gel selectively adhere to *n*Bu gel and *t*Bu gel (Fig. 9.2c) [59]. Even when the hydrophobic guests *n*Bu and *t*Bu were changed to *n*-hexyl and cyclohexyl groups, they also show the selective adhesion to αCD and βCD, respectively [60]. As shown above, the adhesion through host-guest interactions are selective process, and the self-assemblies are controllable by choosing the combinations between CDs and guest molecules with.

As CDs are cyclic oligomers of D-glucopyranose units with a conical chiral cavity, the CD cavities are chiral, and the host-guest interactions have potential ability for chiral sensing. Recently, we have achieved macroscopic chiral recognition using CD hydrogel. The βCD hydrogel was found to assemble with a hydrogel having L-tryptophan moieties as side chains (L-Trp gel) to form an aggregate, but it does not interact with a hydrogel with D-tryptophan (D-Trp gel) [61]. Though chiral recognition with the host-guest interaction is generally difficult because of low selectivity of the chiral host, the design of macromolecular host/guest has enabled multivalent effective chiral recognitions to show the macroscopic phenomena.

9.2.2 Stimuli-Responsive On/Off Switching of Adhesion Between Hydrogels

The affinity between CDs and guest molecules can be changed by external stimuli. We investigated the macroscopic gel adhesion-dissociation systems responding to light [62], solvent [63], metal ion [64], pH [65], and temperature [66]. As shown in Fig. 9.3, a macroscopic gel assembly system was developed by using photoresponsive molecule such as azobenzene (Azo). Azo is a dye molecule with two phenyl groups linked with a *N*=*N* double bond. One of the most intriguing properties of Azo is a photoisomerization between *trans*- and *cis*-isomers [67, 68]. The association constant of an inclusion complex between αCD and *trans*-Azo is 2000 M^{-1}, which is much larger than that of *cis*-Azo (35 M^{-1}) [37, 38, 69]. In addition, the two isomers can be reversibly switched by irradiation with proper wavelength of light. That is, αCD selectively include *trans*-Azo, and the inclusion complex can be dissociated by the isomerization to *cis*-Azo. Here, we prepared hydrogels modified with αCD or *trans*-Azo. When αCD gel and *trans*-Azo

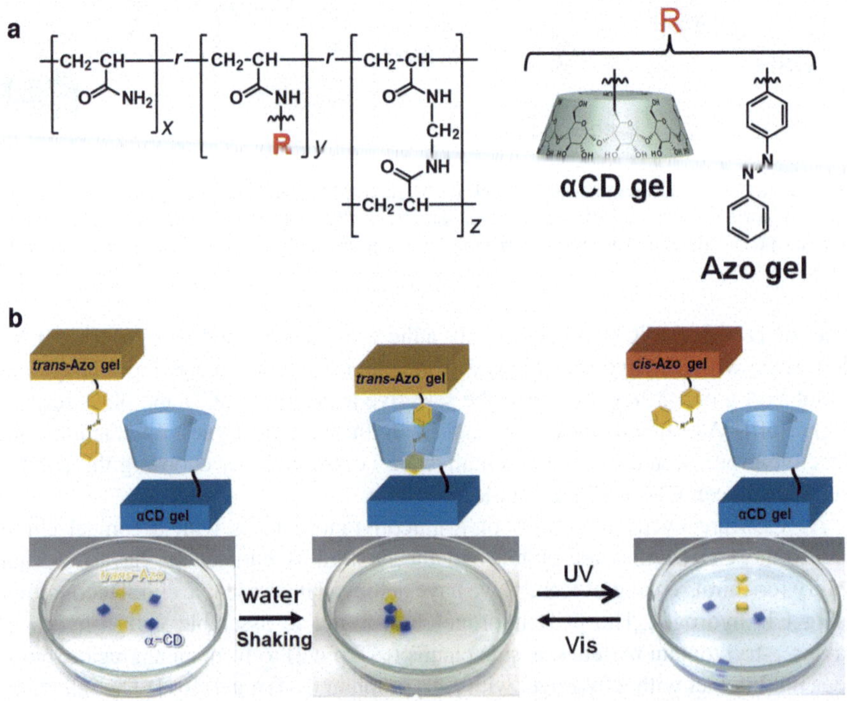

Fig. 9.3 (**a**) Chemical structures of αCD gel and Azo gel. (**b**) Gel assembly and dissociation on photoirradiation. Shaking of the gels led to the assembly of αCD gel/*trans*-Azo gel. Photoirradiation with UV light led to the dissociation gels. The separated αCD gels and Azo gels are found to reassemble on visible light irradiation. Binding of the Azo gels to the αCD gels and dissociation of the combined gels are found to be reversible and reproducible

gel were floated in water and agitated for a minute, αCD hydrogel adhered to *trans*-Azo hydrogel. The assembled gels were found to dissociate by irradiation with UV light, indicating that the *trans* to *cis* photoisomerization of the Azo units in hydrogel decreased the association constant between αCD and Azo, leading to the dissociation of gels. Furthermore, the separated αCD hydrogels and *cis*-Azo hydrogels can reassemble by irradiation of visible light, which triggered the *cis* to *trans* photoisomerization of Azo to be included by αCD [62].

Recombination and switching of hydrogels were realized by using pyrene (Py) as a guest moiety. While Py easily forms dimers or aggregates via intermolecular forces (hydrophobic interaction and π-π stacking) in aqueous media, it remains as an isolated monomer in organic solvents such as dimethyl sulfoxide (DMSO) [70, 71]. -Py-modified hydrogel (Py gel) was prepared as a guest gel in the same manner. In water, the Py hydrogel adhered to a hydrogel having γCD host moieties (γCD gel), because inclusion complexes between γCD and the Py dimer were formed between the gels. On the other hand, changing the solvent with volume fraction of water/DMSO = 1:1 made the Py dimers dissociated, and the Py gels became to adhere to βCD gel. The selectivity of Py gel can be switched by varying the composition of the mix solvent water/DMSO system to show recombination of macroscopic gel assembly [63].

We also achieved metal-ion-responsive adhesion materials. A hydrogel having both βCD and bipyridyl (Bpy) groups as side chains (βCD-Bpy gel) was prepared. βCD-Bpy gel did not adhere to *t*Bu gel. The βCD and Bpy formed an inclusion complex within the same gel, and the βCD could not include further guest molecules not to show the adhesion to *t*Bu gel. However, after immersion in a Cu^{2+} solution, βCD-Bpy hydrogel showed adhesion to *t*Bu hydrogel, indicating that the Bpy moiety left the βCD cavity and formed Cu(Bpy) complex, letting βCD moiety include *t*Bu groups on *t*Bu gel. When the gels were immersed in EDTA to eliminate Cu^{2+} ions from the system, βCD-Bpy hydrogel was found to cease the adhesion to *t*Bu hydrogel again [64].

Assembly triggered by pH has great potentials in biomedical applications. In neutral aqueous solution, hydrogel-tethered dansyl (Dns) groups can adhere to only βCD gels not αCD or γCD gels, because the cavity size of βCD is appropriate for Dns. The assembled βCD gel and Dns gel were found to dissociate by shaking in low pH (≤3.0). In general, inclusion complexes between CD and cation are unstable due to electrostatic interactions. These results indicate that protonation of the Dns moiety made the inclusion complex dissociated to separate the gels in macroscale. The association and dissociation between βCD hydrogel and Dns hydrogel are also repeatable [65].

Among a variety of stimuli, temperature is one of the most important and useful stimulus. The formation of inclusion complexes between CDs and guest compounds strongly depends on temperature [72–74]. Here, we achieved a control of the macroscopic gel assemblies by temperature [66]. By shaking the hydrogels having benzyl moieties (Bz gel) and α, β, and γCD gels in aqueous solution at a room temperature, the Bz gel forms assemblies with only βCD gel. When the temperature was cooled down to 15 °C, Bz gel showed adhesion to both αCD and βCD gel. Further cooling down made Bz gel adhere to all the host gels including γCD gel. By decreasing the temperature, complex formations between CDs and Bz were promoted by entropic gain, letting CDs form the complexes with the Bz moiety. The changes of gel assemblies at various temperatures also showed a good reversibility, suggesting that temperature is a useful stimulus for macroscopic assembly based on molecular recognition.

9.2.3 Macroscopic Self-Assembly Driven by Hydrogen Bond Formations

More sophisticated controls of macroscopic assemblies are available by using dynamic interactions between molecules. There are several reports concerning adhesions between materials via various non-covalent interactions, including dynamic covalent bonds [75], ion-ion interactions [76], and hydrogen bonds [77], among others [78]. Yin and coworkers reported a DNA-based gel assembly system using DNA polymerase-mediated rolling circle amplification (RCA) method on the surface of the gel [79].

We reported macroscopic assemblies and adhesion between materials with hydrogen bond interactions using oligonucleotides or nucleobase pairs [80]. Three types of hydrogels having oligonucleotides as side chains (DNA gels) were prepared by using DNA-tethered monomers **1** (5′-TTTTTCACAGATGAGT-3′ (16-mer)), **1′** (5′-TTTTACTCATCTGTGA-3′ (16-mer), and **2** (5′-TTTTTTTTTTTTTTTT-3′ (16-mer)), respectively. When DNA gels with **1**, **1′**, and **2** were placed to contact each other, adhesive behavior was observed only between those with **1** and **1′**, which have complementary base pairs. The adhesion strength between DNA gels with **1** and **1′** was much larger than that of control samples, indicating that the adhesion of DNA gels was achieved through hybridization of the complementary oligonucleotides with 12 nucleobases. On the basis of these results, we hypothesized that synthetic polymers modified with nucleobases on the polymer side chain could realize selective adhesion by the complementary base pairs. A macroscopic assembling by using organogels modified with adenine (**A**) or thymine (**T**) moieties was carried out in toluene. Although association constants between each single nucleobase are not very high, the **A** and **T** organogels were found to adhere to each other via and to form an alternating assembly. The organogels with the same nucleobase did not form assemblies at all, supporting the complementarity of the gel assembly through hydrogen bond interactions in an organic solvent. As shown above, multivalent interactions on the side chains of polymers are important for integrating and increasing weak interactions of each side chains.

9.2.4 *Macroscopic Self-Assembly by Metal-Ligand Interactions*

We also investigated selective adhesion controlled by metal-ligand interactions [81]. In the adhesion testing, a hydrogel modified with Fe-porphyrin (Fe-Por gel) adhered to a hydrogel modified with $_{\mathrm{L}}$-histidine ($_{\mathrm{L}}$His gel), which acts as a ligand for metal center of Fe-porphyrin. On the other hand, a hydrogel modified with free-base porphyrin (2H-Por gel) did not interact with $_{\mathrm{L}}$His gel. However, addition of $FeCl_3$ to the gels again led to metalation of the porphyrins to show its adhesion property to $_{\mathrm{L}}$His gel.

In biological systems, metal-ligand interactions play important roles. For example, horseradish peroxidase (HRP) exhibits catalytic activity when $_{\mathrm{L}}$-histidine (an active site of apoHRP) and Fe-porphyrin (cofactor) form a complex via metal-ligand interaction. ApoHRP cannot express catalytic activity without Fe-porphyrin [82]. Here, we constructed a system to express catalytic activity by adhesion between apoHRP gel and Fe-Por gel. The apoHRP gel and Fe-porphyrin gel were found to act as a catalyst when they formed a gel assembly. The individual gels did not exhibit catalytic activity when the gel assembly was separated. The catalytic reaction is controllable by changing the gel size or molar content of apoHRP or Fe-Porphyrin moieties in the gels, and these processes are repeatable [83]. These findings are expected to be applicable for various materials, such as drug carriers that release the drug upon gel adhesion.

9.2.5 *Macroscopic Self-Assembly by Boronate-Catechol Interactions*

Boronic acids (BAs) are universally utilized as tools for molecular recognition in the supramolecular chemistry [84–86]. BAs are known to form boronic ester derivatives with various diol compounds [87]. And, the formation and dissociation of the boronic ester derivatives are reversible and controllable by pH change. We developed a hydrogel adhesion system based on the interaction between phenylboronic acid (PB) and catechol (CAT) [88]. In adhesion test, hydrogels modified with PB (PB gel) adhered to hydrogels modified with CAT (CAT gel) to form an assembly in basic aqueous media. The adhesion strength between the gel pieces increased as the mol% of PB and CAT in the hydrogels increased. After changing pH of the aqueous media from basic to acidic, the gel assembly was dissociated. Furthermore, raising pH of the media let the gels reform the assembly. The adhesion and the assembling were found to be controllable by adding competitive saccharide molecules to the system. The adhesive behavior between PB gel and CAT gel varied by the presence of various monosaccharides in basic aqueous media. The adhesion strength between the gels was found to decrease by increasing an association constant of complexation between PB gel and competitive saccharides.

9.2.6 Direct Adhesion of Dissimilar Materials Through Host-Guest Interactions

As manufactures used in our daily life are aggregates of dissimilar materials, adhesion between dissimilar materials is one of the most important issues in materials technology. Although there are some methods for the adhesion of dissimilar materials, such as using adhesives, nails, or wedges, interfacial separation and destruction of the adhesives are severe problems. If the dissimilar materials can be directly adhered without adhesives, that method would be a powerful tool for the adhesion.

We developed direct adhesions of materials using the CD gels and glass substrate modified with guest molecules (Fig. 9.4a). The αCD and βCD gels were placed onto glass surface modified with adamantane (Ad sub) under wet conditions for 24 h. Although αCD gel did not adhere to the Ad sub, βCD gel adhered to the surface, indicating that the adhesion based on host-guest interactions is still available even between dissimilar materials. It was also found that the adhesion strength of the CD gels on the guest substrate is controllable by external stimuli. We showed that adhesion of αCD gel for a glass substrate modified with Azo (Azo sub) was controlled by irradiations of UV and visible lights. The adhesive behavior of αCD gel on Azo sub can be switched by the photoisomerization of Azo moieties (Fig. 9.4b). These results suggest that using host-guest interaction is a feasible method for controlling direct adhesions of dissimilar materials [89].

In recent years, we also achieved adhesions between gel and glass substrate by employing covalent bond formation, such as Sonogashira cross-coupling reaction [90], Suzuki-Miyaura cross-coupling reaction [91], and azide-alkyne cycloaddition reaction [92]. The direct adhesions were found to be stronger than that of conventional adhesives, and organic solvents invading adhesives such as DMSO cannot dissociate the direct adhesions. These results indicate that design of chemical structure on the surface of material is effective for adhesion.

9.2.7 Formation of Hydrogel Self-Assemblies Controlled by Multi-interactions

On the basis of the reports shown above, assemblies of three types of gels by using multiple interactions were investigated (Fig. 9.5). Here, a ferrocene (Fc) derivative was used to prepare a hydrogel modified with Fc (Fc gel). Fc reversibly changes its charge from neutral to cationic in response to chemical stimuli of redox reagents. An association constant (K_a) of Fc with βCD is larger than those of αCD or γCD

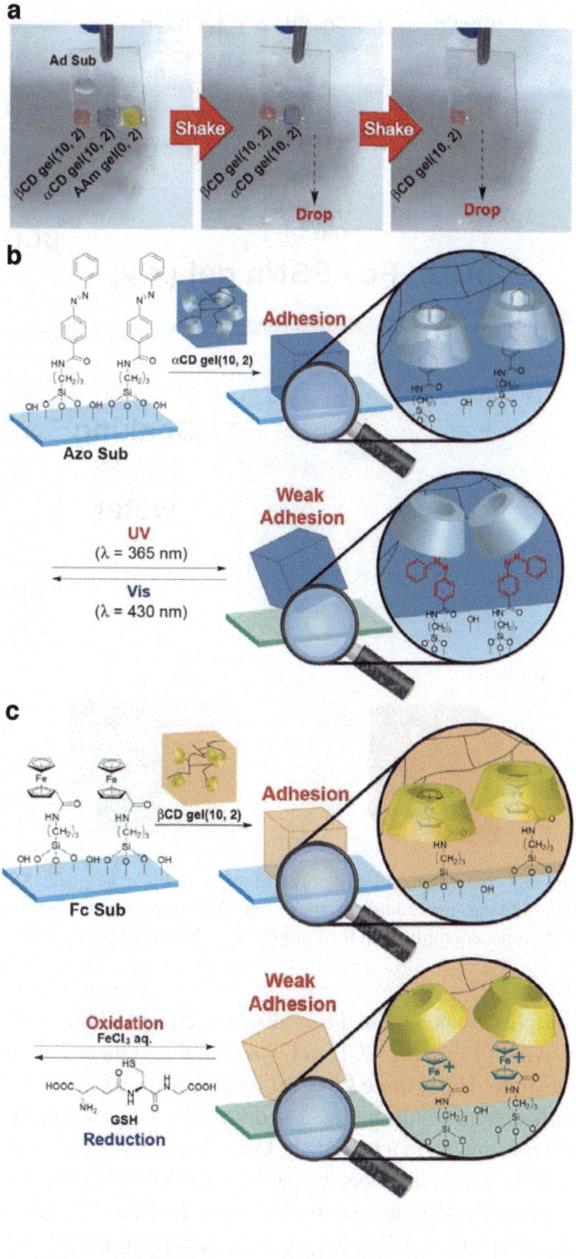

Fig. 9.4 Selective direct adhesion between host-modified gels and glass substrates modified with adamantane guest (**a**). Schematic illustration of the adhesion between the αCD gel and the Azo substrate (**b**) and between the βCD gel and the Fc substrate (**c**). αCD gel adhered to the Azo substrate, but the UV-irradiated Azo substrate had a significantly low adherence property to the αCD gel. βCD gel adhered to the reduced state of the Fc substrate, but the oxidized state of the Fc Sub had a significantly low adherence property to the βCD gel

(Fc/αCD; $K_a = 0.14 \times 10^3\ M^{-1}$, Fc/βCD; $K_a = 17 \times 10^3\ M^{-1}$, Fc/γCD; $K_a = 0.90 \times 10^3\ M^{-1}$) [93, 94]. While βCD shows high affinity for the reduced state of Fc due to its hydrophobic nature, the oxidized state of Fc (Fc^+) exhibits low affinity for βCD because of its electrostatic unstableness. Therefore, an inclusion

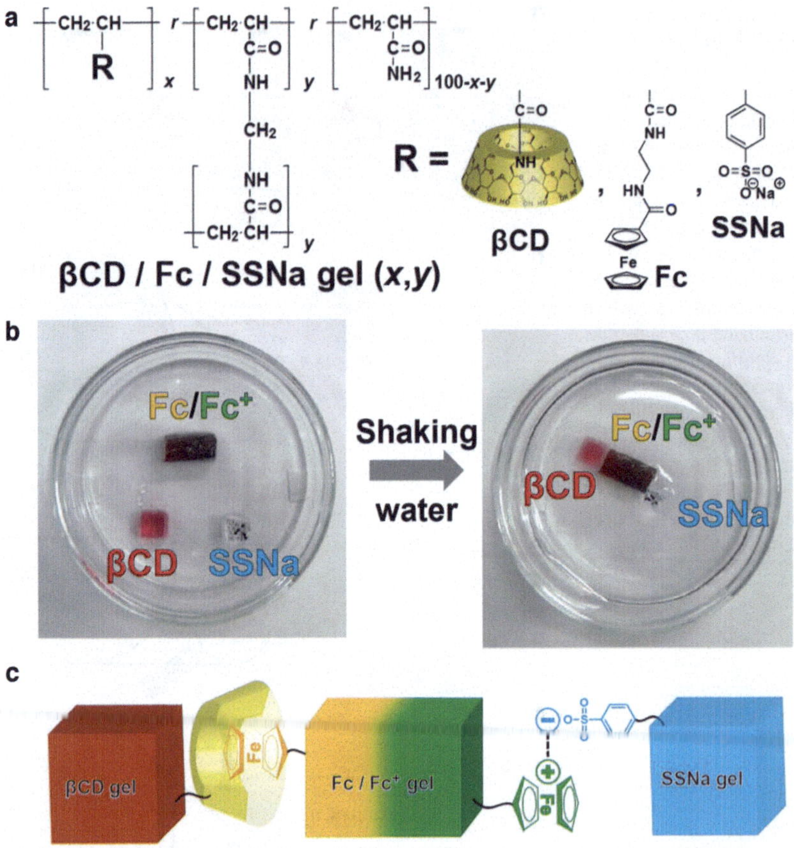

Fig. 9.5 (**a**) Chemical structures of βCD gel, Fc gel, and SSNa gel. (**b**) Photographs of the experimental procedure to make an ABC-type assembly. (**c**) Schematic illustration of the ABC-type assembly with the composition βCD gel-Fc/Fc^+ gel-SSNa gel

complex of βCD and Fc is dissociated by addition of oxidizing reagent [95]. As shown in Fig. 9.4c, βCD gel was found to adhere only to reduced Fc-modified glass substrate. Here, three kinds of hydrogel (βCD gel, Fc gel, and a hydrogel modified with *p*-styrenesulfonic acid sodium salt as an anion (SSNa gel)) were added to an aqueous media in a petri dish. Under reductive conditions, Fc gel selectively adhered to βCD gel, and it does not adhere to the SSNa gel. After addition of an oxidative reagent, Fc residue became Fc^+, and then the Fc^+ gel was separated from βCD gel and adhered to SSNa gel via ionic interactions. We prepared Fc/Fc^+ gel, one half surface of which was oxidized and the other was reduced. When Fc/Fc^+ gel was shaken with the βCD and SSNa gels in the petri dish, the hydrogels formed ABC-type assemblies consisting of βCD gel, Fc/Fc^+ gel, and SSNa gel, aligned by that order. Use of two discrete non-covalent interactions allowed for control of order of the hydrogel assemblies (Fig. 9.5b, c) [96].

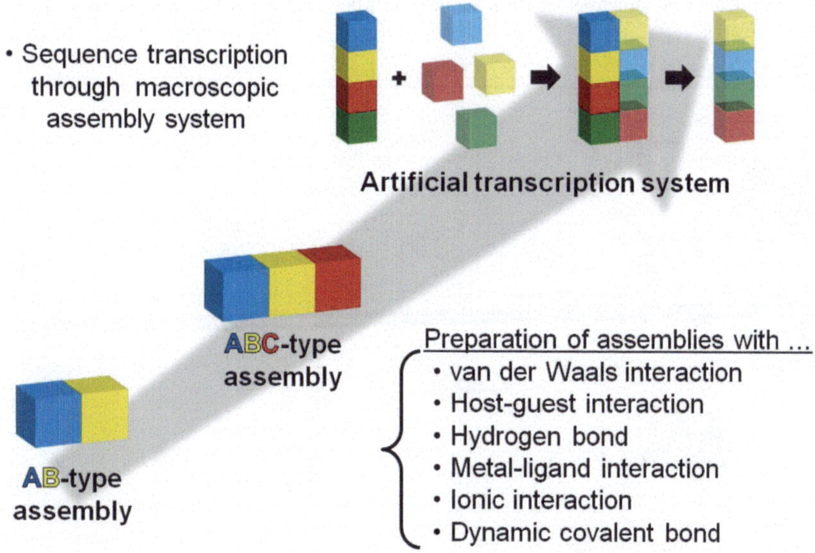

Fig. 9.6 Future prospects of the macroscopic self-assemblies with non-covalent interaction

As described above, several types of chemical interactions can be used to form various macroscopic assemblies. Outer faces of the assemblies still have an ability to recognize other molecules. By using the outer face, a macroscopic assembly system would realize an artificial transcription system in the near future (Fig. 9.6).

9.3 Sol-Gel Switching Materials Through Molecular Recognition

9.3.1 Conventional Photoresponsive Sol-Gel Switching Materials

Gels are divided into two main categories according to type of cross-linking interactions, whether it is a physical interaction [97–99] or a chemical bonding [100–102]. The gels based on physical interactions, stabilized by non-covalent cross-linkers, are easier to tune their physical properties than those with covalent cross-linking bonds. To control the properties of gels, low-molecular-weight physical gels [103–110] and hydrogels formed by host-guest interactions [111–114] have been widely studied and found to exhibit switching elastic properties. The hydrogels formed by host-guest interactions should be more sensitive than the low-molecular-weight physical gels to external stimuli, because of their selective and specific behaviors. Although there are some reports on stimuli-responsive hydrogels with a phase transition, there have been few reports on light- or redox-responsive

Table 9.3 Chemical structures of the host and guest polymers demonstrating stimuli-responsive sol-gel transitions

CD host polymer	Guest polymer	Ref.
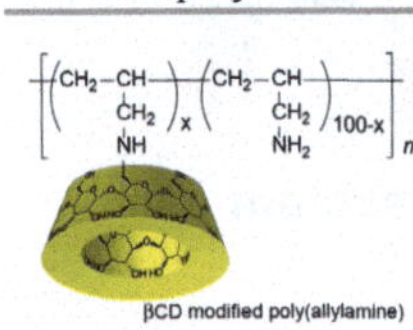	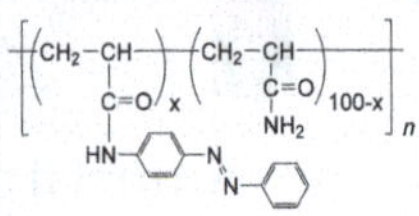	[69]
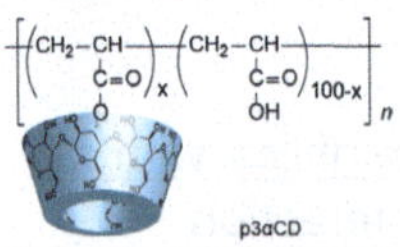	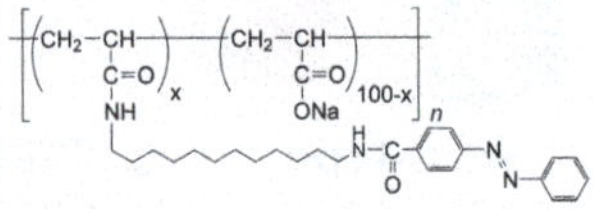	[37]
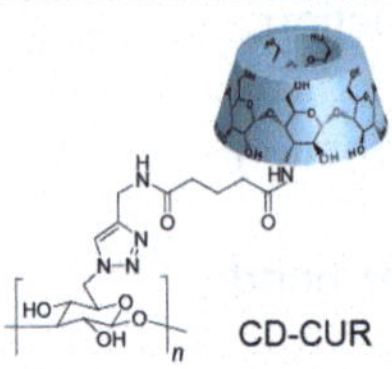	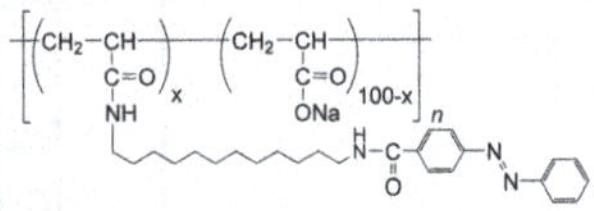	[115]
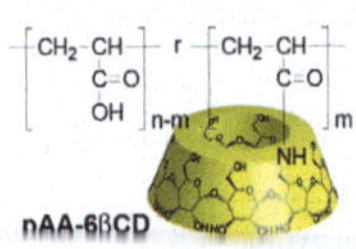	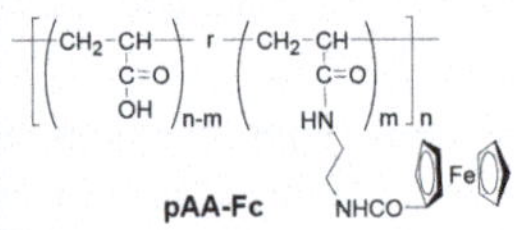	[116]

hydrogels. We developed light- and redox-driven sol-gel switching by using host-guest interactions (Table 9.3).

9.3.2 Photoresponsive Sol-Gel Switching Materials Through Host-Guest Interactions

As described above, Azo undergoes photoisomerization when it is irradiated by UV and visible light [67, 68]. αCD or βCD selectively forms an inclusion complex with the *trans*-isomer, but the *cis*-isomer does not, because the association constant of CDs with *trans*-Azo is higher than that of *cis*-Azo [37, 38, 69].

Utilizing the switchable interaction of host-guest complexation between αCD and Azo, we have investigated the preparation of photoresponsive sol-gel switching materials. By controlling the association and dissociation of the inclusion complex of αCD with Azo, the cross-linking density between polymers can be adjusted, resulting in the sol-gel switching behavior. αCD has been modified to curdlan (β-1,3 glucan, CUR), hereafter called CD-CUR, and used as a host polymer. Azo has been modified to poly(acrylic acid) (pAA-Azo). When CD-CUR and pAA-Azo were mixed with a ratio of 1:1 monomer unit in water, the viscosity of the solution

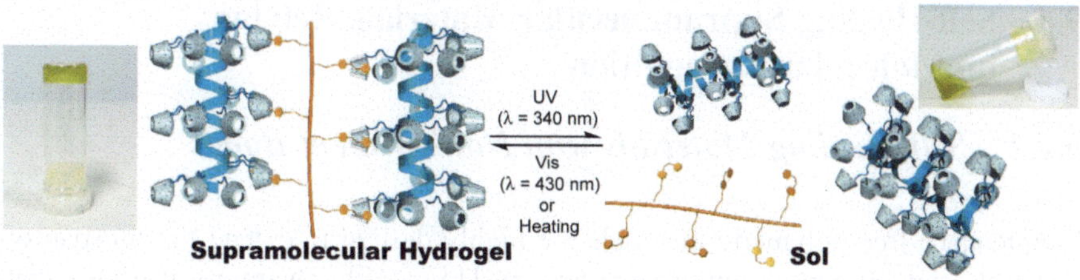

Fig. 9.7 Schematic representation of the interactions of the αCD unit with the azobenzene moieties upon irradiation with UV light (365 nm) and visible light (430 nm) or heating at 60 °C

increased, and the substrates are then transformed into a hydrogel. The viscosity of the CD-CUR/pAA-Azo hydrogel was decreased by irradiation with UV light ($\lambda = 365$ nm) to convert the gel into a sol. Visible light ($\lambda = 430$ nm) or heating (60 °C) generally induces back-isomerization of the *cis*-Azo to the *trans*-Azo (Fig. 9.7). It was found that the viscosity of the CD-CUR/pAA-Azo sol can be recovered by irradiation with visible light. These changes of viscosity of CD-CUR/pAA-Azo can be repeatedly triggered by irradiations with UV or visible light [115].

9.3.3 Redox-Responsive Sol-Gel Switching Materials Through Host-Guest Interactions

The photoresponsive sol-gel switching in the precious chapter employs Azo as a stimuli-responsive unit. Here, stimuli-responsive host-guest system was developed by introducing suitable functional units into the guest polymer. The desired stimuli-responsive function can be incorporated into the units. To construct redox-responsive sol-gel switching materials, Fc has been used as a redox-responsive guest molecule. As shown above, K_a of βCD with Fc is higher than that of αCD or γCD [55]. By utilizing the property of redox-responsive complex formation, gel-sol transition systems have been realized through complex formation between βCD and Fc on the side chains on poly(acrylic acid) [116].

A mixture of pAA modified with βCD (pAA-βCD) and pAA modified with Fc (pAA-Fc) was found to form a self-standing hydrogel. Treating the hydrogel with sodium hypochlorite as an oxidant changed state of the gel to sol, because βCD cannot include the ferrocenium cation (Fc^+). Treatment of the sol with a reducing agent [glutathione (GSH)] restores the gel. Additionally, the gel-sol transition can be induced via an electrochemical redox reaction. Thus, breaking and formation of the cross-linking points in pAA-βCD/pAA-Fc materials produces redox-responsive sol-gel switching systems.

9.4 Self-Healing Supramolecular Materials Relying on Molecular Recognition

9.4.1 Self-Healing Materials with Non-covalent Bonds

Maintenance-free polymeric materials are highly desired to extend the lifespan of materials and promote a sustainable society [117–119]. Therefore, methods that impart self-healing abilities and toughness to materials have attracted much attention over the years. There are two principal approaches for design of the self-healing materials, that is, "physical" and "chemical" approaches.

The physical self-healing materials are designed by utilizing capsules filled with healing agents [120, 121] or with functionalized stress relaxation properties. Slide-ring materials are polymers cross-linked with polyrotaxane (PRx) [49, 50], which are composed of ring molecules threaded on an axle polymers with bulky stoppers at both ends. They show high flexibility, entropic elasticity, and physical self-healing ability [122–124]. These properties are a result of the sliding motion of the ring molecule along the axle. In this regard, self-healing materials using PRx are mainly limited to hydrogels because of PRx's solubility. For example, preparation of PRx elastomer via bulk polymerization is difficult, because PRx with αCD and poly (ethylene glycol) (PEG) is not dissolved in conventional vinyl monomers. Recently, we synthesize a new type of acrylated PRx with αCD and PEG as a polymerizable hydrophobic PRx, which can be copolymerized with some abundant acrylate without polymerization solvent. The obtained materials show high stress relaxation and deformation hysteresis, namely, self-healing property. These results have demonstrated that the topologically cross-linked structure composed of polyrotaxanes plays an important role in mechanical properties even in bulk state [125].

The chemical self-healing materials use reversible chemical reactions [126–129], hydrogen bonds [130], electrostatic interactions [131, 132], coordination bonds [133], π-π stacking [134], and host-guest interactions as cross-linking. On the basis of their reversibility and stimuli-responsiveness, these non-covalent bonds are a relatively easy way to introduce desired functions into materials.

We have developed self-healing supramolecular hydrogels based on host-guest interactions on the side chains of water-soluble polymers (Fig. 9.8). There are two effective approaches to prepare supramolecular self-healing materials through host-guest interactions: (1) from a mixture of host and guest polymers (Fig. 9.8b) and (2) from polymerization of host and guest monomers (Fig. 9.8c and Table 9.4).

9.4.2 Self-Healing Hydrogels Formed by Host and Guest Polymers

Supramolecular materials (pAA-βCD/pAA-Fc) have been synthesized from host and guest polymers (Fig. 9.9). The self-standing materials are stabilized by host-guest

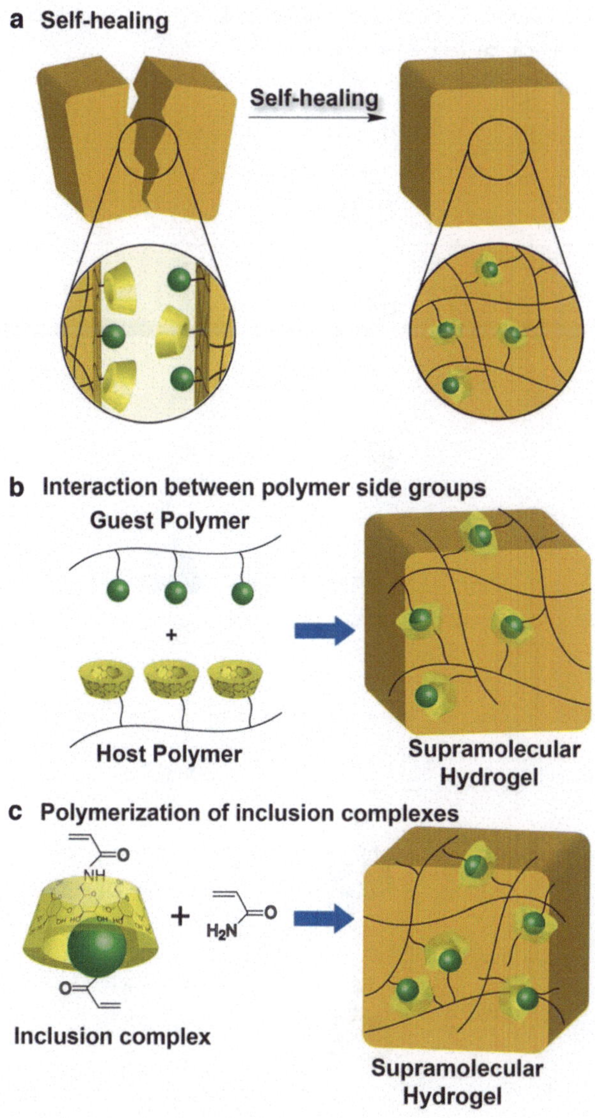

Fig. 9.8 Preparation of self-healing materials through host-guest interactions. (**a**) Proposed image of the self-healing materials. (**b**) The preparation of the self-healing material from host polymer and guest polymer. (**c**) The preparation of the self-healing material from polymerization of inclusion complex between host monomer and guest monomer

interactions. After the pAA-βCD/pAA-Fc material was cut in two pieces with a razor, these two parts can adhere again. The contact surface was found to disappear over the course of several hours (Fig. 9.9b). And, 85% of the initial strength of pAA-βCD/pAA-Fc was recovered after 24 h. When the material is ruptured, the

Table 9.4 Chemical structures of self-healing materials through host-guest interactions

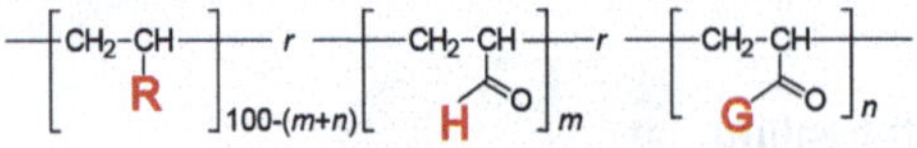

Main-chain polymer (R)	Host (H)	Guest (G)	Ref.
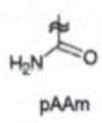		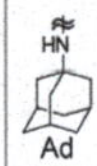	[135–137]
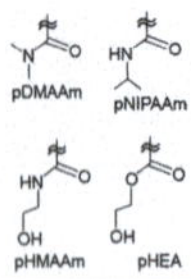		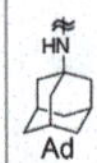	[137]
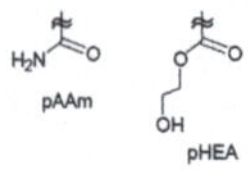		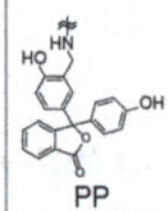	[139]
	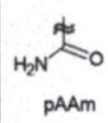	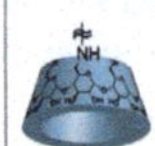	[135, 136]
			[138]
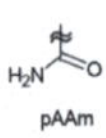	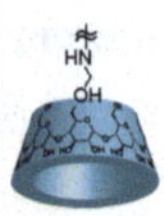	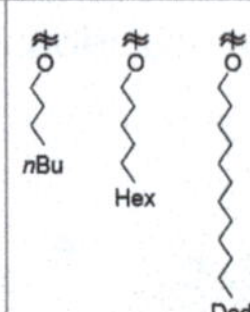	[138]
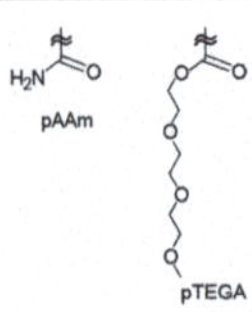	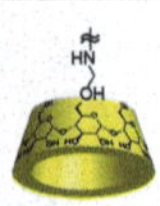	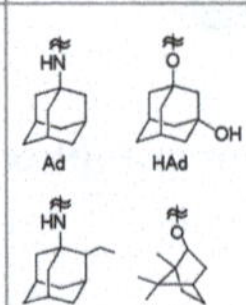	[138]

polymer network was broken at the host-guest cross-linking points. As these cross-linking points are reversible, the material shows readhesion and self-healing only by contact the ruptured surfaces.

Treatment of the cut surface with the oxidizing agent (sodium hypochlorite) prevented the pieces from readhering, because the oxidized Fc^{+} moieties cannot be included by βCD. However, treatment with a reducing agent induced readherence to form a single piece. Thus, the redox-responsiveness of the adhesion proves that the self-healing is processed by host-guest interactions (Fig. 9.9c) [116].

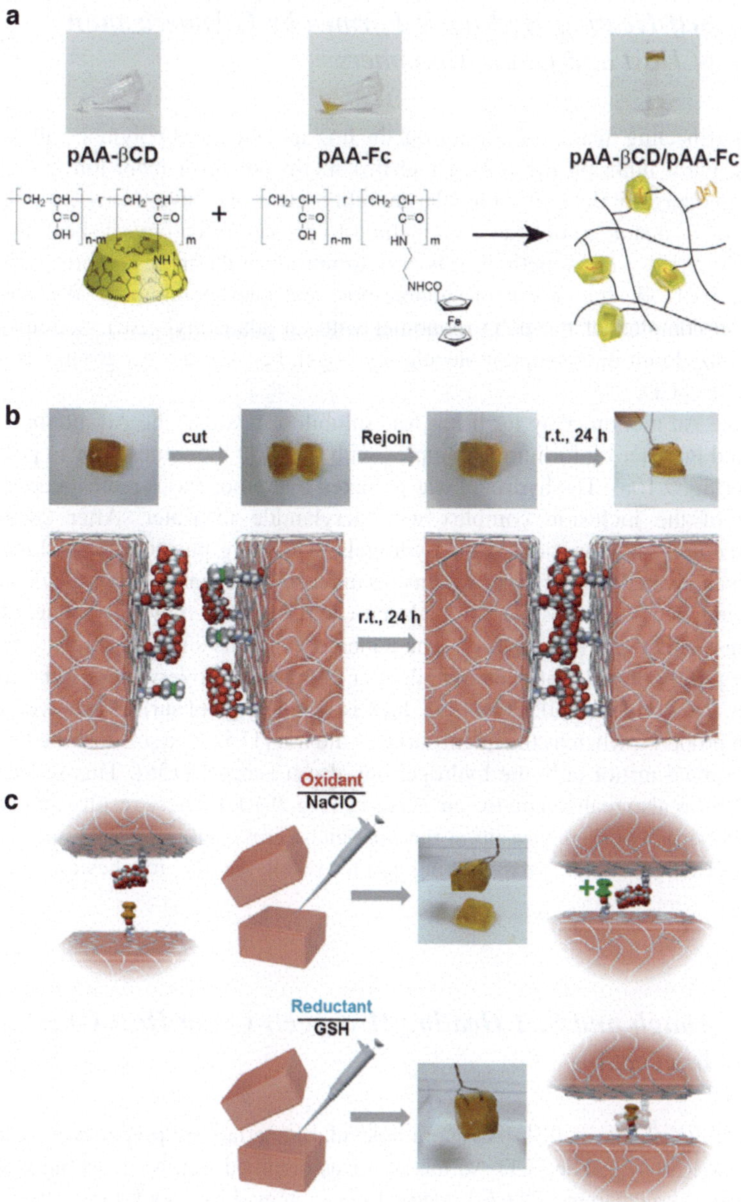

Fig. 9.9 (**a**) Redox-responsive self-healing hydrogels. (**b**) Self-healing of the hydrogel composed of host polymer modified with βCD and guest polymer modified with ferrocene. (**c**) Redox-responsive switching of self-healing ability. When the cut surface was coated by oxidant (NaClO), no self-healing was occurred. Subsequently, the cut surface was restored when the oxidized surface was treated with reductant (glutathione (GSH))

9.4.3 Self-Healing Hydrogels Formed by Polymerization of Host and Guest Monomers

In the self-healing material formed from the host and the guest polymers, all the host and the guest units on the polymer chains might not form inclusion complexes, owing to the restrictions of molecular mobility. Here, we have attempted polymerization of inclusion complexes as a method to prepare effective self-healing materials (Fig. 9.10). In this method, it is easy to introduce the inclusion complex itself into materials via the choice of suitable host and guest monomers. We chose an inclusion complex of the βCD monomer with an adamantyl (Ad) monomer as a preorganized unit in the supramolecular hydrogel, because the Ad group has a high affinity for βCD.

As the Ad monomer by itself has low solubility in water, the Ad monomer was dissolved in water by forming a complex with the βCD monomer prior to polymerization (Fig. 9.10a). The hydrogel was prepared by homogenous radical copolymerization of the inclusion complex with acrylamide in water. After 24 h, the homogenous solutions yielded the hydrogels, which are purified by washing with DMSO and water several times to remove unreacted reagents.

When two cut surfaces were brought into contact, the two pieces readhered, thus forming a single gel (Fig. 9.10b). The adhered materials showed almost complete recovery of the initial material strength after 24 h. The recovery ratio of the rupture strength increased with adhesive time. Interestingly, only cut surfaces showed a self-healing property, whereas the uncut surfaces did not [135]. The self-healing property was observed in not only the hydrogel but also a xerogel [136]. This self-healing property was also realized on the cut surfaces (Fig. 9.10c). These results support that the polymer network is torn up at the host-guest cross-linking points and that the re-complexation of the cross-linking points results in the readhesion and self-healing.

9.4.4 Tough and Self-Healing Hydrogels Using Host-Guest Interaction

The previous chapter's βCD-Ad supramolecular materials are prepared at a concentration of 1 mol/kg. The βCD-Ad material was obtained as a white turbid material. However, a transparent βCD-Ad material was obtained by copolymerization of the monomers with concentration of 2 mol/kg. The transparent βCD-Ad material exhibits high flexibility, toughness, and self-healing capability through host-guest interactions (Fig. 9.11). The material shows a self-healing property in both wet and dry states owing to the reversible nature of the host-guest interactions. The polymerization can took place on a glass substrate by curing under light to obtain transparent βCD-Ad-coated glass. The coated film exhibits the self-healing property regarding scratch damage. The reversible host-guest interactions, dispersion of

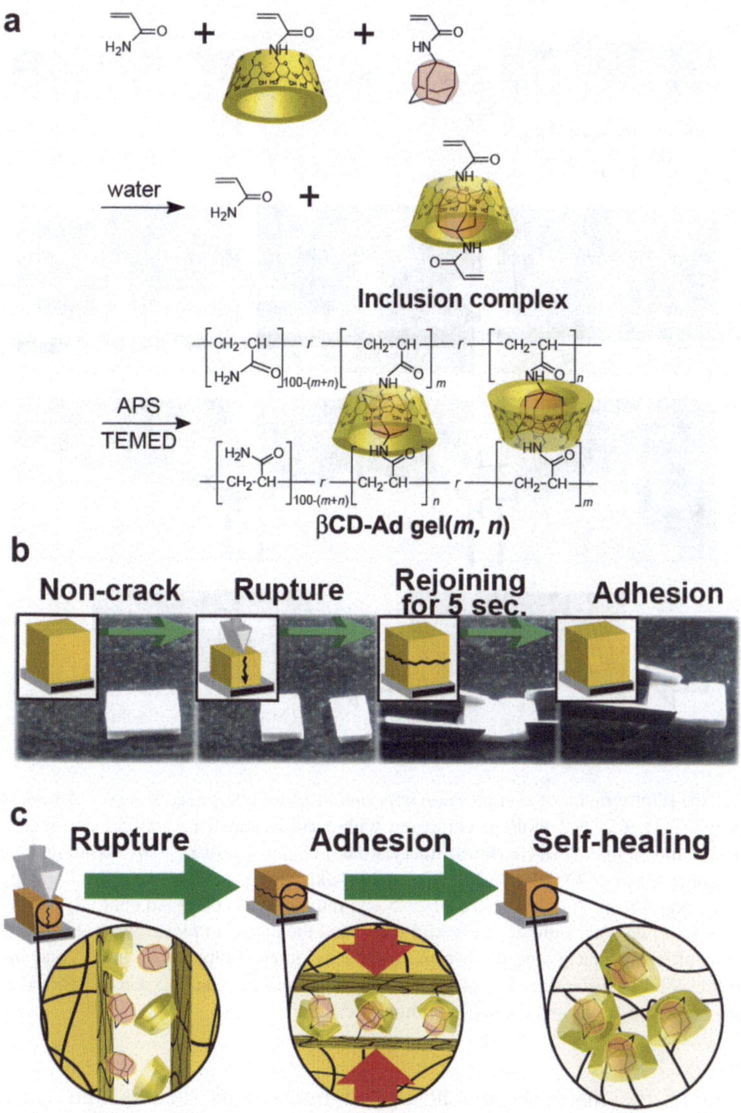

Fig. 9.10 Self-healing hydrogels prepared from βCD and Ad monomers. (**a**) Copolymerization of the βCD/Ad inclusion complex with acrylamide monomer in aqueous solutions. (**b**) Adhesive behavior of the βCD-Ad hydrogel. (**c**) Schematic illustration of the rapid and complete self-healing between cut surfaces

stress, and adhesiveness at the molecular levels were found to be responsible for these toughness and self-healing properties [137].

The attractiveness of using CDs as a host is the variety of guest molecules. CDs can include linear alkyl groups, spherical alkyl groups, aromatic groups, ferrocene, fullerene, and so on. Now, we have studied the relationship between the mechanical

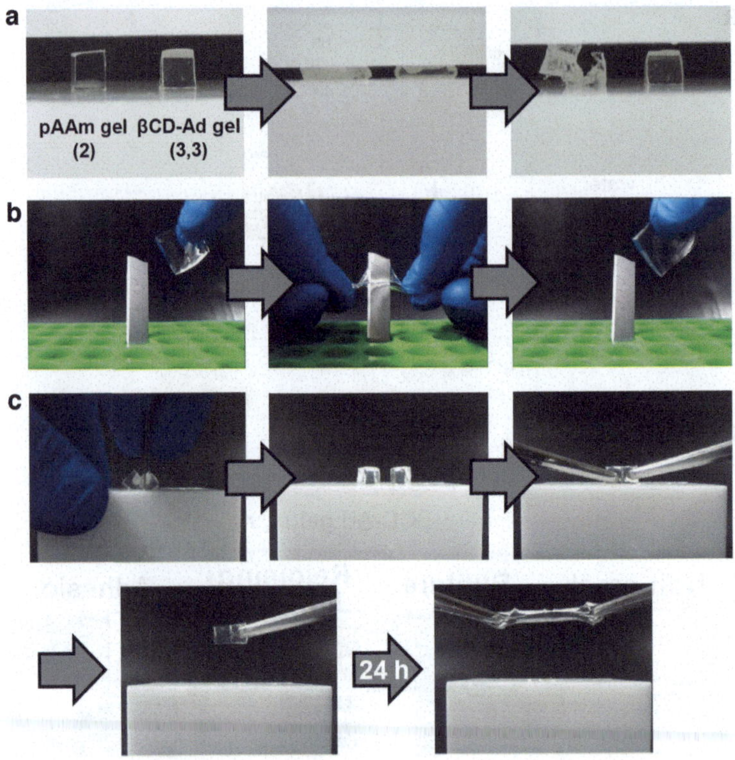

Fig. 9.11 (**a**) Photographs of compression experiment of a cube (size, 5 × 5 × 5 mm^3)-shaped βCD-Ad gel (3,3) (C_m = 2 mol/kg) compared with a cube-shaped pAAm gel chemically cross-linked with 2 mol% of *N,N'*-methylenebis(acrylamide) (C_m = 2 mol/kg). (**b**) Photographs of stab-resistant properties of βCD-Ad gel (2,2) (C_m = 2 mol/kg) using a cutter blade. (**c**) Photographs of self-healing experiment. A cuboid (size, 10 × 5 × 5 mm^3)-shaped βCD-Ad gel (3,3) (C_m = 2 mol/kg) was cut into two pieces under a powerful force, and then two cut pieces were attached together again. Soon after the attachment, the two pieces adhered to each other to be lifted up against their own weight. After still standing for 24 h under humid condition, the two pieces adhered strongly enough to be pulled from opposite sides without breakage

properties of the materials and host-guest interactions, that is, the association constants of CDs with guest molecules and molecular structures of the guest molecules [138]. In the case of covalently cross-linked hydrogels, increasing the rupture stress decreases the rupture strain. Interestingly, in the case of our host-guest gels, an increase in the association constants of CDs with guests slightly increases the rupture stress of the gel. In contrast, the rupture strain of the host-guest gel shows no significant changes. Note that, although the association constant of αCD with dodecyl guest (Dod) is lower than βCD with isobornyl guest (Ibr) (Dod/αCD, K_a = 990 M^{-1}; Ibr/βCD, K_a = 1300 M^{-1}), both the rupture stress and strain of αCD-Dod gel are higher than those of βCD-Ibr gel. This is probably because of the difference of molecular structure of the guest molecules. Ibr is the spherical alkyl guest, while Dod is the linear alkyl guest. Thus, αCD unit is allowed to form *pseudo*-rotaxane with the Dod on the polymer side chain. When a load is applied to the hydrogel, the αCD unit would slide along the alkyl chain leading the stress distribution. Since the stress can be efficiently dispersed in αCD-Dod hydrogel, the gel has been toughened.

9.4.5 Multifunctional Stimuli-Responsive Supramolecular Materials with Coloring and Self-Healing Properties

Multifunctionality is currently a hot topic in science and engineering circles and is gathering more and more attentions. However, fabricating polymeric materials with multiple functions requires multiple intricate reaction steps; the development of efficient synthetic methodologies for multifunctional polymeric materials has been in high demand. To produce multifunctional polymeric materials by simple method, we employ phenolphthalein (PP) as a guest molecule, because PP exhibits purple color under basic aqueous solution; however an inclusion complex with CD becomes colorless. Thus, the βCD-PP hydrogel exhibits a reproducible color change based on formation and dissociation of inclusion complex between βCD and PP at heating or adding competitive molecules in basic condition. In addition, the βCD-PP hydrogel shows self-healing property because of βCD and PP grafted on the side chain [139].

9.4.6 Self-Healing Materials Formed by Polyrotaxanes with Reversible Bonds

The previous self-healing materials, which used βCD and guest molecules, have achieved healing properties through chemical interactions. Combination of the reversibility and the mobility of the cross-linking points was found to play an important role in achieving efficient self-healing. We proposed self-healing materials with a combination of chemical and physical interactions (Fig. 9.12a). In this work, a self-healing material based on dynamic covalent was reported. The boronic acid-diol interaction was used as reversible bonds, and the cross-linking points were tethered to sliding rings on a linear polymer chain linked with polyrotaxane structure (Fig. 9.12b). The mobility of ring molecules along the axle polymer in polyrotaxane allows sliding nature of the cross-linking points to give the resultant material with rapid and efficient self-healing properties. A scratch-curable top coating on this hard material was prepared by the method of light curing. In a coating consisting of the self-healing material, a scratch damage recovered its initial state under semidry conditions (Fig. 9.12a) [140].

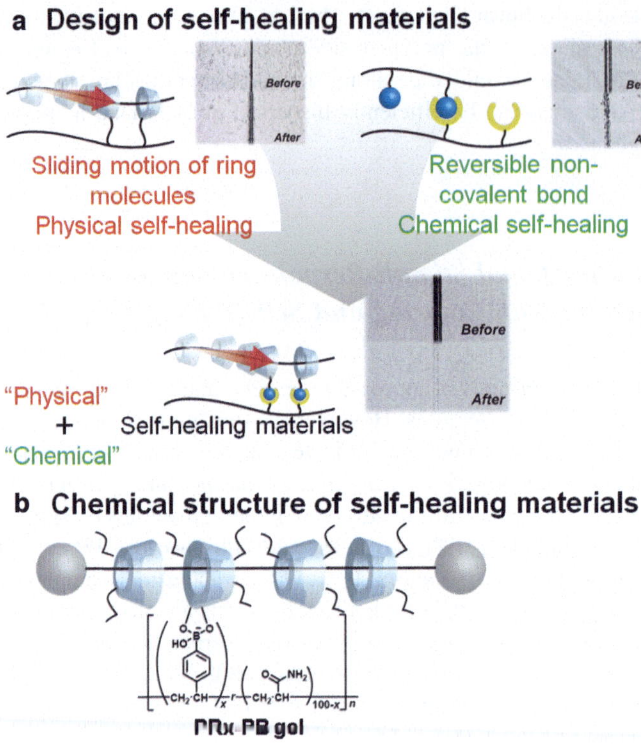

Fig. 9.12 (**a**) Design of self-healing materials with polyrotaxane (PRx) based on αCD functionalized by physical and chemical interactions. (**b**) Chemical structure of PRx-PB gel as self-healing materials

9.5 Supramolecular Actuator Controlled by Molecular Recognition

9.5.1 *Conventional Actuators*

Actuators are an important class of materials with applications in diverse fields such as medicine, physics, materials science, and materials engineering. The motion of actuators is reminiscent of those of living creatures and artificial muscles. Functions such as deformation, transformation, and pressing would be realized through movement of actuators. One of the topics in the research fields of actuators is enhancing energy conversion efficiency for input energies (electric, thermal, change, and photo energies) [141–143]. Many attempts have been made to realize controlling the actuators by using input energies or external stimuli. Deformation of organic or inorganic molecules, electrostrictive effect, and piezoelectricity were employed to create actuators that can be controlled by input energy [144–147]. Polymeric actuators (polymer gels [148–151], liquid crystalline elastomers [36, 152–162],

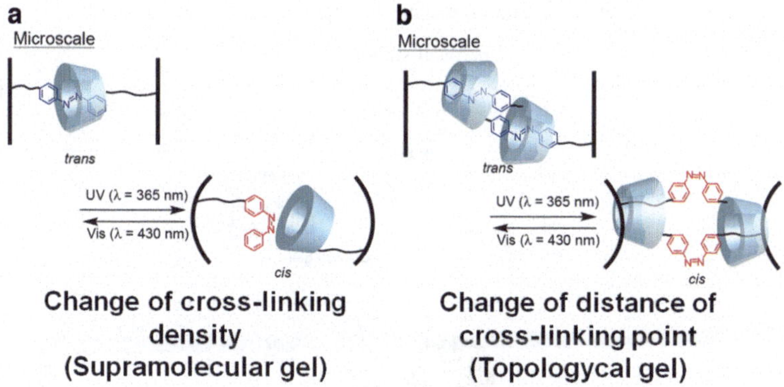

Fig. 9.13 Design of supramolecular actuator from (**a**) supramolecular gel controlled by change of cross-linking density and (**b**) topological gel controlled by change of length between cross-linking point

conjugated polymers [163], carbon nanotubes [164–167]) attracted much attention from researchers, owing to their lightness, high amount of displacement, and high energy efficiency, all of which may be used to realize applications beyond biological movements, for example, in "soft robotics."

To prepare stimuli-responsive materials using host-guest interactions, two approaches are available. One is by changing the density of the host-guest cross-linking (Fig. 9.13a), and the other is by utilizing host-guest molecular machines as cross-linking points (Fig. 9.13b). In the first approach, we can control the actuating by the complex formation between host and guest molecules at the polymer side chains. The association and dissociation of inclusion complexes on the polymer side chains result in contraction and expansion motions due to changes in the cross-linking density under water. In the second case, actuation is controlled by changing the end-to-end length of the [c2]daisy chain molecular machine driven by external stimuli. Using these two methods, the stimuli-responsive actuators have been investigated (Tables 9.5 and 9.6).

9.5.2 Photoresponsive Supramolecular Actuator

Self-healing supramolecular materials (βCD-Ad materials) do not have cross-linking of covalent bond, which is important for the effective self-healing property. Furthermore, without the covalent cross-linking point, the polymer chains would be spatially separated by the dissociation of the inclusion complex on the polymer chains to result in the sol-gel switching property. To prepare self-standing gel even when the host-guest bonds are dissociated, the covalent cross-linking is also required for the supramolecular actuator. Here, we introduce host-guest supramolecular actuators showing photoresponsive contraction-expansion. αCD and Azo units were introduced into the covalently cross-linked acrylamide hydrogel. This hydrogel (αCD-Azo

Table 9.5 Chemical structures of stimuli-responsive supramolecular gel actuators

External stimuli	Chemical structures	Ref.
Photo (UV and Vis lights)	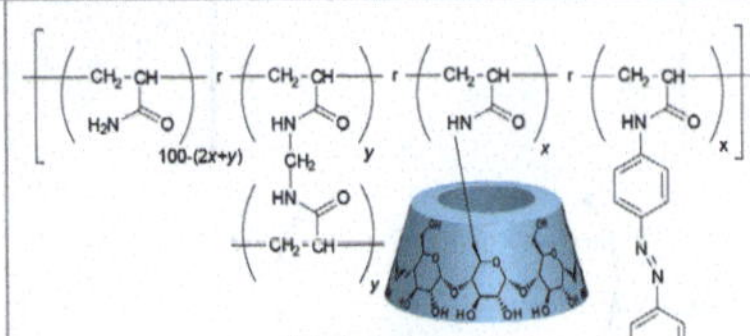 αCD-Azo gel(*x*, *y*)	[168]
Redox	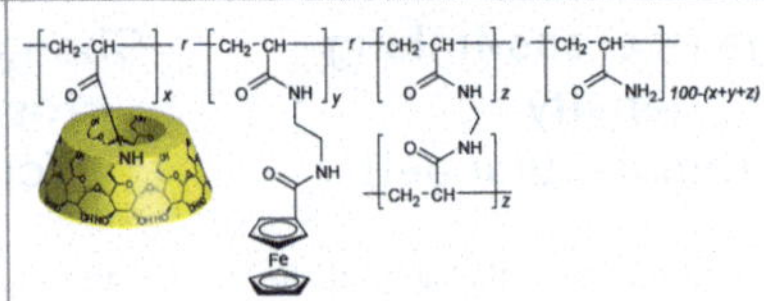 βCD-Fc gel(*x*,*y*,*z*)	[169]
Redox	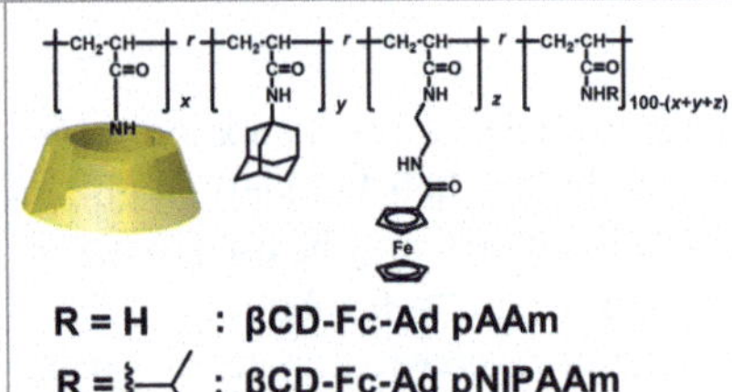	[170]

hydrogel) has both the covalent bonds and the host-guest supramolecular cross-linking. It was found that UV irradiation to αCD-Azo hydrogels in water increased the weight and volume of the hydrogel, whereas irradiating with visible light restored the hydrogel's initial volume. The contraction-expansion of the αCD-Azo hydrogel were driven by the formation and dissociation of the inclusion complex between αCD and Azo units. The plate-shaped αCD-Azo gel was found to bend in the same direction as the incident light. During the bending, the hydrogel increased its volume at UV-irradiated side, whereas the volume at the opposite side remained constant. Thus, the strain deformation of the αCD-Azo hydrogels led to the flex behavior triggered by UV light (Fig. 9.14a). On the other hand, the αCD-Azo hydrogels irradiated by visible light restored its original shape, as Azo was isomerized to *trans-* form to be included by αCD. This actuation can show a clear back-and-forth motion that depends on the wavelength, regardless of the irradiation history [168].

9.5.3 Redox-Responsive Supramolecular Actuator

Supramolecular actuators are able to respond to other stimuli. By using the design of the αCD-Azo hydrogel's actuator, we prepared various actuators that respond to desired stimuli. To realize a redox-responsive supramolecular actuator, a hydrogel

Table 9.6 Chemical structures of stimuli-responsive topological gel actuators

External stimuli	Chemical structures	Ref.
Photo (UV and Vis lights)	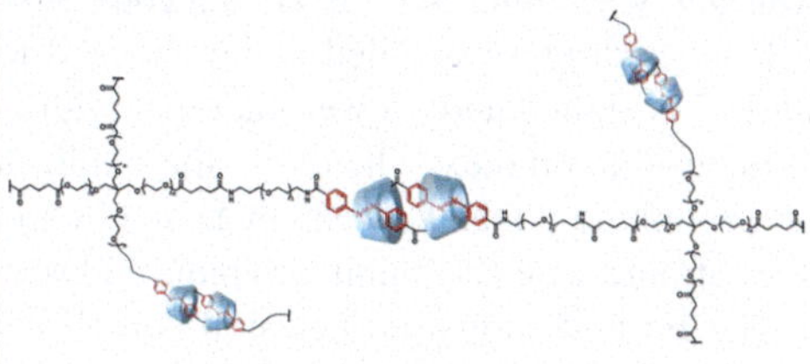 [c2]AzoCD$_2$ gel	[171]
Photo (UV and Vis lights)	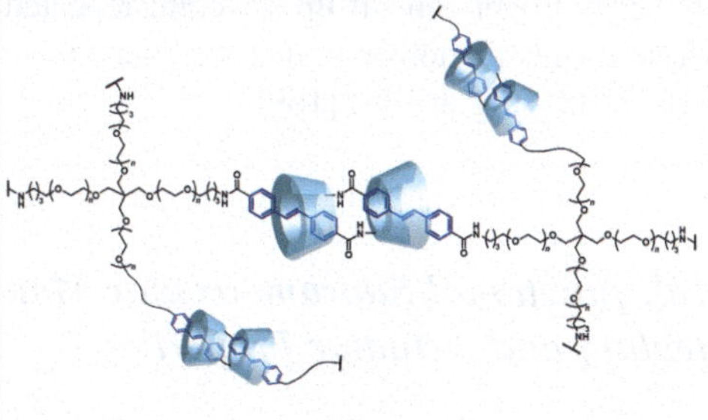 [c2]StiCD$_2$ gel	–
Photo (UV and Vis lights)	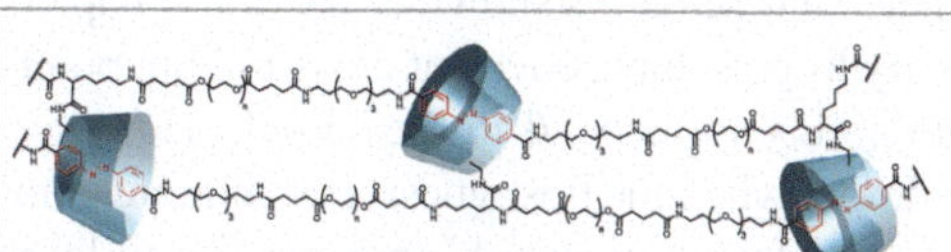 αCD-Azo hydrogel and xerogel	[172]

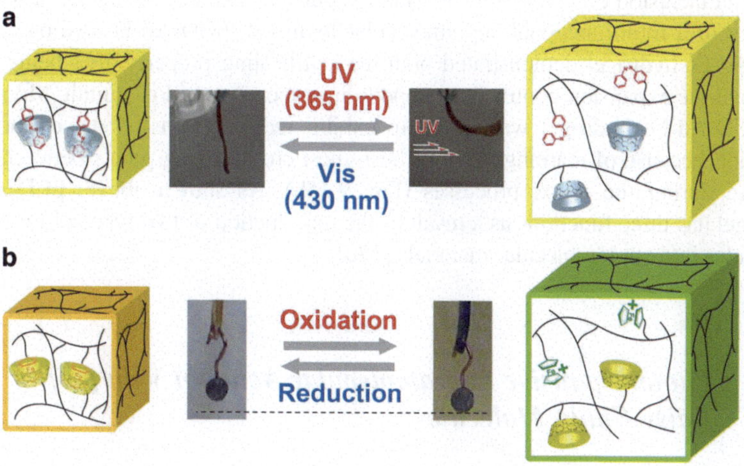

Fig. 9.14 Schematic illustrations of (**a**) photoresponsive and (**b**) redox-responsive supramolecular actuators

containing βCD as a host molecule and Fc as a guest molecule has been introduced into the covalently cross-linked acrylamide hydrogel (βCD-Fc hydrogel). When βCD-Fc hydrogel was immersed in an aqueous solution of oxidizing reagent, oxidized βCD-Fc hydrogel expanded as Fc changed to Fc^+. Association constant of βCD with Fc^+ is significantly lower than that with Fc, thus resulting in dissociation of the inclusion complex. Therefore, the oxidized βCD-Fc hydrogel lost the host-guest cross-linking points between βCD and Fc to show the expanding behavior. When the cationic state (Fc^+) unit was reduced back to the neutral state (Fc), the gel shrunk. The Fc unit forms an inclusion complex with the βCD unit, and the reduced βCD-Fc restored the cross-linking points to restore the original shape.

The βCD-Fc hydrogel can lift up a weight repeatedly in response to the redox reagents. These results demonstrate that the chemical energy can be converted into the mechanical work (Fig. 9.14b) [169].

9.5.4 Multifunctional Supramolecular Materials with Self-Healing and Actuator Properties

The supramolecular actuators shown in previous chapters, namely, the αCD-Azo and the βCD-Fc hydrogels, have covalent cross-linking points. The covalent cross-linking itself has no ability to respond to the stimuli. However, it is required to form self-standing hydrogel when the host-guest cross-linking points are dissociated. If polymer chains in supramolecular materials are cross-linked by multiple non-covalent bonds, multifunctional supramolecular materials would be realized without covalently cross-linking. We have used two different types of host-guest inclusion complexes of βCD: one is an inclusion complex with Ad, and the other is with Fc. Figure 9.15a shows a strategy of a multifunctional supramolecular hydrogel (βCD-Ad-Fc hydrogel). The βCD-Ad-Fc hydrogel demonstrated both the self-healing property after being damaged and the expansion-contraction property in response to redox stimuli. Moreover, the βCD-Ad-Fc hydrogel was found to exhibit redox-responsive shape-morphing behavior, because of rearrangement of host-guest cross-linking points between βCD and Ad during the redox processes (Fig. 9.15b). As shown above, βCD-Ad-Fc hydrogel has three functions as a result of the introduction of two types of functional molecules into supramolecular materials [170].

9.5.5 Photoresponsive Supramolecular Actuator with [c2] Daisy Chain Molecule

Figure 9.13 shows that the supramolecular actuators have been designed by two approaches. The materials in previous chapters are categorized as supramolecular gels, which function by controlling the complex formation between the host and

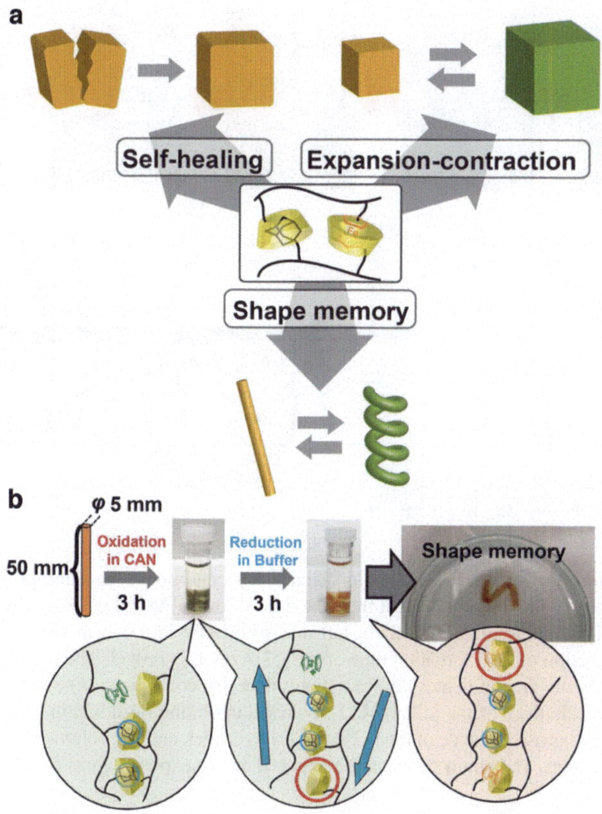

Fig. 9.15 Schematic illustrations of (**a**) redox-responsive multifunctional materials and (**b**) redox-responsive shape memory behavior

guest units on the polymer side chain (Fig. 9.13a). The other approach is use of a topological cross-linking, in which some kind of rotaxane complexes are employed as supramolecular cross-linking points. The cross-linking of rotaxanes can slide along the polymer chain driven by external stimuli. Biological muscles consume chemical energy through ATP dephosphorylation to perform macroscopic mechanical work. Myosin and actin filaments in muscle cells show sliding motions leading to contraction and expansion. Mechanical motion of the topological cross-linking network is reminiscent of a contracting muscle fibril.

Now, we developed photoresponsive wet- and dry-type topological gels ([c2]AzoCD$_2$ hydrogel and [c2]AzoCD$_2$ xerogel) as shown in Table 9.6. The [c2]AzoCD$_2$ hydrogel and xerogel were prepared via polycondensation between four-armed poly(ethylene glycol) and a [c2]daisy chain, which is a double-threaded rotaxane of αCD and Azo as photoresponsive molecular machine (Fig. 9.16). The obtained [c2]AzoCD$_2$ hydrogel and xerogel are topologically cross-linked materials, and they are not covalently cross-linked. Light-induced actuation arises from the

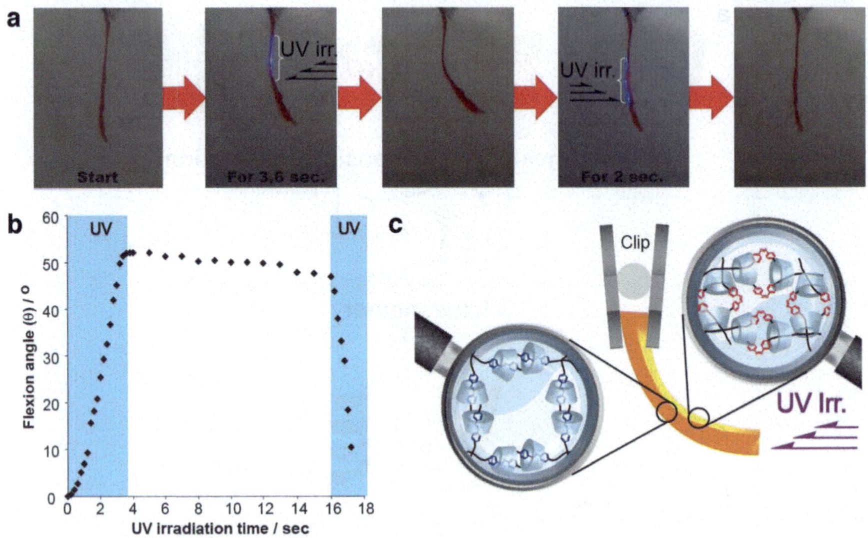

Fig. 9.16 Photoresponsive xerogel actuators under dry conditions. (**a**) Photographs of the [c2]AzoCD$_2$ xerogel irradiated with UV light from the right side. [c2]AzoCD$_2$ xerogel bends to the right but continuous UV light irradiation from the left side restores the initial form. (**b**) Plots of the flexion angle (θ) as a function of the irradiation time of the [c2]AzoCD$_2$ xerogel. The blue areas denote UV irradiation. The white area indicates no irradiation under dry conditions. (**c**) Schematic illustration of the bending mechanism of the [c2]AzoCD$_2$ hydrogel upon photoirradiation. UV light decreases the volume of the exposed surface of the [c2]AzoCD$_2$ hydrogel, and the volume of the non-exposed side remains constant. The strain between the exposed and unexposed areas results in flexion

sliding motion of the [c2]daisy chain unit (Fig. 9.16c), and the strip-shaped [c2]AzoCD$_2$ hydrogel bends toward the source of UV irradiation. The bending direction of the [c2]AzoCD$_2$ hydrogel is opposite to that of the host-guest-type actuators such as αCD-Azo hydrogel, because the [c2]daisy chain molecule shrinks by the photoirradiation.

Surprisingly, response of the [c2]AzoCD$_2$ xerogel was found to be very fast to show bending of 7° per second (Fig. 9.16a, b) in dry condition. This is 10,800 times faster than the [c2]AzoCD$_2$ hydrogel (7° for 3 h). This type of dry actuator does not depend on diffusion process of the media, which takes longer time than any other chemical or physical processes. The dry actuator [c2]AzoCD$_2$ xerogel is supposed to be released from the slow diffusion process of water to show the fast response. The [c2]AzoCD$_2$ xerogel was also found to lift up an object by irradiation with UV to perform the conversion of photo energy into mechanical work [171].

For further rapid responding, we choose stilbene (Sti) instead of Azo dye molecule, because the isomerization from *trans*-isomer to *cis*-isomer in [c2]daisy chain [c2]StiCD$_2$ is 60 times faster than [c2]AzoCD$_2$. Consequently, an initial bending speed of [c2]StiCD$_2$ hydrogel was so fast to show 2.6 degree/s, which is 600 times faster than that of [c2]AzoCD$_2$ hydrogel. Moreover, the bending speed of [c2]

$StiCD_2$ xerogel showed higher value of 12 degree/s. As shown above, using [c2] daisy chain molecular machines is a feasible method to prepare rapid responding artificial actuator.

9.5.6 *Photoresponsive Supramolecular Actuator with [2] Rotaxane*

As described in the previous chapter, use of the rotaxane complexes as cross-linking point is an effective way to prepare rapidly responding artificial actuator. However, preparation of the [c2]daisy chains (Fig. 9.17a) requires multiple intricate steps. Development of efficient synthetic method for artificial actuators is still in high demand. Here, we designed a new type of polymeric actuator based on a [2]rotaxane (Fig. 9.17b), which is made of [2]rotaxanes and polymer chains. We prepared an actuator αCD-Azo hydrogel by polycondensation of a *pseudo*[2]rotaxane based on lysine-modified αCD (Lys-αCD) and diamino-oligo(ethylene glycol)-modified Azo (Am_2Azo). The resulting polymer is topologically cross-linked by movable links with a rotaxane structure (Fig. 9.18a) [172]. Interestingly, rupture strain of αCD-Azo hydrogels showed 2800%. In addition, rupture energy of this material was 50 times higher than that of a covalently cross-linked control gel. The cross-linking of [2] rotaxane structure allows sliding motion of αCD along the PEG chains. And, the

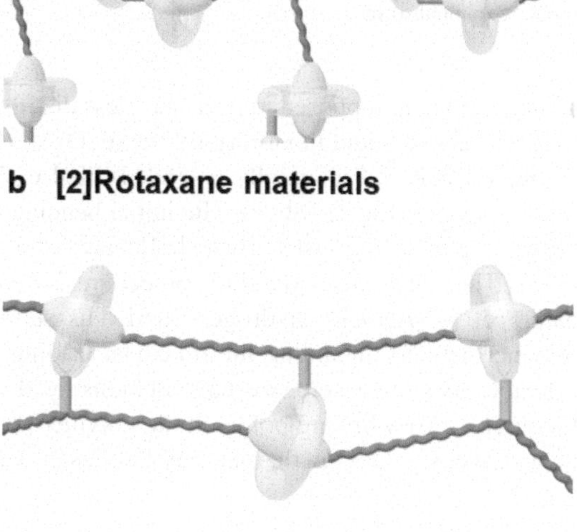

Fig. 9.17 Schematic diagram of (**a**) [c2]daisy chain materials and (**b**) [2] rotaxane materials

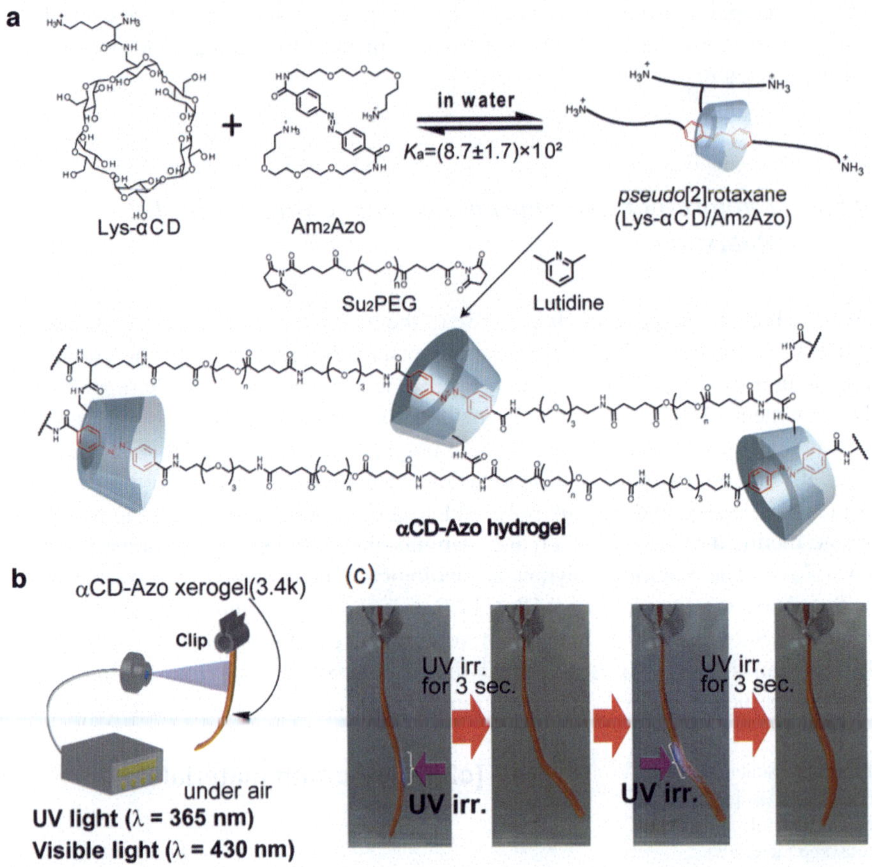

Fig. 9.18 (**a**) Preparation of the αCD-Azo hydrogel. (**b**) Experimental equipment and irradiation experimental setup of the αCD-Azo xerogel in air. (**c**) Light irradiation from the right side of the αCD-Azo xerogel bends to the right side. Subsequently irradiation with UV light from the left side restore the initial form

sliding motion is supposed to result in stress distribution within the polymer network to give toughness and flexibility for the αCD-Azo hydrogels against external stress.

Interestingly, an uniaxially extended dried αCD-Azo gel was found to show a faster response (Fig. 9.18b, c). The initial bending speed of αCD-Azo xerogel is 7.5 degree/s, which is faster than [c2]daisy chain actuator [c2]AzoCD$_2$ xerogel [171]. The uniaxially extending procedure is realized by the tough and flexible nature of [c2]AzoCD$_2$ hydrogels, and it is supposed to allow alignment of the polymer network including the rotaxane, leading to the fast response of the dried actuator. As shown here, we have demonstrated that the [2]rotaxane structures as topological cross-linking points in the polymer network are useful to prepare tough photoresponsive polymeric material with faster actuation.

9.6 Conclusion

The supramolecular materials utilizing molecular recognition have enabled various functions as described in this review. Non-covalent bonds are a powerful tool for preparations of functional supramolecular materials. These functions and properties are easily to be introduced by using supramolecular interactions such as host-guest complexations, which show selective and specific behaviors. They can be controlled reversibly, owing to the nature of non-covalent bonds. This review has mainly described host-guest interactions of polymeric materials using CDs as host molecules to realize attractive functions, such as macroscopic self-assembles, self-healing materials, sol-gel switching systems, and actuators. These functionalities were not achieved by conventional covalently cross-linked materials.

Although diverse molecular recognition systems with CDs have produced the fascinate materials with interesting properties and functions, these supramolecular materials have yet to be completely understood in terms of macromolecular science, molecular dynamics, surface science, and statistical thermodynamics. For example, supramolecular cross-linking including host-guest interaction is a generally weak bonding. However, some of the materials demonstrated simultaneously acquire both toughness and flexibility, which are not achieved by conventional materials at the same time. Additionally, the supramolecular materials also have self-healing properties. These behaviors might be owing to the stress- or pressure-distributing property of the supramolecular networks with the weak and reforming bonds. We are now investigating the mechanisms behind these functions.

While some of supramolecular materials has been inspired from biological systems, the topologically interlocked materials such as the movable cross-linking are not found in the nature. They are also available for molecular designs to enhance mechanical properties or to create new functions. In the near future, the supramolecular materials based on host-guest interactions would advance materials science and engineering.

Acknowledgments This work was financially supported by the ImPACT Program of the Council for Science, Technology and Innovation (Cabinet Office, Government of Japan), a Grant-in-Aid for Scientific Research (B) (No. 26288062 and 18H02035) from MEXT of Japan, the Research Grant Program of the Asahi Glass Foundation (2015).

References

1. R.A. Mooney, R. Landick, RNA polymerase unveiled. Cell **98**, 687–690 (1999)
2. R. Kovall, B.W. Matthews, Toroidal structure of lambda-exonuclease. Science **277**, 1824–1827 (1997)
3. M.A. Trakselis, S.C. Alley, E. Abel-Santos, S.J. Benkovic, Creating a dynamic picture of the sliding clamp during T4 DNA polymerase holoenzyme assembly by using fluorescence resonance energy transfer. Proc. Natl. Acad. Sci. U. S. A. **98**, 8368–8375 (2001)

4. S.J. Benkovic, A.M. Valentine, F. Salinas, Replisome-mediated DNA replication. Annu. Rev. Biochem. **70**, 181–208 (2001)
5. W.A. Breyer, B.W. Matthews, A structural basis for processivity. Protein Sci. **10**, 1699–1711 (2001)
6. E.T. Kool, Recognition of DNA, RNA, and proteins by circular oligonucleotides. Acc. Chem. Res. **31**, 502–510 (1998)
7. B. Alberts, A. Johnson, J. Lewis, M. Raff, K. Robert, P. Walter, *Molecular Biology of the Cell*, 5th edn. (Garland Science, New York, 2008)
8. H. Yin, M.D. Wang, K. Svoboda, K. Landridk, S.M. Block, J. Gelles, Transcription against an applied force. Science **270**, 1653–1657 (1995)
9. N. Hirokawa, Kinesin and dynein superfamily proteins and the mechanism of organelle transport. Science **279**, 519–526 (1998)
10. R.D. Vale, R.A. Milligan, The way things move: looking under the hood of molecular motor proteins. Science **288**, 88–95 (2000)
11. B.M. Gumbiner, Cell adhesion: the molecular basis of tissue architecture and morphogenesis. Cell **84**, 345–357 (1996)
12. K.M. Yamada, S.S. Yamada, I. Pastan, Cell surface protein partially restores morphology, adhesiveness, and contact inhibition of movement to transformed fibroblasts. Proc. Natl. Acad. Sci. U. S. A. **73**, 1217–1221 (1976)
13. C.J. Pedersen, The discovery of crown ethers (Noble Lecture). Angew. Chem. Int. Ed. **27**, 1021–1027 (1988)
14. D.J. Cram, The design of molecular hosts, guests, and their complexes (Nobel Lecture). Angew. Chem. Int. Ed. **27**, 1009–1112 (1988)
15. S.H. Gellman, Introduction: molecular recognition. Chem. Rev. **97**, 1231–1232 (1997)
16. J.-M. Lehn, *Supramolecular Chemistry* (Wiley-VCH, Weinheim, 1995)
17. J.-M. Lehn, Supramolecular chemistry—scope and perspectives molecules, supermolecules, and molecular devices (Nobel Lecture). Angew. Chem. Int. Ed. **27**, 89–112 (1988)
18. W.L. Mock, *Comprehensive Supramolecular Chemistry*, vol 2 (Pergamon, Oxford, 1996)
19. L. Brunsveld, B.J.B. Folmer, E.W. Meijer, R.P. Sijbesma, Supramolecular polymers. Chem. Rev. **101**, 4071–4098 (2001)
20. A. Ciferri, *Supramolecular Polymers*, 2nd edn. (CRC Press, Florida, 2005)
21. P.R. Ashton, R. Ballardini, V. Balzani, I. Baxter, A. Credi, M.C.T. Fyfe, M.T. Gandolfi, M. Gomez-Lopez, M.-V. Martinez-Diaz, A. Piersanti, N. Spencer, J.F. Stoddart, M. Venturi, A.J.P. White, D. Williams, Acid−base controllable molecular shuttles. J. Am. Chem. Soc. **120**, 11932–11942 (1998)
22. M.C. Jimenez-Molero, C. Dietrich-Buchecker, Sauvage: chemically induced contraction and stretching of a linear rotaxane dimer. Chem. Eur. J. **8**, 1456–1466 (2002)
23. J.D. Badjic, V. Balzani, A. Credi, S. Silvi, J.F. Stoddart, A molecular elevator. Science **303**, 1845–1849 (2004)
24. A. Mirzoian, A.E. Kaifer, Reactive pseudorotaxanes: inclusion complexation of reduced viologens by the hosts β-cyclodextrin and heptakis(2,6-di-o-methyl)-β-cyclodextrin. Chem. Eur. J. **3**, 1052–1057 (1997)
25. N. Armaroli, V. Balzani, J.-P. Collin, P. Gavina, J.-P. Sauvage, B. Ventura, Rotaxanes incorporating two different coordinating units in their thread: synthesis and electrochemically and photochemically induced molecular motions. J. Am. Chem. Soc. **121**, 4397–4408 (1999)
26. A. Altieri, F.G. Gatti, E.R. Kay, D.A. Leigh, D. Martel, F. Paolucci, A.M.Z. Slawin, J.K.Y. Wong, Electrochemically switchable hydrogen-bonded molecular shuttles. J. Am. Chem. Soc. **125**, 8644–8654 (2003)
27. Y. Liu, A.H. Flood, J.F. Stoddart, Thermally and electrochemically controllable self-complexing molecular switches. J. Am. Chem. Soc. **126**, 9150–9151 (2004)
28. O. Lukin, A. Godt, F. Vögtle, Residual topological isomerism of intertwined molecules. Chem. Eur. J. **10**, 1878–1883 (2004)

29. N. Koumura, R.W.J. Zijlstra, R.A. Van Delden, N. Harada, B.L. Feringa, Light-driven monodirectional molecular rotor. Nature **401**, 152–155 (1999)
30. A.M. Brouwer, C. Frochot, F.G. Gatti, D.A. Leigh, L. Mottier, F. Paolucci, S. Roffia, G.W.H. Wurpel, Photoinduction of fast, reversible translational motion in a hydrogen-bonded molecular shuttle. Science **291**, 2124–2128 (2001)
31. S. Shinkai, T. Nakaji, Y. Nishida, T. Ogawa, O. Manabe, Photoresponsive crown ethers. 1. Cis-trans isomerism of azobenzene as a tool to enforce conformational changes of crown ethers and polymers. J. Am. Chem. Soc. **102**, 5860–5865 (1980)
32. H. Murakami, A. Kawabuchi, K. Kotoo, M. Kunitake, N. Nakashima, A light-driven molecular shuttle based on a rotaxane. J. Am. Chem. Soc. **119**, 7605–7606 (1997)
33. T. Muraoka, K. Kinbara, Y. Kobayashi, T. Aida, Light-driven open−close motion of chiral molecular scissors. J. Am. Chem. Soc. **125**, 5612–5613 (2003)
34. M. Irie, Photoresponsive polymers. Adv. Polym. Sci. **94**, 27–67 (1990)
35. Y. Inoue, P. Kuad, Y. Okumura, Y. Takashima, H. Yamaguchi, A. Harada, Thermal and photochemical switching of conformation of poly(ethylene glycol)-substituted cyclodextrin with an azobenzene group at the chain end. J. Am. Chem. Soc. **129**, 6396–6397 (2007)
36. Y. Yu, M. Nakano, T. Ikeda, Photomechanics: directed bending of a polymer film by light. Nature **425**, 145 (2003)
37. I. Tomatsu, A. Hashidzume, A. Harada, Contrast viscosity changes upon photoirradiation for mixtures of poly(acrylic acid)-based α-Cyclodextrin and azobenzene polymers. J. Am. Chem. Soc. **128**, 2226–2227 (2006)
38. I. Tomatsu, A. Hashidzume, A. Harada, Cyclodextrin-based side-chain polyrotaxane with unidirectional inclusion in aqueous media. Angew. Chem. Int. Ed. **45**, 4605–4608 (2006)
39. M.L. Bender, M. Komiyama, *Cyclodextrin Chemistry, Reactivity and Structure Concepts in Organic Chemistry* (Springer, Berlin, 1978)
40. J. Szejtli, *Cyclodextrins and Their Inclusion Complexes* (Akadémiai Kiadó, Budapest, 1982)
41. J. Szejtli, T. Osa, *Comprehensive Supramolecular Chemistry* (Pergamon, Oxford, 1996)
42. C.J. Easton, S.F. Lincoln, *Modified Cyclodextrins: Scaffolds and Templates for Supramolecular Chemistry* (Imperial College Press, London, 1999)
43. R Breslow, Artificial enzymes and enzyme models, in *Advances in Enzymology and Related Areas of Molecular Biology*, ed. by A. Meister, (Wiley, New York, 1986), pp. 1–60
44. M Komiyama, H Shigekawa, Cyclodextrins as enzyme models, in *Comprehensive Supramolecular Chemistry*, ed. by J. Szejtli, T. Osa, (Pergamon, Oxford, 1996), pp. 401–422
45. B. Ronald, *Artificial Enzymes* (Wiley-VCH Verlag GmbH & Co. KGaA, Weinheim, 2005)
46. A. Harada, Y. Takashima, M. Nakahata, Supramolecular polymeric materials via cyclodextrin-guest interactions. Acc. Chem. Res. **47**, 2128–2140 (2014)
47. M. Furue, A. Harada, S. Nozakura, Preparation of cyclodextrin-containing polymers and their catalysis in ester-hydrolysis. J. Polym. Sci. Polym. Lett. Ed. **13**, 357–360 (1975)
48. A. Harada, H. Adachi, Y. Kawaguchi, M. Kamachi, Recognition of alkyl groups on a polymer chain by cyclodextrins. Macromolecules **30**, 5181–5182 (1997)
49. A. Harada, M. Kamachi, Complex formation between poly(ethylene glycol) and α-cyclodextrin. Macromolecules **23**, 2821–2823 (1990)
50. A. Harada, J. Li, M. Kamachi, The molecular necklace: a rotaxane containing many threaded α-cyclodextrins. Nature **356**, 325–327 (1992)
51. F.M. Raymo, J.F. Stoddart, Interlocked macromolecules. Chem. Rev. **99**, 1643–1663 (1999)
52. G. Wenz, B.-H. Han, A. Mller, Cyclodextrin rotaxanes and polyrotaxanes. Chem. Rev. **106**, 782–817 (2006)
53. A. Harada, A. Hashidzume, H. Yamaguchi, Y. Takashima, Polymeric rotaxanes. Chem. Rev. **109**, 5974–6023 (2009)
54. A.W. Bosmana, R.P. Sijbesmab, E.W. Meijer, Supramolecular polymers at work. Mater. Today **7**, 34–39 (2004)
55. T.F.A. de Greef, E.W. Meijer, Materials science: supramolecular polymers. Nature **453**, 171–173 (2008)

56. E. Krieg, M. Bastings, M. C, P. Besenius, B. Rybtchinski, Supramolecular polymers in aqueous media. Chem. Rev. **116**, 2414–2477 (2016)
57. P.M. Gasper, B. Fuglestad, E.A. Komives, P.R.L. Markwick, J.A. McCammon, Allosteric networks in thrombin distinguish procoagulant vs. anticoagulant activities. Proc. Natl. Acad. Sci. U. S. A. **109**, 21216–21222 (2012)
58. A. Ghosh, S. Vishveshwara, Variations in clique and community patterns in protein structures during allosteric communication: Investigation of dynamically equilibrated structures of methionyl tRNA synthetase complexes. Biochemistry **47**, 11398–11407 (2008)
59. A. Harada, R. Kobayashi, Y. Takashima, A. Hashidzume, H. Yamaguchi, Macroscopic self-assembly through molecular recognition. Nat. Chem. **3**, 34 (2011)
60. H. Yamaguchi, R. Kobayashi, Y. Takashima, A. Hashidzume, A. Harada, Self-assembly of gels through molecular recognition of cyclodextrins: shape selectivity for linear and cyclic guest molecules. Macromolecules **44**, 2395–2399 (2011)
61. Y. Zheng, Y. Kobayashi, T. Sekine, Y. Takashima, A. Hashidzume, H. Yamaguchi, A. Harada, Visible chiral discrimination via macroscopic selective assembly. Commun. Chem. **1**, 4 (2018)
62. H. Yamaguchi, Y. Kobayashi, R. Kobayashi, Y. Takashima, A. Hashidzume, A. Harada, Photoswitchable gel assembly based on molecular recognition. Nat. Commun. **3**, 603 (2012)
63. Y. Zheng, A. Hashidzume, Y. Takashima, H. Yamaguchi, A. Harada, Switching of macroscopic molecular recognition selectivity using a mixed solvent system. Nat. Commun. **3**, 831 (2012)
64. T. Nakamura, Y. Takashima, A. Hashidzume, H. Yamaguchi, A. Harada, A metal-ion-responsive adhesive material via switching of molecular recognition properties. Nat. Commun. **5**, 4622 (2014)
65. Y. Zheng, A. Hashidzume, A. Harada, pH-responsive self-assembly by molecular recognition on a macroscopic scale. Macromol. Rapid Commun. **34**, 1062–1066 (2013)
66. Y. Zheng, A. Hashidzume, Y. Takashima, H. Yamaguchi, A. Harada, Switching of macroscopic molecular recognition selectivity using a mixed solvent system. ACS Macro Lett. **1**, 1083–1085 (2012)
67. B.L. Feringa, *Molecular Switches* (Wiley-VCH Verlag GmbH, Hoboken, 2001)
68. M. Irie, Diarylethenes for memories and switches. Chem. Rev. **100**, 1685–1716 (2000)
69. Y. Takashima, T. Nakayama, M. Miyauchi, Y. Kawaguchi, H. Yamaguchi, A. Harada, Complex formation and gelation between copolymers containing pendant azobenzene groups and cyclodextrin polymers. Chem. Lett. **33**, 890–891 (2004)
70. C.D. Geddes, J.R. Lalowicz, *Advanced Concepts in Fluorescence Sensing: Macromolecular Sensing* (Springer Science+Business Media, New York, 1991)
71. L. Mohanambe, S. Vasudevan, Aromatic molecules in restricted geometries: pyrene excimer formation in an anchored bilayer. J. Phys. Chem. B **110**, 14345–14354 (2006)
72. A. Harada, *Supramolecular Polymer Chemistry* (WileyVCH, Weinheim, 2012)
73. M.V. Rekharsky, Y. Inoue, Chem. Rev. **98**, 1875–1917 (1998)
74. P. Brocos, N. Diaz-Vergara, X. Banquy, S. Pérez-Casas, M. Costas, A. Piñeiro, Similarities and differences between cyclodextrin-sodium dodecyl sulfate host-guest complexes of different stoichiometries: molecular dynamics simulations at several temperatures. J. Phys. Chem. B **114**, 12455–12467 (2010)
75. E. Liang, H. Zhou, X. Ding, Z. Zheng, Y. Penga, Fabrication of a rhythmic assembly system based on reversible formation of dynamic covalent bonds in a chemical oscillator. Chem. Commun. **49**, 5384–5386 (2013)
76. T. Asoh, A. Kikuchi, Electrophoretic adhesion of stimuli-responsive hydrogels. Chem. Commun. **46**, 7793–7795 (2010)
77. C.A. Anderson, A.R. Jones, E.M. Briggs, E.J. Novitsky, D.W. Kuykendall, N.R. Sottos, S.C. Zimmerman, High-affinity DNA base analogs as supramolecular, nanoscale promoters of macroscopic adhesion. J. Am. Chem. Soc. **135**, 7288–7295 (2013)

78. Y. Ahn, Y. Jang, N. Selvapalam, G. Yun, K. Kim, Supramolecular velcro for reversible underwater adhesion. Angew. Chem. Int. Ed. **52**, 3140–3144 (2013)
79. H. Qi, M. Ghodousi, Y. Du, C. Grun, H. Bae, P. Yin, A. Khademhosseini, DNA-directed self-assembly of shape-controlled hydrogels. Nat. Commun. **4**, 2275 (2013)
80. M. Nakahata, Y. Takashima, A. Hashidzume, A. Harada, Macroscopic self-assembly based on complementary interaction between nucleobase pairs. Chem. Eur. J. **21**, 2770–2774 (2015)
81. Y. Kobayashi, Y. Takashima, A. Hashidzume, H. Yamaguchi, A. Harada, Reversible self-assembly of gels through metal-ligand interactions. Sci. Rep. **3**, 1243 (2013)
82. H.B. Dunford, *Peroxidase* (Wiley-VCH, Weinheim, 1999)
83. Y. Kobayashi, Y. Takashima, A. Hashidzume, H. Yamaguchi, A. Harada, Manual control of catalytic reactions: reactions by an apoenzyme gel and a cofactor gel. Sci. Rep. **5**, 16254 (2015)
84. D.G. Hall, *Boronic Acids* (Wiley-VCH, Weinheim, 2005)
85. T.D. James, M.D. Phillips, S. Shinkai, *Boronic Acids in Saccharide Recognition* (RSC Publishing, Cambridge, 2006)
86. S.T. Bull, M.G. Davidson, J.M.H. van den Elsen, J.S. Fossey, A.T.A. Jenkins, Y.-B. Jiang, Y. Kubo, F. Marken, K. Sakurai, J. Zhao, T.D. James, Exploiting the reversible covalent bonding of boronic acids: recognition, sensing, and assembly. Acc. Chem. Res. **46**, 312–326 (2012)
87. J.P. Lorand, J.O. Edwards, Polyol complexes and structure of the benzeneboronate ion. J. Org. Chem. **24**, 769–774 (1959)
88. M. Nakahata, S. Mori, Y. Takashima, A. Hashidzume, H. Yamaguchi, A. Harada, pH- and sugar-responsive gel assemblies based on boronate-catechol interactions. ACS Macro Lett. **3**, 337–340 (2014)
89. Y. Takashima, T. Sahara, T. Sekine, T. Kakuta, M. Nakahata, M. Otsubo, Y. Kobayashi, A. Harada, Supramolecular adhesives to hard surfaces: adhesion between host hydrogels and guest glass substrates through molecular recognition. Macromol. Rapid Commun. **35**, 1646–1652 (2014)
90. T. Sekine, T. Kakuta, T. Nakamura, Y. Kobayashi, Y. Takashima, A. Harada, A macroscopic reaction: direct covalent bond formation between materials using a Suzuki-Miyaura cross-coupling reaction. Sci. Rep. **4**, 6348 (2014)
91. T. Sekine, Y. Takashima, A. Harada, Direct covalent bond formation between materials using copper(I)-catalyzed azide alkyne cycloaddition reactions. RSC Adv. **5**, 56130–56135 (2015)
92. T. Sekine, Y. Takashima, H. Yamaguchi, A. Harada, Direct adhesion of dissimilar materials using Sonogashira cross-coupling reaction. Chem. Lett. **45**, 1250 (2016)
93. A. Harada, S. Takahashi, Preparation and properties of cyclodextrin inclusion compounds of organometallic complexes. Ferrocene inclusion compounds. J. Incl. Phenom. **2**, 791–798 (1984)
94. J.S. Wu, K. Toda, A. Tanaka, I. Sanemasa, Association constants of ferrocene with cyclodextrins in aqueous medium determined by solubility measurements of ferrocene. Bull. Chem. Soc. Jpn. **71**, 1615–1618 (1998)
95. A.U. Moozyckine, J.L. Bookham, M.E. Deary, D.M. Davies, Structure and stability of cyclodextrin inclusion complexes with the ferrocenium cation in aqueous solution: H NMR studies. J. Chem. Soc. Perkin Trans. **2**, 1858–1862 (2001)
96. M. Nakahata, Y. Takashima, A. Harada, Redox-responsive macroscopic gel assembly based on discrete dual interactions. Angew. Chem. Int. Ed. **53**, 3617–3621 (2014)
97. C.E. Beaty, W.M. Saltzman, Controlled growth factor delivery induces differential neurite outgrowth in three-dimensional cell cultures. J. Control. Release **24**, 15 (1993)
98. S. Kim, K.E. Healy, Synthesis and characterization of injectable poly(N-isopropylacrylamide-co-acrylic acid) hydrogels with proteolytically degradable cross-links. Biomacromolecules **4**, 1214 (2003)
99. M.J. Mahoney, K.S. Anseth, Three-dimensional growth and function of neural tissue in degradable polyethylene glycol hydrogels. Biomaterials **27**, 2265–2274 (2006)

100. W.A. Petka, J.L. Harden, K.P. McGrath, D. Wirtz, D.A. Tirrell, Reversible hydrogels from self-assembling artificial proteins. Science **281**, 389–392 (1998)
101. W. Shen, R.G.H. Lammertink, J.K. Sakata, J.A. Kornfield, D.A. Tirrell, Assembly of an artificial protein hydrogel through leucine zipper aggregation and disulfide bond formation. Macromolecules **38**, 3909 (2005)
102. W.E. Hennink, C.F. van Nostrum, Novel crosslinking methods to design hydrogels. Adv. Drug Deliv. Rev. **54**, 13–36 (2002)
103. C. Amiel, B.J. Sébille, New associating polymer systems involving water soluble β-cyclodextrin polymers. Incl. Phenom. Mol. Recognit. Chem. **25**, 61–67 (1996)
104. C. Amiel, L. Moine, A. Sandier, W. Brown, C. David, F. Hauss, E. Renard, M. Gosselet, B. Sebille, *Macromolecular Assemblies Generated by Inclusion Complexes Between Amphipathic Polymers and β-Cyclodextrin Polymers in Aqueous Media in Stimuli-Responsive Water Soluble and Amphiphilic Polymers* (Oxford University Press, Washington DC, 2001)
105. C. Amiel, B. Sebille, Association between amphiphilic poly(ethylene oxide) and β-cyclodextrin polymers: aggregation and phase separation. Adv. Colloid Interf. Sci. **79**, 105 (1999)
106. G. Pouliquen, C. Amiel, C. Tribet, Photoresponsive viscosity and host-guest association in aqueous mixtures of poly-cyclodextrin with azobenzene-modified poly(acrylic)acid. J. Phys. Chem. B **111**, 5587–5595 (2007)
107. M. Weickenmeier, G. Wenz, J. Huff, Association thickener by host guest interaction of a β-cyclodextrin polymer and a polymer with hydrophobic side-groups. Macromol. Rapid Commun. **18**, 1117–1123 (1997)
108. V. Baurckbuchler, A.-L. Kjniksen, C. Galant, R. Lund, C. Amiel, K.D. Knudsen, B. Nyström, Rheological and structural characterization of the interactions between cyclodextrin compounds and hydrophobically modified alginate. Biomacromolecules **7**, 1871–1878 (2006)
109. V. Wintgens, S. Daoud-Mahammed, R. Gref, L. Bouteiller, C. Amiel, Aqueous polysaccharide associations mediated by β-cyclodextrin polymers. Biomacromolecules **9**, 1434–1442 (2008)
110. F.v.d. Manakker, M.v.d. Pot, T. Vermonden, C.F.v. Nostrum, W.E. Hennink, Self-assembling hydrogels based on β-cyclodextrin/cholesterol inclusion complexes. Macromolecules **41**, 1766–1773 (2008)
111. X. Guo, J. Wang, L. Li, D. Pham, P. Clements, S.F. Lincoln, B.L. May, Q. Chen, L. Zheng, R.K. Prud'homme, Tailoring polymeric hydrogels through cyclodextrin host-guest complexation. Macromol. Rapid Commun. **31**, 300–304 (2010)
112. J. Wang, D.T. Pham, X. Guo, L. Li, S.F. Lincoln, Z. Luo, H. Ke, L. Zheng, R.K. Prud'homme, Polymeric networks assembled by adamantyl and β-cyclodextrin substituted poly(acrylate)s: host-guest interactions, and the effects of ionic strength and extent of substitution. Ind. Eng. Chem. Res. **49**, 609–612 (2010)
113. O.W. Kretschmann, S. Choi, M. Miyauchi, I. Tomatsu, A. Harada, H. Ritter, Switchable hydrogels obtained by supramolecular cross-linking of adamantyl-containing LCST copolymers with cyclodextrin dimers. Angew. Chem. Int. Ed. **45**, 4361–4365 (2006)
114. C. Koopmans, H. Ritter, Formation of physical hydrogels via host−guest interactions of β-cyclodextrin polymers and copolymers bearing adamantyl groups. Macromolecules **41**, 7418–7422 (2008)
115. S. Tamesue, Y. Takashima, H. Yamaguchi, S. Shinkai, A. Harada, Photoswitchable supramolecular hydrogels formed by cyclodextrins and azobenzene polymers. Angew. Chem. Int. Ed. **49**, 7461–7464 (2010)
116. M. Nakahata, Y. Takashima, H. Yamaguchi, A. Harada, Redox-responsive self-healing materials formed from host-guest polymers. Nat. Commun. **2**, 511 (2011)
117. W.H. Binder, *Self-Healing Polymers: From Principles to Applications* (Wiley-VCH, Hoboken, 2013)
118. S. Burattini, B.W. Greenland, D. Chappell, H.M. Colquhoun, W. Hayes, Healable polymeric materials: a tutorial review. Chem. Soc. Rev. **39**, 1973–1985 (2010)

119. Y. Yang, M.W. Urban, Self-healing polymeric materials. Chem. Soc. Rev. **42**, 7446–7467 (2013)
120. S.R. White, N.R. Sottos, P.H. Geubelle, J.S. Moore, M.R. Keesler, S.R. Sriram, E.N. Brown, S. Viswanathan, Autonomic healing of polymer composites. Nature **409**, 794–797 (2001)
121. S.R. White, J.S. Moore, N.R. Sottos, B.P. Krull, W.A. Santa Cruz, R.C.R. Gergely, Restoration of large damage volumes in polymers. Science **344**, 620–623 (2014)
122. Y. Okumura, K. Ito, The polyrotaxane gel: a topological gel by figure-of-eight cross-links. Adv. Mater. **13**, 485–487 (2001)
123. T. Karino, Y. Okumura, C. Zhao, T. Kataoka, K. Ito, M. Shibayama, SANS studies on deformation mechanism of slide-ring gel. Macromolecules **38**, 6161–6167 (2005)
124. K. Mayumi, K. Ito, K. Kato, *Polyrotaxane and Slide-Ring Materials* (RSC Publishing, London, 2015)
125. K. Koyanagi, Y. Takashima, H. Yamaguchi, A. Harada, Movable cross-linked polymeric materials from bulk polymerization of reactive polyrotaxane cross-linker with acrylate monomers. Macromolecules **50**, 5695–5700 (2017)
126. X. Chen, M.A. Dam, K. Ono, A. Mal, H. Shen, S.R. Nutt, K. Sheran, F. Wudl, A thermally re-mendable cross-linked polymeric material. Science **295**, 1698–1702 (2002)
127. B. Ghosh, M.W. Urban, Self-repairing oxetane-substituted chitosan polyurethane networks. Science **323**, 1458–1460 (2009)
128. Y. Amamoto, J. Kamada, H. Otsuka, A. Takahara, K. Matyjaszewski, Repeatable photoinduced self-healing of covalently cross-linked polymers through reshuffling of trithiocarbonate units. Angew. Chem. Int. Ed. **50**, 1660–1663 (2011)
129. K. Imato, M. Nishihara, T. Kanehara, Y. Amamoto, A. Takahara, H. Otsuka, Self-healing of chemical gels cross-linked by diarylbibenzofuranone-based trigger-free dynamic covalent bonds at room temperature. Angew. Chem. Int. Ed. **51**, 1138–1142 (2012)
130. P. Cordier, F. Tournilhac, C. Soulie-Ziakovic, L. Leibler, Self-healing and thermoreversible rubber from supramolecular assembly. Nature **451**, 977–980 (2008)
131. Q. Wang, J.L. Mynar, M. Yoshida, E. Lee, M. Lee, K. Okuro, K. Kinbara, T. Aida, High-water-content mouldable hydrogels by mixing clay and a dendritic molecular binder. Nature **463**, 339–343 (2010)
132. T.L. Sun, T. Kurokawa, S. Kuroda, A.B. Ihsan, T. Akasaki, K. Sato, M.A. Haque, T. Nakajima, J.P. Gong, Physical hydrogels composed of polyampholytes demonstrate high toughness and viscoelasticity. Nat. Mater. **12**, 932–937 (2013)
133. M. Burnworth, L. Tang, J.R. Kumpfer, A.J. Duncan, F.L. Beyer, G.L. Fiore, S.J. Rowan, C. Weder, Optically healable supramolecular polymers. Nature **472**, 334–337 (2011)
134. S. Burattini, B.W. Greenland, D.H. Merino, W. Weng, J. Seppala, H.M. Colquhoun, W. Hayes, M.E. Mackay, I.W. Hamley, S.J. Rowan, A healable supramolecular polymer blend based on aromatic π-π stacking and hydrogen-bonding interactions. J. Am. Chem. Soc. **132**, 12051–12058 (2010)
135. T. Kakuta, Y. Takashima, M. Nakahata, M. Otsubo, H. Yamaguchi, A. Harada, Preorganized hydrogel: self-healing properties of supramolecular hydrogels formed by polymerization of host-guest-monomers that contain cyclodextrins and hydrophobic guest groups. Adv. Mater. **25**, 2849–2853 (2013)
136. T. Kakuta, Y. Takashima, T. Sano, T. Nakamura, Y. Kobayashi, H. Yamaguchi, A. Harada, Adhesion between semihard polymer materials containing cyclodextrin and adamantane based on host-guest interaction. Macromolecules **48**, 732–738 (2015)
137. M. Nakahata, Y. Takashima, A. Harada, Highly flexible, tough, and self-healing supramolecular polymeric materials using host-guest interaction. Macromol. Rapid Commun. **37**, 86–92 (2016)
138. Y. Takashima, Y. Sawa, K. Iwaso, M. Nakahata, H. Yamaguchi, A. Harada, Supramolecular materials cross-linked by host-guest inclusion complexes: the effect of side chain molecules on mechanical properties. Macromolecules **50**, 86–92 (2017)

139. Y. Takashima, K. Yonekura, K. Koyanagi, K. Iwaso, M. Nakahata, H. Yamaguchi, A. Harada, Multifunctional stimuli-responsive supramolecular materials with stretching, coloring, and self-healing properties functionalized via host-guest interactions. Macromolecules **50**, 4414–4150 (2017)
140. M. Nakahata, S. Mori, Y. Takashima, H. Yamaguchi, A. Harada, Self-healing materials formed by cross-linked polyrotaxanes with reversible bonds. Chem **1**, 766–775 (2016)
141. M.V. Gandhi, B.S. Thompson, *Smart Materials and Structures* (Chapman & Hall, London, 1992)
142. W.U. Marek, *Handbook of Stimuli-Responsive Materials* (Wiley-VCH Verlag GmbH & Co. KGaA, Weinheim, 2011)
143. S. Minko, *Responsive Polymer Materials: Design and Applications* (Wiley-Blackwell, Hoboken, 2006)
144. T. Sekitani, M. Takamiy, Y. Noguchi, S. Nakano, Y. Kato, T. Sakurai, T. Someya, A large-area wireless power-transmission sheet using printed organic transistors and plastic MEMS switches. Nat. Mater. **6**, 413–417 (2007)
145. S. Kobatake, S. Takami, H. Muto, T. Ishikawa, M. Irie, Rapid and reversible shape changes of molecular crystals on photoirradiation. Nature **446**, 778–781 (2007)
146. F. Terao, M. Morimoto, M. Irie, Light-driven molecular-crystal actuators: rapid and reversible bending of rodlike mixed crystals of diarylethene derivatives. Angew. Chem. Int. Ed. **51**, 901–904 (2012)
147. A. Coskun, M. Banaszak, R.D. Astumian, J.F. Stoddart, B.A. Grzybowski, Great expectations: can artificial molecular machines deliver on their promise? Chem. Soc. Rev. **41**, 19–30 (2012)
148. Y. Osada, H. Okuzaki, H. Hori, A polymer gel with electrically driven motility. Nature **355**, 242–244 (1992)
149. Y. Osada, A. Matsuda, Shape memory in hydrogels. Nature **376**, 219 (1995)
150. D.J. Beebe, J.S. Moore, J.M. Bauer, Q. Yu, R.H. Liu, C. Devadoss, B.-H. Jo, Functional hydrogel structures for autonomous flow control inside microfluidic channels. Nature **404**, 588–590 (2000)
151. A. Sidorenko, T. Krupenkin, A. Taylor, P. Fratzl, J. Aizenberg, Reversible switching of hydrogel-actuated nanostructures into complex micropatterns. Science **315**, 487–490 (2007)
152. T. Ikeda, J. Mamiya, Y. Yu, Photomechanics of liquid-crystalline elastomers and other polymers. Angew. Chem. Int. Ed. **46**, 506–528 (2007)
153. C. Ohm, M. Brehmer, R. Zentel, Liquid crystalline elastomers as actuators and sensors. Adv. Mater. **22**, 3366–3387 (2010)
154. G.S. Kumar, D.C. Neckers, Photochemistry of azobenzene-containing polymers. Chem. Rev. **89**, 1915–1925 (1989)
155. T. Ikeda, O. Tsutsumi, Optical switching and image storage by means of azobenzene liquid-crystal films. Science **268**, 1873–1875 (1995)
156. T. Hugel, N.B. Holland, A. Cattani, L. Moroder, M. Seitz, H.E. Gaub, Single-molecule optomechanical cycle. Science **296**, 1103–1106 (2002)
157. M. Camacho-Lopez, H. Finkelmann, P. Palffy-Muhoray, M. Shelley, Fast liquid-crystal elastomer swims into the dark. Nat. Mater. **3**, 307–310 (2004)
158. C.L. van Oosten, C.W.M. Bastiaansen, D.J. Broer, Printed artificial cilia from liquid-crystal network actuators modularly driven by light. Nat. Mater. **8**, 677–682 (2009)
159. Z. Sekkat, J. Wood, W. Knoll, Reorientation mechanism of azobenzenes within the trans → cis photoisomerization. J. Phys. Chem. **99**, 17226–17234 (1995)
160. T.J. White, S.V. Serak, N.V. Tabiryan, R.A. Vaia, T.J. Bunning, Polarization-controlled, photodriven bending in monodomain liquid crystal elastomer cantilevers. J. Mater. Chem. **19**, 1080–1085 (2009)
161. N. Hosono, T. Kajitani, T. Fukushima, K. Ito, S. Sasaki, M. Takata, T. Aida, Large-area three-dimensional molecular ordering of a polymer brush by one-step processing. Science **330**, 808–811 (2010)

162. K.D. Harris, C.W.M. Bastiaansen, J. Lub, D.J. Broer, Self-assembled polymer films for controlled agent-driven motion. Nano Lett. **5**, 1857–1860 (2005)
163. W. Lu, A.G. Fadeev, F. Qi, E. Smela, B.R. Mattes, J. Ding, G.M. Spinks, J. Mazurkiewicz, D. Zhou, G.G. Wallace, D.R. Macfarlane, S.A. Forsyth, M. Forsyth, Use of ionic liquids for π-conjugated polymer electrochemical devices. Science **297**, 983–987 (2002)
164. R.H. Baughman, C. Cui, A.A. Zakhidov, Z. Iqbal, J.N. Barisci, G.M. Spinks, G.G. Wallace, A. Mazzoldi, D.D. Rossi, A.G. Rinzler, O. Jaschinski, S. Roth, M. Kertesz, Carbon nanotube actuators. Science **284**, 1340–1344 (1999)
165. P. Kim, C.M. Lieber, Nanotube nanotweezers. Science **286**, 2148–2150 (1999)
166. Y. Zhang, S. Iijima, Elastic response of carbon nanotube bundles to visible light. Phys. Rev. Lett. **82**, 3472–3475 (1999)
167. G.M. Spinks, G.G. Wallace, L.S. Fifield, L.R. Dalton, A. Mazzoldi, D.D. Rossi, I.I. Khayrullin, R.H. Baughman, Pneumatic carbon nanotube actuators. Adv. Mater. **14**, 1728–1732 (2002)
168. Y. Takashima, S. Hatanaka, M. Otsubo, M. Nakahata, T. Kakuta, A. Hashidzume, H. Yamaguchi, A. Harada, Expansion-contraction of photoresponsive artificial muscle regulated by host-guest interactions. Nat. Commun. **3**, 1270 (2012)
169. M. Nakahata, Y. Takashima, A. Hashidzume, A. Harada, Redox-generated mechanical motion of a supramolecular polymeric actuator based on host-guest interactions. Angew. Chem. Int. Ed. **52**, 5731–5735 (2013)
170. K. Miyamae, M. Nakahata, Y. Takashima, A. Harada, Self-healing, expansion-contraction, and shape-memory properties of preorganized supramolecular hydrogel through host-guest interactions. Angew. Chem. Int. Ed. **54**, 8984–8987 (2015)
171. K. Iwaso, Y. Takashima, A. Harada, Fast response dry-type artificial molecular muscles with [c2]daisy chains. Nat. Chem. **8**, 625–632 (2016)
172. Y. Takashima, Y. Hayashi, M. Osaki, F. Kaneko, H. Yamaguchi, A. Harada, A photoresponsive polymeric actuator topologically cross-linked by movable units based on a [2]rotaxane. Macromolecules **56**, 4688–4693 (2018)

Chapter 10
Photo-responsive Dynamic Molecular Catalyst for Spatiotemporal Control of Chemical Reactions

Tatsushi Imahori

Abstract Stimuli-responsive molecular catalyst enables multiple control of chemical reaction(s) to demonstrate highly sophisticated chemical transformations. Among them, photo-responsive dynamic molecular catalyst that changes the structure by photo-stimuli and switches the catalytic functions has recently attracted much interest due to the potential of photo-stimuli as clean and diverse noninvasive stimuli. This chapter focuses on the photo-responsive dynamic catalysts that reversibly switch the catalytic function on the basis of the dynamic conformation changes induced by external photo-stimuli. Recent progresses in the field of photo-responsive dynamic catalysts are highlighted. Although the progress is still small in present stage, this type of catalysts could open up new possibilities in various fields of scientific researches as well as synthetic chemistry.

Keywords Photo-responsive catalysts · Photoswitchable catalysts · Dynamic molecular space · Spatiotemporal control of reaction

10.1 Introduction

Molecular catalyst has remarkably progressed and plays an important role in modern organic synthetic chemistry. Many catalysts that demonstrate highly efficient and selective reactions through effective activation of substrate and/or reactant and stabilization and regulation of transition state by the finely constructed reaction space have been developed. However, development of molecular catalyst is still insufficient. Conventional molecular catalysts generally have a fixed catalytic function and basically control reactions by the same mechanism, which means that they cannot display multiple catalytic functions to demonstrate multiple control of reaction(s). Multiple control of reaction(s) could demonstrate more sophisticated chemical transformations than single reaction control. For instance, a catalyst that

T. Imahori (✉)
Department of Industrial Chemistry, Tokyo University of Science, Shinjyuku-ku, Tokyo, Japan
e-mail: imahori@rs.tus.ac.jp

S. Shirakawa (ed.), *Designed Molecular Space in Material Science and Catalysis*,
https://doi.org/10.1007/978-981-13-1256-4_10

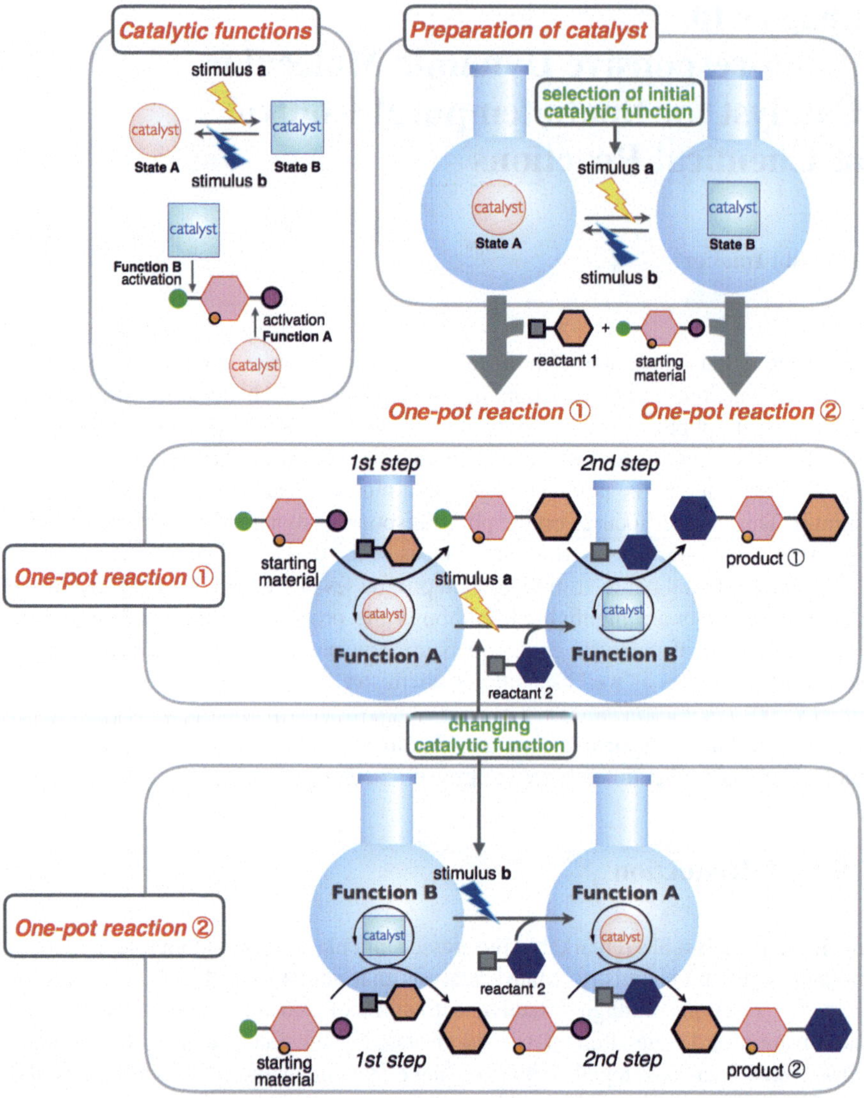

Fig. 10.1 One-pot multistep (two steps) synthesis with a switchable catalyst

switches multiple catalytic functions in a stepwise manner and promotes different reactions by each function can demonstrate environmentally benign one-pot multistep synthesis (Fig. 10.1) [1–3]. Conventional catalysts with a fixed function cannot demonstrate this type of temporal control of reactions. In addition to the conventional reaction control by catalytic reaction space improving catalytic activity and reaction selectivity, temporal control of reaction should be accomplished by molecular catalyst. Spatiotemporal control of reaction by molecular catalyst is a next goal of catalyst development.

To demonstrate the spatiotemporal control of chemical reaction, stimuli-responsive catalysts that switch the catalytic function by external stimuli have attracted much attention in recent years, and some types of stimuli-responsive catalysts have been developed [4–6]. Among them, stimuli-responsive dynamic catalyst that changes the structure in response to a selected stimulus and switches the catalytic function is promising. Stimuli-responsive dynamic catalyst can switch not only catalytic activity, but also selectivity or reaction type. Moreover stimuli-responsive dynamic catalyst reversibly switches the functions based on reversible structural changes, which enable sophisticated temporal control of reactions [7–10]. Light [11, 12], chemical [9, 13–20], pH [21–24], redox [25], and so on have been utilized as the stimuli to induce the structural changes. Especially, photo-responsive dynamic catalysts have been extensively developed due to the advantages. As well as being potentially clean, light can readily provide diverse noninvasive stimuli by selecting the wavelength, which allows precise control of the functions of catalysts [6, 26]. In this chapter, recent progresses of the photo-responsive dynamic catalysts toward spatiotemporal control of chemical reactions are summarized.

10.2 Photo-responsive Dynamic Catalyst

In photo-responsive dynamic catalysts, photochromic molecules that reversibly change the structure in response to photo-stimuli are introduced to induce the dynamic feature. Azobenzene [27], stilbene [28], and *cis*-1,2-diarylethene [29] are generally utilized as the photochromic molecules (Fig. 10.2). The catalytic reaction space adjusts the structure and switch catalytic functions based on the photo-responsive structural changes of the introduced photochromic molecule.

Most of the reported photo-responsive dynamic catalysts can be classified into the following three types:

1. Photo-responsive dynamic cooperative catalyst
2. Photo-responsive dynamic catalyst switching steric environment
3. Photo-responsive dynamic catalyst switching electronic feature

10.2.1 Photo-responsive Dynamic Cooperative Catalyst

In the early stage of the development of photo-responsive dynamic cooperative catalysts, photoswitchable templates based on molecular recognition were studied [30, 31].

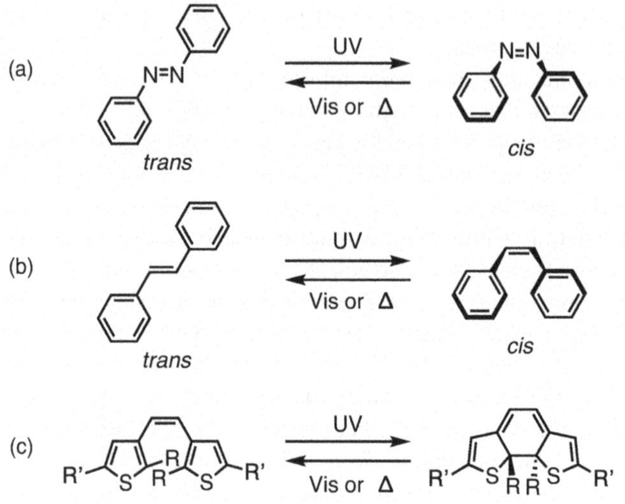

Fig. 10.2 Photochromic molecules

Rebek et al. reported a photo-responsive cooperative catalyst utilizing adenine receptors [30]. An azobenzene core is introduced to the catalyst **1** to induce reversible overall conformational changes based on photo-isomerization of azobenzene [27], and two imide-based adenine receptors are connected to the core to capture reactants containing an adenine unit (Fig. 10.3). Irradiation of 366 nm light induces isomerization of the *trans*-azobenzene core to the *cis*-form to afford a 1:1 mixture of *cis-/trans*-azobenzene isomers of the catalyst (*cis*-**1** and *trans*-**1**) at the photostationary state. Visible light (>420 nm) leads to the reverse isomerization, and the ratio of the isomers reaches to *cis*-**1**:*trans*-**1** = 30:70. This isomerization modulates the catalytic activity in the amide formation of amine **2** and *p*-nitrophenyl ester **3**, which contain an adenine unit, respectively (Scheme 10.1). Although the *trans*-azobenzene catalyst hardly accelerated the amide-forming reaction, the 1:1 mixture of *cis-/trans*-azobenzene isomers caused nearly tenfold enhancement in the reaction rate. The *cis*-azobenzene form of the catalyst (*cis*-**1**) closely locates the two adenine recognition sites and brings amine **2** and *p*-nitrophenyl ester **3** in proximity to each other, in which an ideal template for the amide formation would be constructed. In the *trans*-azobenzene catalyst (*trans*-**1**), the two adenine recognition sites are separately placed, and it is difficult to induce their cooperative function. The addition of competitive binders with an adenine unit decreased the catalytic activity of *cis*-**1**. These results support the assumed template mechanism. The produced amide containing two adenine units also binds to the catalyst and impedes the reaction. The product inhibition is a problem of this photo-responsive template.

Cacciapaglia and co-workers reported a similar photo-responsive catalysis with a photoswitchable dynamic template utilizing metal coordinations [31]. The template **5** consists of two crown ether units with a barium cation each and an azobenzene core to control the spatial arrangement of the crown ether units by photo-irradiation

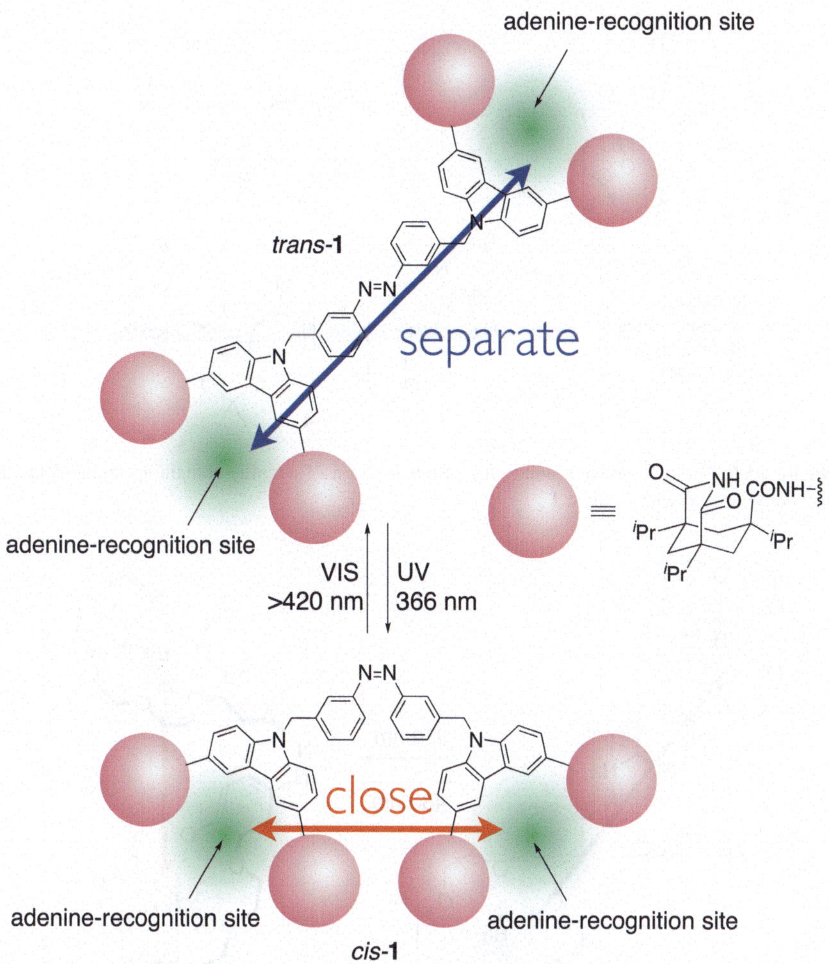

Fig. 10.3 Photo-responsive template utilizing adenine receptors

(Fig. 10.4). UV light (370 nm) irradiation isomerizes the *trans*-azobenzene core to the *cis*-azobenzene form, and the ratio of the isomers reaches to *cis*-azobenzene form (*cis*-**5**):*trans*-azobenene form (*trans*-**5**) = 95:5 at the photostationary state. In the twisted *cis*-azobenzene form, the crown ether units are arranged close to each other. In contrast, visible light (480 nm) induces the reverse isomerization to give the *trans*-azobenzene template as the major isomer (*cis*-**5**: *trans*-**5** = 19: 81). The *trans*-azobenzene form has a spread structure, in which the crown ether units are located apart from each other. This photoswitchable template functions as a reversible photo-responsive catalyst for ethanolysis of anilide **6** containing a *para*-carboxylate group (Scheme 10.2). Although the *trans*-azobenzene form showed only a low catalytic activity, the *cis*-azobenzene catalyst substantially accelerated the reaction. The close two crown ether units in the *cis*-azobenzene form would provide a suitable

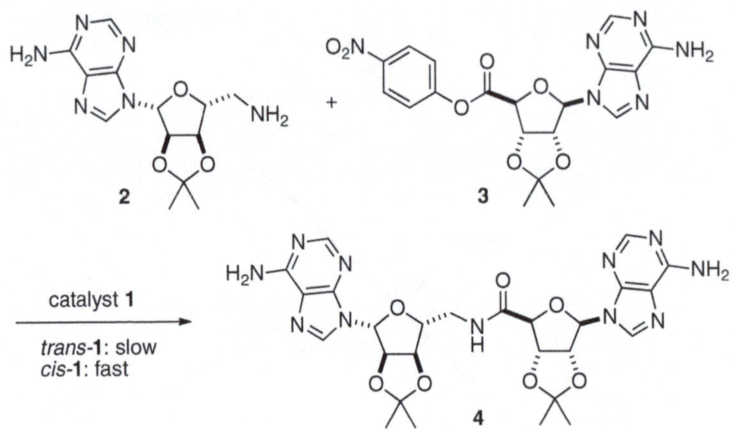

Scheme 10.1 Photo-responsive amide formation reaction of adenine-containing amine **2** and *p*-nitrophenyl ester **3** with catalyst **1**

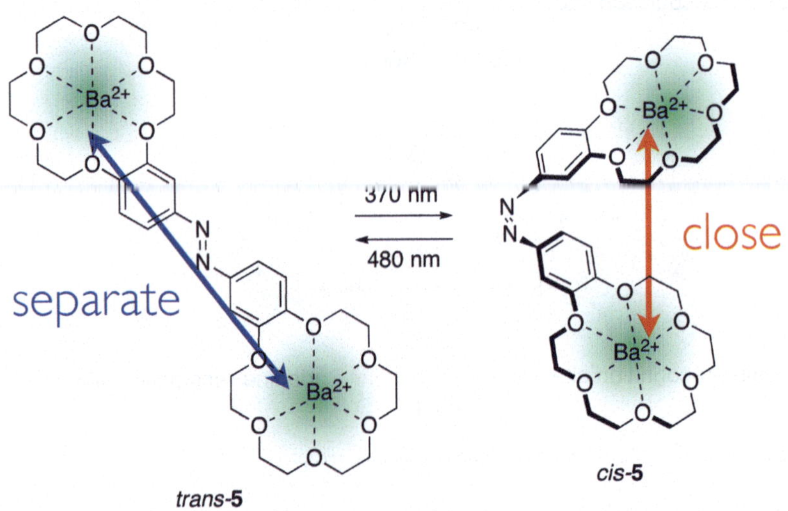

Fig. 10.4 Photo-responsive template utilizing metal coordinations

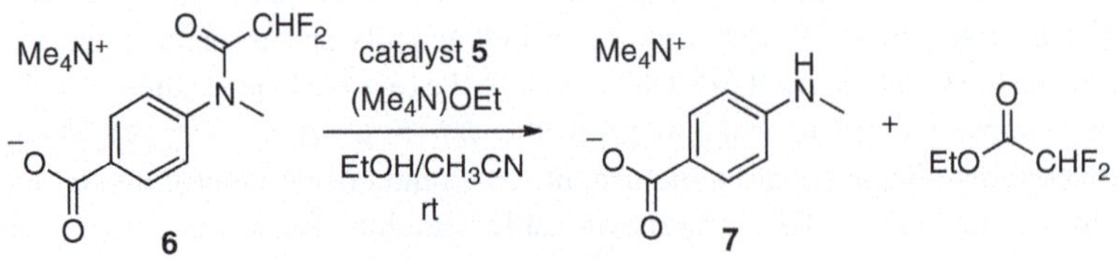

Scheme 10.2 Photo-responsive ethanolysis of anilide **6** with catalyst **5**

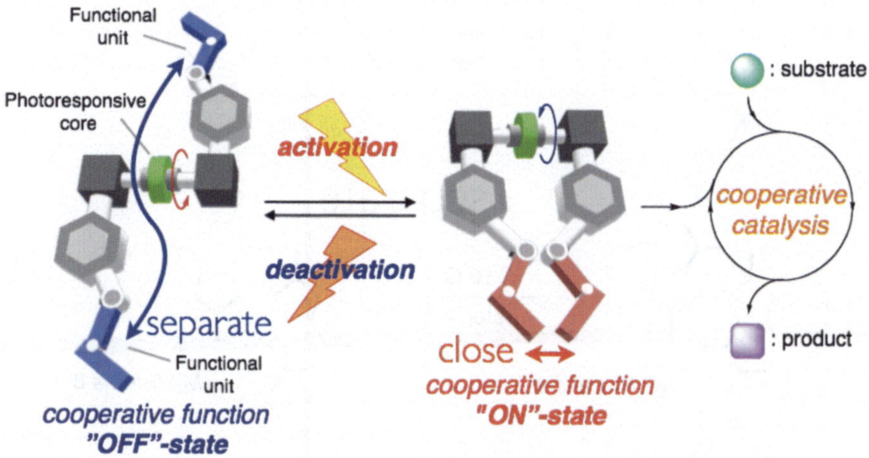

Fig. 10.5 Concept of the photo-responsive bifunctional organocatalysts based on conformational changes

reaction space (suitable reaction template) for the ethanolysis, in which one barium center functions as a binding site for a carboxylate group on the anilide substrate and the other places a nucleophilic ethoxide ion near the amide group of the substrate (reaction site) to promote the reaction. Two-point binding of the anilide substrate to the two barium centers of the *cis*-azobenzene catalyst at the carboxylate group and the amide group also seems plausible as the alternative catalytic mechanism. The amide group would be effectively activated by a Lewis acidic barium center through the two-point binding. The reaction with the *cis*-azobenzene catalyst proceeded approximately three times faster than the reaction with the *trans*-azobenzene catalyst. The product inhibition takes place in this catalyst, too.

The photoswitchable templates based on molecular recognition are problematic in terms of product inhibition. To overcome this drawback, photo-responsive dynamic cooperative catalysts utilizing simple weak interactions to activate and/or regulate reactants have been developed. As a cooperative catalyst utilizing simple weak interactions, bifunctional organocatalysts have attracted much interest [32–36]. Catalytic activities of various bifunctional organocatalysts are attributed to the cooperative function of the incorporated two functional groups (units) inducing weak interactions. Bifunctional organocatalysts have been applied to the photo-responsive cooperative catalysts based on dynamic conformational changes (Fig.10.5) [37–41].

Feringa et al. reported a photo-responsive dynamic acid-base combined catalyst **8** [37]. The authors have developed unidirectional rotary molecular motors based on photo- and thermal isomerization of chiral stilbene derivatives [42–44]. The stimuli-responsive chiral molecular motor was introduced to a bifunctional organocatalyst to gain switchable catalytic activity and stereoselectivity by changing the spatial arrangement of the functional units. The photo-responsive catalyst **8** consists of a thiourea unit as an acid functionality (hydrogen-bonding donor), a *N*,*N*-dimethyl-4-aminopyridine (DMAP) unit as a Brønsted base functionality, and a chiral stilbene

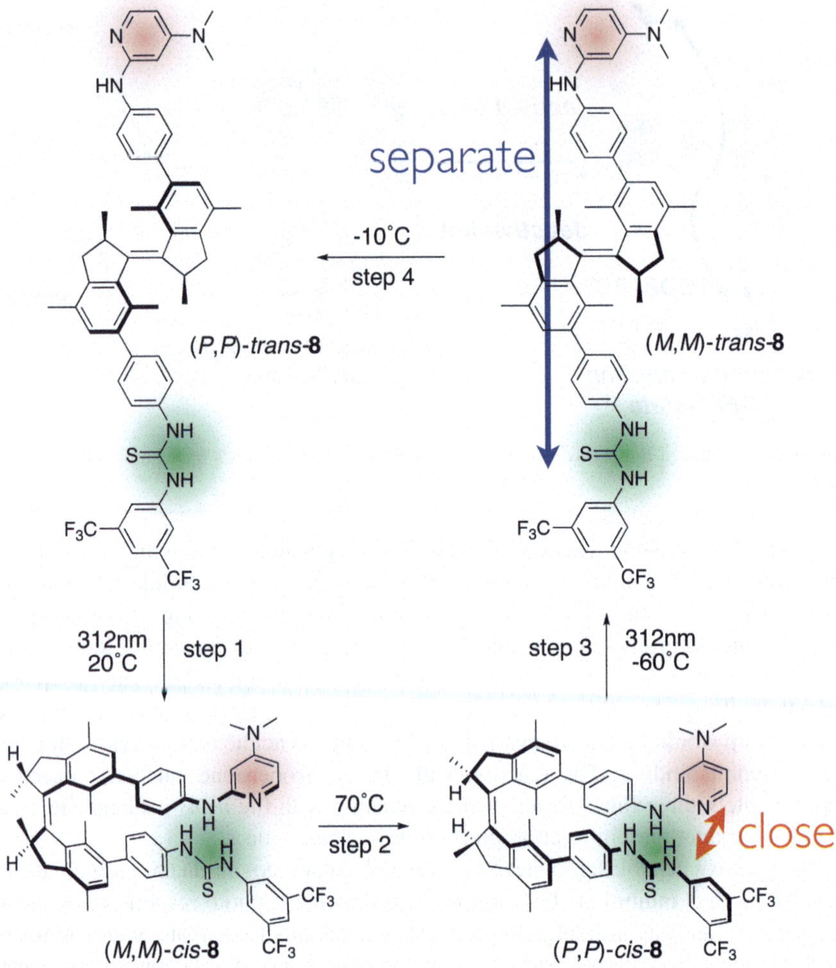

Fig. 10.6 Rotary molecular motor-based stimuli-responsive bifunctional organocatalyst

core as the unidirectional rotary motor (Fig. 10.6). A thiourea unit and a DMAP unit can cooperatively function in a bifunctional organocatalyst to promote Michael reaction [45]. The molecular motor rotates the functional units in unidirectional fashion and changes their spatial arrangement. Irradiation of 312 nm light induces isomerization of the thermally stable *trans*-stilbene derivative, (*P*,*P*)-*trans*-**8** to the *cis*-stilbene isomer ((*M*,*M*)-*cis*-**8**) at 20 °C, in which the two functionalities are brought into close proximity (Fig. 10.6, step 1). This *cis*-stilbene isomer thermally inverses the helix on *cis*-stilbene to provide (*P*,*P*)-*cis*-**8** at 70 °C, and the functionalities remain in close proximity (Fig. 10.6, step 2). Then irradiation of 312 nm light at −60 °C isomerizes (*P*,*P*)-*cis*-**8** to the *trans*-stilbene isomer ((*M*,*M*)-*trans*-**8**)

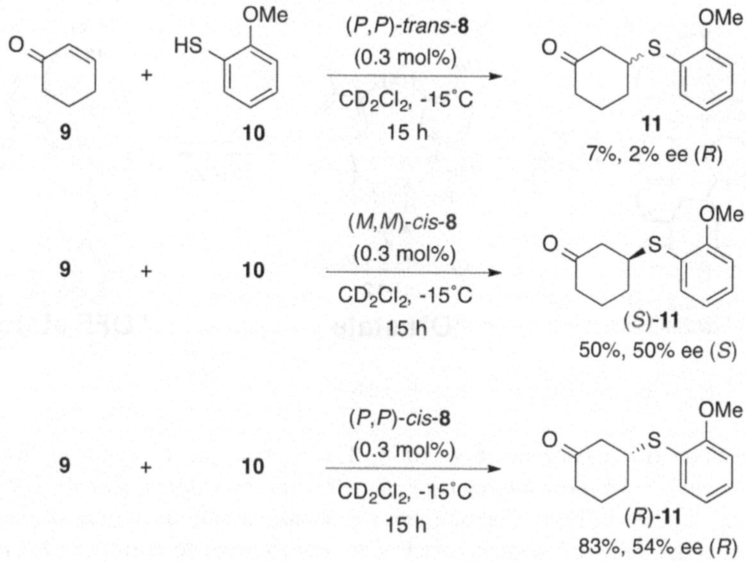

Scheme 10.3 Stimuli-responsive control over the activity and stereoselectivity of catalyst **8** in thio-Michael reaction

(Fig. 10.6, step 3). Finally, (*M*,*M*)-*trans*-**8** thermally inverses the helix to afford original (*P*,*P*)-*trans*-**8** at −10 °C (Fig. 10.6, step 4). In these *trans*-stilbene isomers ((*M*,*M*)-*trans*-**8** and (*P*,*P*)-*trans*-**8**), the acid and base functionalities are located separately. These stepwise conformational changes modulate the catalytic activity and stereoselectivity in the thio-Michael reaction [46] between 2-cyclohexen-1-one (**9**) and thiophenol **10** (Scheme 10.3). The *trans*-stilbene catalyst ((*P*,*P*)-*trans*-**8**), in which cooperative function of the separate acid and base functionalities seems unlikely, showed low catalytic activity with no stereoselectivity (7%, *e.r.* (*S*:*R*) = 49:51). On the other hand, the *cis*-stilbene isomers ((*M*,*M*)-*cis*-**8** and (*P*,*P*)-*cis*-**8**) showed higher catalytic activities. (*M*,*M*)-*cis*-**8** promoted the Michael addition in 50% yield with substantial enantiomeric excess (*e.r.* (*S*:*R*) = 75:25) in the identical conditions. (*P*,*P*)-*cis*-**8** provided a more rapid reaction (83% yield) and interestingly showed opposite enantioselectivity (*e.r.* (*S*:*R*) = 23:77). In these *cis*-stilbene isomers, the closely placed acid and base functionalities would function cooperatively to promote the reaction, and the helicity of the motor unit (*cis*-stilbene unit) controls the stereochemistry of the reaction. Stimuli-responsive activity and selectivity control of a catalyst via dynamic conformational changes was demonstrated for the first time.

A short time later, our group reported a related photo-responsive bifunctional organocatalyst based on dynamic conformational changes [38]. An azobenzene core was introduced to a bis(trityl alcohol) catalyst, in which cooperative function of the two acidic trityl alcohol units acquires the catalytic activity [47], to modulate the cooperative function by the overall conformational changes based on photo-

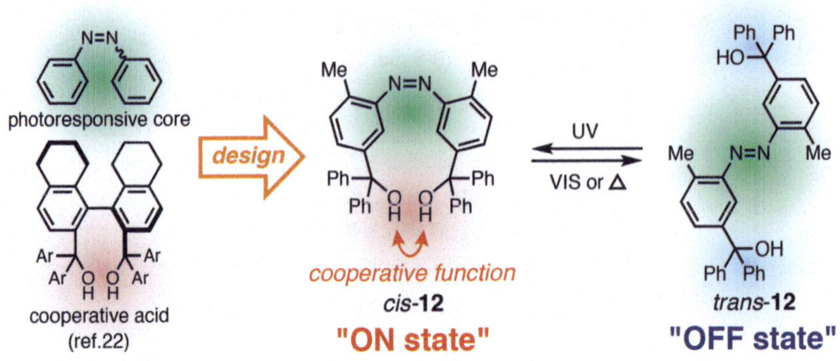

Fig. 10.7 Photo-responsive bis(trityl alcohol) catalyst **12**

isomerization of azobenzene (Fig. 10.7). The azobenzene-incorporated bis(trityl alcohol) catalyst **12** reversibly changes the structure by external stimuli. UV light (365 nm) irradiation isomerizes the *trans*-azobenzene core to the *cis*-azobenzene form, and the ratio of isomers reaches to *cis*-azobenzene form (*cis*-**12**):*trans*-azobenzene form (*trans*-**12**) = 95:5 at the photostationary state. In the *cis*-azobenzene catalyst, the two trityl alcohol units are located in proximity to each other. Visible light (>420 nm) induces the reverse isomerization to give an almost 1:1 mixture of *cis-/trans*-azobenzene isomers of the catalyst (*cis*-**12**:*trans*-**12** = 52:48). This isomerization from *cis*-azobenzene catalyst to the *trans*-azobenzene isomer can be completed via thermal isomerization (*cis*-**12**:*trans*-**12** = <1:>99, 150 °C, 24 h). The *trans*-azobenzene form places the two trityl alcohol units apart from each other. This photoswitchable bis(trityl alcohol) **12** functions as a reversible photo-responsive catalyst for Morita-Baylis-Hillman reaction [48, 49] between **13** and **14** (Scheme 10.4). Although the *trans*-azobenzene form showed low catalytic activity, the *cis*-azobenzene catalyst substantially accelerated the reaction. The proximal trityl alcohol units in the *cis*-azobenzene catalyst would cooperatively function to promote the reaction. The difference in the catalytic activities of the both isomeric catalysts was observed up to approximately 86-fold.

Photo-responsive dynamic cooperative catalyst that demonstrates switchable cooperative deactivation has also been developed. Pericàs and co-workers reported an azobenzene-based photo-responsive thiourea catalyst with a nitro group (Fig. 10.8) focusing on interaction of thiourea with nitro group [50]. Since thiourea can interact with nitro group through hydrogen bondings, the dynamic thiourea catalyst **16** switches interaction with the introduced nitro group on the basis of the photo-responsive structural changes. In the *cis*-azobenzene isomer (*cis*-**16**), the nitro group is placed close to the thiourea unit and would interact with the thiourea unit to decrease the catalytic activity. On the other hand, the nitro group is placed apart from the thiourea unit in the *trans*-azobenzene isomer (*trans*-**16**). The free thiourea unit of the *trans*-azobenzene catalyst would display good catalytic activity. The dynamic thiourea catalyst **16** reversibly changes the structure by external stimuli. UV light (365 nm) irradiation for the *trans*-azobenzene catalyst induces isomerization to the

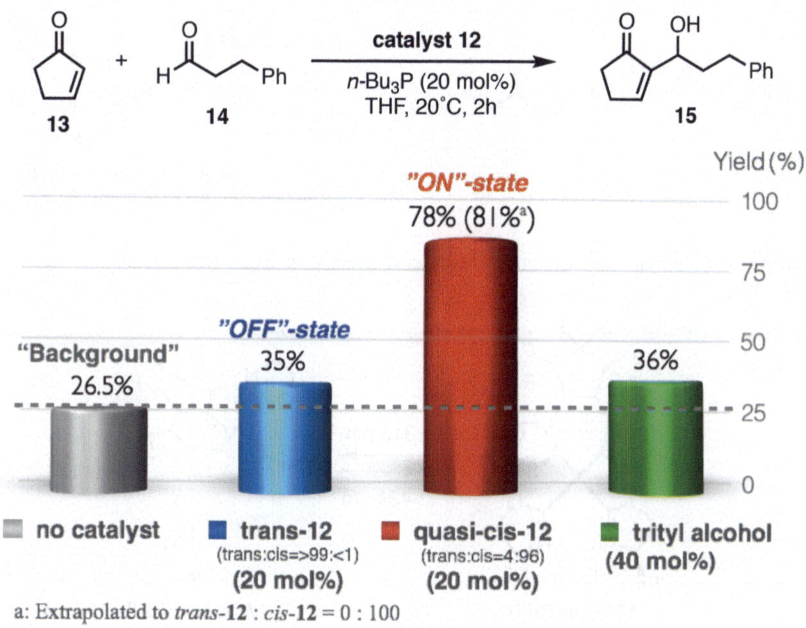

Scheme 10.4 Photo-responsive Morita-Baylis-Hillman reaction with catalyst **12**

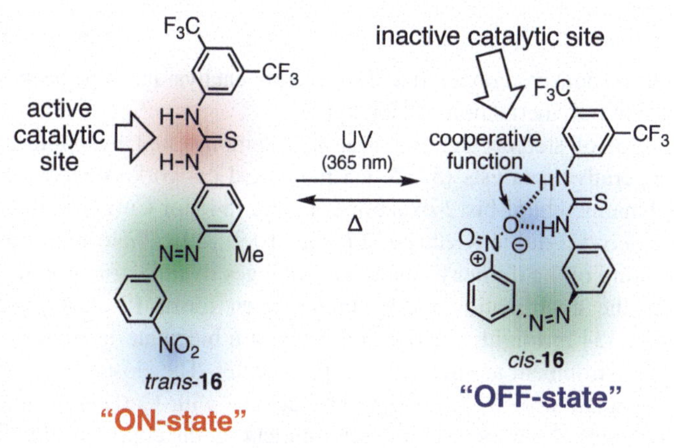

Fig. 10.8 Photo-responsive cooperative deactivation catalyst **16**

cis-azobenzene form, and the ratio of isomers reaches to *cis*-**16**:*trans*-**16** = 56:44 at the photostationary state. Visible light (450 nm) irradiation isomerizes the *cis*-azobenzene catalyst to the *trans*-azobenzene form almost completely (*cis*-**16**:*trans*-**16** = 1:99). The catalyst **16** actually demonstrated switchable catalytic activity in nitro-Michael reaction (Scheme 10.5).

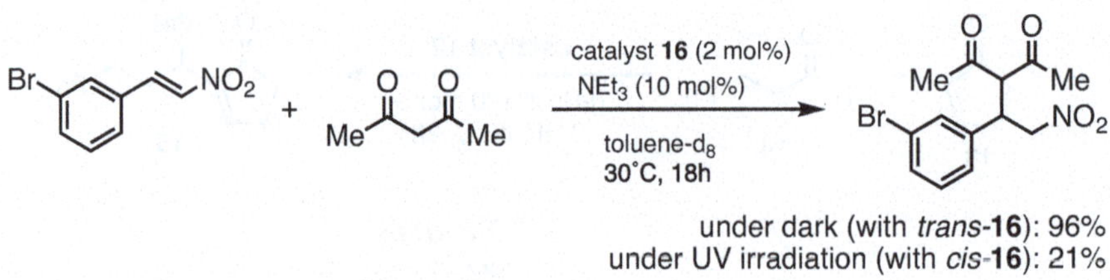

Scheme 10.5 Photo-responsive nitro-Michael reaction with catalyst **16**

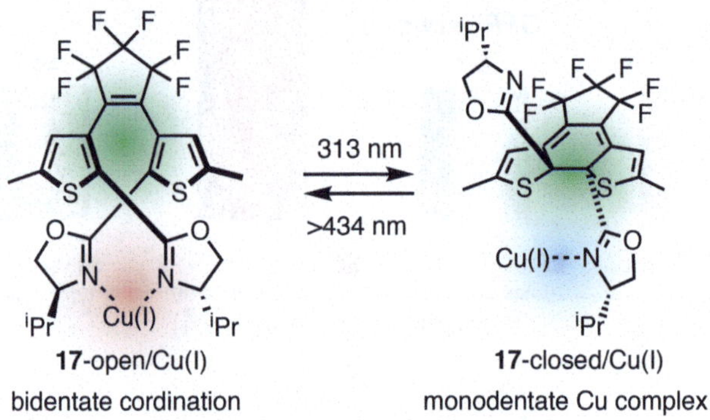

Fig. 10.9 Photo-responsive dynamic bis(2-oxazoline) ligand **17**

This photo-responsive cooperative deactivation catalyst has also been applied to photoswitchable polymerization of lactide [51].

Photo-control of stereoselectivity was also demonstrated by photo-responsive dynamic cooperative catalysts [52, 53]. Branda and co-workers reported a photo-responsive dynamic chiral bis(2-oxazoline) ligand [54] for Cu(I) catalyst based on photochromic *cis*-1,2-dithienylethene (**17**, Fig. 10.9) [29]. Photo-induced ring closure/ring opening of the dithienylethene unit changes overall structure of the ligand and switches the coordination mode. In the open form, **17**-open generated by irradiation of >434 nm light, Cu(I) coordinates in a bidentate fashion to construct a rigid chiral environment around active copper center. The closed form (**17**-closed) induced by irradiation of 313 nm light coordinates with Cu(I) in a monodentate fashion, which provides a loose chiral environment on the copper center. These **17**/Cu(I) complexes promoted cyclopropanation of styrene (**18**) with ethyl diazoacetate (**19**) [55, 56] in different stereoselectivities (Scheme 10.6). The reaction with **17**-open/Cu(I) provided *trans*- and *cis*-cyclopropane **20** with moderate enantioselectivities (*cis*-**20**, 50% ee; *trans*-**20,** 30% ee). On the other hand, the complex with **17**-closed afforded *cis*- and *trans*-**20** with lower enantioselectivities (*cis*-**20,** 5% ee; *trans*-**20,** 5% ee). These stereochemical outcomes probably reflect the chiral environments around active copper center modulated by the photo-induced structural changes of ligand. The drawback of this system is the limited photoswitchability. The photo-cyclization of 2,2′-disubstituted dithienylethenes is

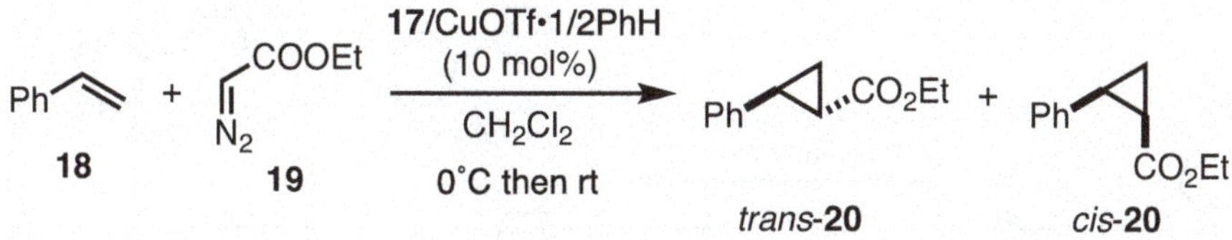

catalyst	ratio		
17-open/Cu(I)	55 (30% ee)	:	45 (50% ee)
17-closed/Cu(I)	63 (5% ee)	:	37 (5% ee)

Scheme 10.6 Photo-control of stereoselectivity in Cu-catalyzed cyclopropanation of styrene with ligand **17**

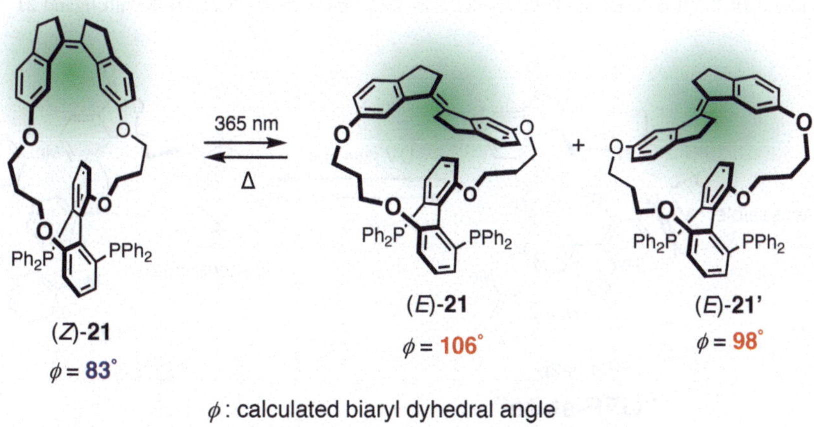

Fig. 10.10 Photo-responsive dynamic axially chiral biaryl bis(phosphine) ligand **21**

impeded by the steric hindrance, and only a minor amount of closed isomer was obtained. Furthermore, reversibility of the photo-cyclization/photo-ring opening was hampered by the complexation with Cu(I).

Craig and co-workers also reported a photoswitchable dynamic chiral ligand for Pd catalysts [53], a macrocyclic axially chiral biaryl bis(phosphine) ligand **21** that incorporate a photo-responsive stilbene unit (Fig. 10.10). Photo-isomerization of the stilbene unit allows photochemical manipulation of the biaryl geometry and changes the stereochemical outcomes in Pd-catalyzed reactions. UV light (365 nm) irradiation to the (*Z*)-stilbene isomer ((*Z*)-**21**) induces reversible isomerization and provides 68:23:9 mixture of three isomers, (*Z*)-**21**, (*E*)-**21**, and (*E*)-**21**′ with different calculated biaryl dihedral angles (ϕ) at the photostationary state. The isomers (*Z*)-**21** and (*E*)-**21** displayed different stereoselectivity in asymmetric Heck reaction (Scheme 10.7) and asymmetric allylic alkylation.

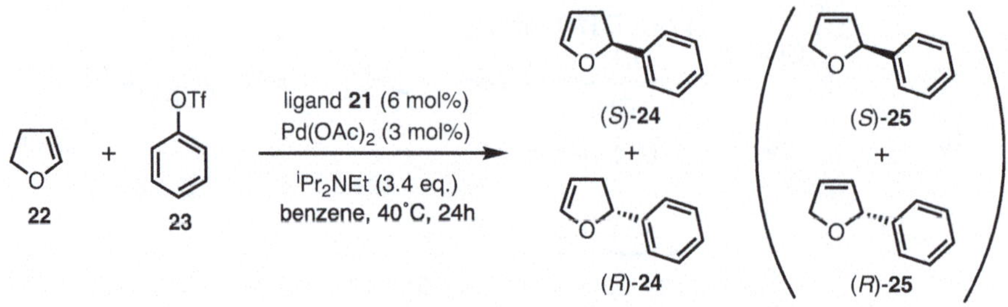

Ligand	conversion (%)	**24 : 25**	ee of (*S*)-**24** (%)
(*Z*)-**21**	55	97 : 3	96
(*E*)-**21**	95	98 : 2	79

Scheme 10.7 Photo-control of stereoselectivity in asymmetric Heck reaction with ligand **21**

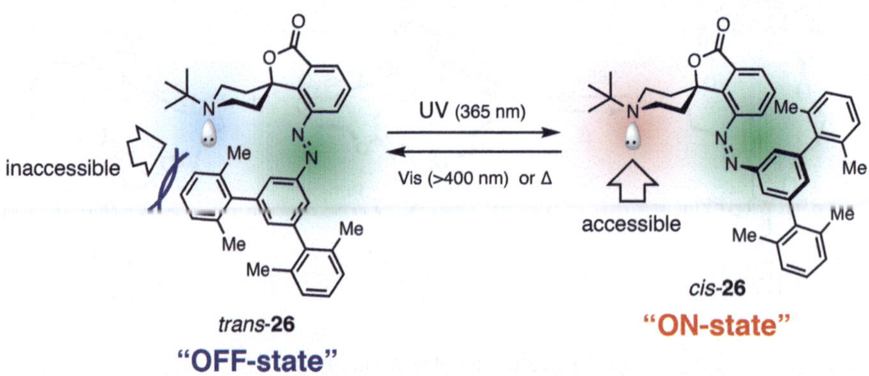

Fig. 10.11 Photo-responsive dynamic Brønsted base catalyst **26**

10.2.2 Photo-responsive Dynamic Catalyst Switching Steric Environment

In the second type of photo-responsive dynamic catalysts, steric environment around the catalytic center is switchable in response to photo-stimuli. Since access to the catalytic center is greatly affected by the surrounding steric environment, catalytic functions can be regulated by switching the steric environment around the catalytic center.

Hecht et al. reported a photo-responsive dynamic Brønsted base catalyst that switches the steric environment around the catalytic center, 3°-amine center, based on photo-isomerization of azobenzene (Fig. 10.11) [57, 58]. The azobenzene-based catalyst **26** reversibly changes the structure by photo-stimuli. Irradiation of 365 nm light induces isomerization of the *trans*-azobenzene catalyst (*trans*-**26**) to the *cis*-

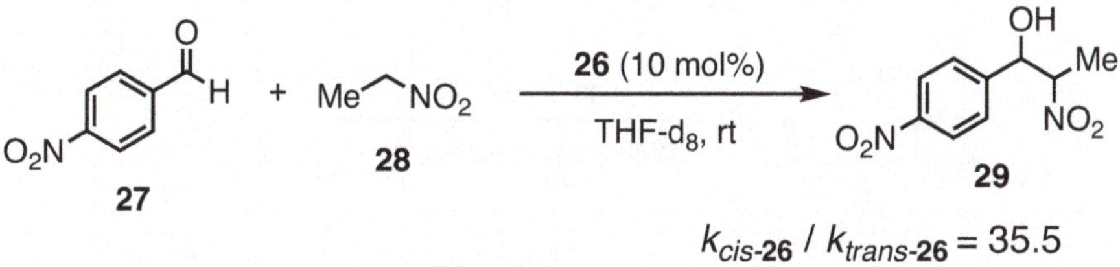

Scheme 10.8 Photo-responsive Henry reaction with catalyst **26**

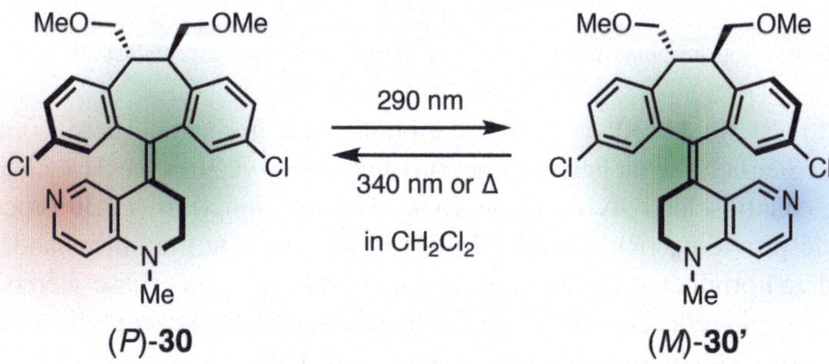

Fig. 10.12 Photo-responsive dynamic DMAP catalyst **30** switching helical chirality

azobenzene isomer (*cis*-**26**). The ratio of isomers reaches to *cis*-**26**:*trans*-**26** = >95:5 at the photostationary state. In the *cis*-azobenzene form, steric bulkiness around the basic catalytic center is removed, and the catalytic center becomes sterically accessible. Visible light (>400 nm) irradiation induces the reverse isomerization (*cis*-**26**:*trans*-**26** = 10:90 at the photostationary state). In the *trans*-azobenzene form, the basic catalytic center is shielded by surrounding structure, which would inhibit access to the catalytic center. This photoswitchable catalyst **26** demonstrated functions as a reversible photo-responsive Brønsted base catalyst in Henry reaction between **27** and **28** (Scheme 10.8). Although the *trans*-azobenzene form showed low catalytic activity, the *cis*-azobenzene catalyst substantially accelerated the reaction. The difference in the catalytic activities of the both isomeric catalysts was observed up to approximately 35.5-fold.

Photo-control of stereoselectivity by photo-responsive dynamic catalysts switching steric environment was also demonstrated. Chen and co-workers reported a stilbazole-based chiral DMAP [59] derivative as a photo-responsive dynamic chiral nucleophilic catalyst (**30**, Fig. 10.12) [60]. The stilbazole-based DMAP derivative has a chiral dibenzosuberane unit, and the chirality induces a helical chirality in the stilbazole core including a DMAP unit. Since photo-isomerization of the stilbazole core flips the helical chirality, chiral environment of the DMAP unit is switchable in response to photo-stimuli. The DMAP derivative with (*P*)-helical chirality ((*P*)-**30**) almost completely isomerizes to the derivative with (*M*)-helical chirality ((*M*)-**30'**) by 290 nm light ((*P*)-**30**:(*M*)-**30'** = <1:>99 at the photostationary state). The reverse isomerization is induced by 340 nm light, and the ratio of isomers

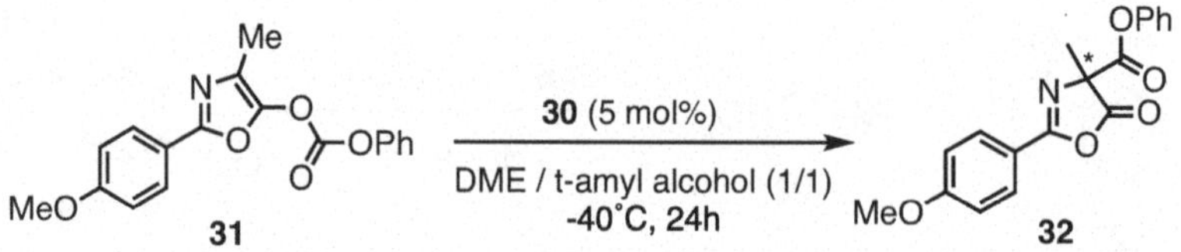

catalyst	yield (%)	ee (%) (config.)
(*P*)-**30**	84	87 (*R*)
(*M*)-**30'**	81	91 (*S*)

Scheme 10.9 Photo-responsive Steglich rearrangement reaction with catalyst **30**

reaches to (*P*)-**30**:(*M*)-**30'** = 91:9 at the photostationary state. These isomers each promoted Steglich rearrangement reaction of *O*-carboxylazlactone **31** with an almost opposite enantioselectivity (Scheme 10.9). The reaction with (*P*)-**30** proceeded in 84% yield providing (*R*)-product ((*R*)-**32**) in 87% ee. On the other hand, (*M*)-**30'** provided (*S*)-product ((*S*)-**32**) in 81% yield and 91% ee. These stereochemical outcomes would reflect the chiral environments around the nucleophilic active center modulated by the photo-responsive structural changes.

10.2.3 Photo-responsive Dynamic Catalyst Switching Electronic Feature

As the third type of photo-responsive dynamic catalysts, photo-responsive dynamic catalysts switching electronic environment of the catalytic center have been studied. Since photo-isomerization of *cis*-1,2-diarylethene involves rearrangement of some covalent bonds, electronic features of the isomers should be different from each other. By focusing on this feature of *cis*-1,2-diarylethene, some photo-responsive dynamic carbenes as catalyst or ligand for metal catalysts have been developed [61–63].

Bielawski et al. developed a *cis*-1,2-diarylethene-based carbene catalyst **33** for transesterification reaction (Fig. 10.13, Scheme 10.10) [61]. The carbene catalyst reversibly changes the structure by photo-stimuli. Irradiation of 313 nm light induces isomerization of the open form of **33** (**33**-open) to the closed form (**33**-closed). In the open form, more electron-donating unsaturated carbine is formed. Visible light (>500 nm) irradiation induces the reverse isomerization (**33**-closed to **33**-open). In the closed form, less electron-donating saturated carbine is formed. This photoswitchable carbene **33** demonstrated photo-responsive catalytic activity in transesterification reaction between **34** and **35** (Scheme 10.10). The open form showed approximately 12.5 times higher catalytic activity. The photoswitching of the electron-donating ability would change the Brønsted basicity and modulate the catalytic activity.

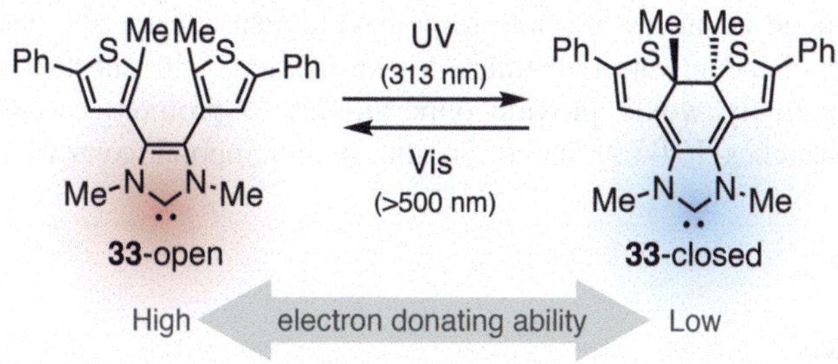

Fig. 10.13 Photo-responsive dynamic carbene catalyst **33**

Scheme 10.10 Photo-responsive transesterification reaction with catalyst **33**

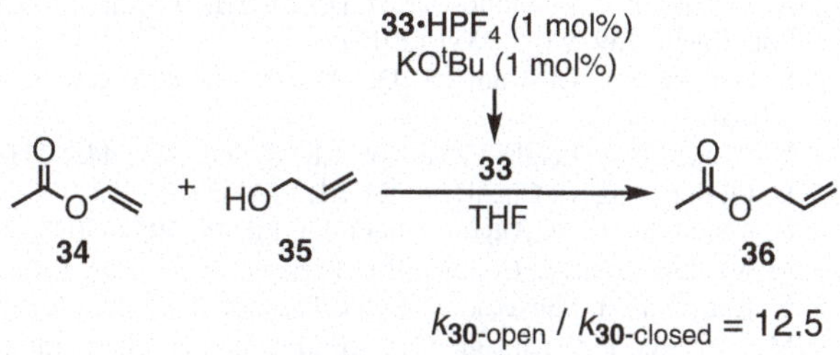

Applications of similar type of photoswitchable dynamic carbenes to ligands for metal catalysts have also been developed. Bielawski et al. demonstrated photo-responsive Rh-catalyzed hydroboration of alkenes and alkynes using **33** as the ligand for the Rh catalyst [62]. The photoswitchable electron-donating ability of **33** switched the catalytic activity of the Rh catalyst. The photo-responsive dynamic carbene ligand that switches reaction course of Ru-catalyzed olefin metathesis, ring-closing metathesis versus ring-opening metathesis polymerization, has also been developed [63].

10.3 Conclusion

In this chapter, recent progresses in the field of photo-responsive dynamic catalysts are highlighted. Some catalysts that switch the catalytic functions in response to photo-stimuli have been developed, and spatiotemporal controls of chemical reaction(s) have been demonstrated with the dynamic catalysts. However, the photoswitchable catalytic functions in the developed photo-responsive dynamic catalysts are limited. Most of the photo-responsive dynamic catalysts switch the catalytic activity. To demonstrate versatile spatiotemporal controls of chemical reaction(s), switching of other catalytic functions, such as selectivity or reaction type, should be accomplished. The field of photo-responsive dynamic catalysts is still immature.

The continued advances in photo-responsive dynamic catalysts would demonstrate multiregulated advanced chemical transformations with substantial practicalities and eventually would provide opportunities to promote various fields of scientific researches [64–69] based on the spatiotemporal control of chemical reactions.

References

1. S. Semwal, J. Choudhury, Angew. Chem. Int. Ed. **56**, 5556–5560 (2017)
2. K. Eichstaedt, J. Jaramillo-Garcia, D.A. Leigh, V. Marcos, S. Pisano, T.A. Singleton, J. Am. Soc. Chem. **139**, 9376–9381 (2017)
3. S. Gaikwad, A. Goswami, S. De, M. Schmittel, Angew. Chem. Int. Ed. **55**, 10512–10517 (2016)
4. V. Blanco, D.A. Leigh, V. Marcos, Chem. Soc. Rev. **44**, 5341–5370 (2015)
5. U. Lüning, Angew. Chem. Int. Ed. **51**, 8163–8165 (2012)
6. R.S. Stoll, S. Hecht, Angew. Chem. Int. Ed. **49**, 5054–5075 (2010)
7. M. Vlatković, B.S.L. Collons, B.L. Feringa, Chem. Eur. J. **22**, 17080–17111 (2016)
8. N. Kumagai, M. Shibasaki, Cat. Sci. Technol. **3**, 41–57 (2013)
9. M.J. Wiester, P.A. Ulmann, C.A. Mirkin, Angew. Chem. Int. Ed. **50**, 114–137 (2011)
10. D.A. Leigh, V. Marcos, M.R. Wilson, ACS Catal. **4**, 4490–4497 (2014)
11. B.M. Neilson, C.W. Bielawski, ACS Catal. **3**, 1874–1885 (2013)
12. T. Imahori, S. Kurihara, Chem. Lett. **43**, 1524–1531 (2014)
13. Y. Yoshinaga, T. Yamamoto, M. Suginome, ACS Macro Lett. **6**, 705–710 (2017)
14. T. Yamamoto, R. Murakami, M. Suginome, J. Am. Soc. Chem. **139**, 2557–2560 (2017)
15. Y. Akai, T. Yamamoto, Y. Nagata, T. Ohmura, M. Suginome, J. Am. Soc. Chem. **134**, 11092–11095 (2012)
16. T. Yamamoto, T. Yamada, Y. Nagata, M. Suginome, J. Am. Soc. Chem. **132**, 7899–7901 (2010)
17. S. De, S. Pramanik, M. Schmittel, Angew. Chem. Int. Ed. **53**, 14255–14259 (2014)
18. S. De, S. Pramanik, M. Schmittel, Dalton Trans. **43**, 10977–10982 (2014)
19. C.M. McGuirk, J. Mendez-Arroyo, A.M. Lifschitz, C.A. Mirkin, J. Am. Soc. Chem. **136**, 16594–16601 (2014)
20. C.M. McGuirk, C.L. Stern, C.A. Mirkin, J. Am. Soc. Chem. **136**, 4689–4696 (2014)
21. G. De Bo, D.A. Leigh, C.T. McTernan, S. Wang, Chem. Sci. **8**, 7077–7081 (2017)
22. J. Beswick, V. Blanco, G. De Bo, D.A. Leigh, U. Lewandowska, B. Lewandoski, K. Mishiro, Chem. Sci. **6**, 140–143 (2015)
23. V. Blanco, D.A. Leigh, U. Lewandowska, B. Lewandoski, V. Marcos, J. Am. Soc. Chem. **136**, 15775–15780 (2014)
24. V. Blanco, D.A. Leigh, V. Marcos, J.A. Morales-Serma, A.L. Nussbaumer, J. Am. Soc. Chem. **136**, 4905–4908 (2014)
25. S. Mortezaei, N.R. Catarineu, J.W. Canary, J. Am. Soc. Chem. **134**, 8054–8057 (2012)
26. R. Göstl, A. Senf, S. Hecht, Chem. Soc. Rev. **43**, 1982–1996 (2014)
27. E. Merino, M. Ribagorda, Beilstein J. Org. Chem. **8**, 1071–1090 (2012)
28. D.H. Waldeck, Chem. Rev. **91**, 415–436 (1991)
29. M. Irie, Chem. Rev. **100**, 1685–1716 (2000)
30. F. Würthner, J. Rebek Jr., Angew. Chem. Int. Ed. **34**, 446–448 (1995)
31. R. Cacciapaglia, S. Di Stefano, L. Mandolini, J. Am. Soc. Chem. **125**, 2224–2227 (2003)
32. T. Ikariya, M. Shibasaki (eds.), *Bifunctional Molecular Catalysis in Topics in Organometallic Chemistry*, vol 37 (Springer, Berlin/Heidelberg, 2011)
33. M. Shibasaki, M. Kanai, K. Funabashi, Chem. Commun., 1989–1999 (2002)

34. J. Park, S. Hong, Chem. Soc. Rev. **41**, 6931–6943 (2012)
35. L. Stegbauer, F. Sladojevich, D.J. Dixon, Chem. Sci. **3**, 942–958 (2012)
36. S. Matsunaga, M. Shibasaki, Chem. Commun. **50**, 1044–1057 (2014)
37. J. Wamg, B.L. Feringa, Science **331**, 1429–1432 (2011)
38. T. Imahori, R. Yamaguchi, S. Kurihara, Chem. Eur. J. **18**, 10802–10807 (2012)
39. M. Vlatkovic, L. Bernardi, E. Otten, B.L. Feringa, Chem. Commun. **50**, 7773–7775 (2014)
40. D. Zhao, T.M. Neubauer, B.L. Feringa, Nat. Commun. **6**, 6642 (2015)
41. M. Samanta, V.S.R. Krishna, S. Bandyopadhyay, Chem. Commun. **50**, 10577–10579 (2014)
42. N. Koumura, R.W.J. Zijlstra, R.A. van Delden, N. Harada, B.L. Feringa, Nature **401**, 152–155 (1999)
43. R. Eelkeme, M.M. Pollard, J. Vicario, N. Katsonis, B.S. Ramon, C.W.M. Bastiaansen, D.J. Broer, B.L. Feringa, Nature **440**, 163 (2006)
44. M. Schliwa (ed.), *Molecular Motors* (Wiley-VCH, Weinheim, 2003)
45. C. Rabalakos, W.D. Wulff, J. Am. Chem. Soc. **130**, 13524–13525 (2008)
46. N.K. Rana, S. Selvakumar, V.K. Singh, J. Org. Chem. **75**, 2089–2091 (2010)
47. A.K. Unni, N. Takenaka, H. Yamamoto, V.H. Rawal, J. Am. Chem. Soc. **127**, 1336–1337 (2005)
48. Y.M.A. Yamada, S. Ikegami, Tetrahedron Lett. **41**, 2165–2169 (2000)
49. N.T. McDougal, S.E. Schaus, J. Am. Chem. Soc. **125**, 12094–12095 (2003)
50. L. Osorio-Planes, C. Rodríguez-Escrich, M.A. Pericàs, Org. Lett. **16**, 1704–1707 (2014)
51. Z. Dai, Y. Cui, C. Chen, J. Wu, Chem. Commun. **52**, 8826–8829 (2016)
52. D. Sud, T.B. Norsten, N.R. Branda, Angew. Chem. Int. Ed. **44**, 2019–2021 (2005)
53. Z.S. Kean, S. Akbulatov, Y. Tian, R.A. Widenhoefer, R. Boulatov, S.L. Craig, Angew. Chem. Int. Ed. **53**, 14508–14511 (2014)
54. H.A. McManus, P.J. Guiry, Chem. Rev. **104**, 4151–4202 (2004)
55. R.E. Lowenthal, A. Abiko, S. Masamune, Tetrahedron Lett. **31**, 6005–6008 (1990)
56. D.A. Evans, K.A. Woerpel, M.M. Hinman, M.M. Faul, J. Am. Chem. Soc. **113**, 726–728 (1991)
57. M.V. Peters, R.S. Stoll, A. Kühn, S. Hecht, Angew. Chem. Int. Ed. **47**, 5968–5972 (2008)
58. R.S. Stoll, M.V. Peters, A. Kühn, S. Heiles, R. Goddard, M. Bühl, C.M. Thiele, S. Hecht, J. Am. Soc. Chem. **131**, 357–367 (2009)
59. D.J. Berry, C.V. DiGiovanna, S.S. Metrick, R. Murugan, ARKIVOC **i**, 201–226 (2001)
60. C.T. Chen, C.C. Tsai, P.K. Tsou, G.T. Huang, C.H. Yu, Chem. Sci. **8**, 524–529 (2017)
61. B.M. Neilson, C.W. Bielawski, J. Am. Soc. Chem. **134**, 12693–12699 (2012)
62. B.M. Neilson, C.W. Bielawski, Organometallics **32**, 3121–3128 (2013)
63. A.J. Teator, H. Shao, G. Lu, P. Liu, C.W. Bielawski, Organometallics **36**, 490–497 (2017)
64. R.A. Weitekamp, H.A. Atwater, R.H. Grubbs, J. Am. Chem. Soc. **135**, 16817–16820 (2013)
65. D. Wang, K. Wurst, W. Knolle, U. Decker, L. Prager, S. Naumov, M.R. Buchmeiser, Angew. Chem. Int. Ed. **47**, 3267–3270 (2008)
66. S.A. MacDonald, C.G. Willson, J.M.J. Fréchet, Acc. Chem. Res. **27**, 151–158 (1994)
67. B.M. Neilson, C.W. Bielawski, Chem. Commun. **49**, 5453–5455 (2003)
68. A.J. Teator, Y. Tian, M. Chen, J.K. Lee, C.W. Bielawski, Angew. Chem. Int. Ed. **54**, 11559–11563 (2015)
69. W.A. Velema, W. Szymanski, B.L. Feringa, J. Am. Chem. Soc. **136**, 2178–2191 (2014)

Chapter 11
Mimicking Integrated Functions of "Molecular Space" in Biological Systems by Using Crystalline Cavities Consisting of Short Peptides

Ryosuke Miyake

Abstract The intelligent use of "molecular space" is an important feature of biological functions, such as selective or efficient catalysis, molecular transfer, accurate molecular recognition, etc. Peptides provide various types of "molecular space" due to the varietyof functional groups and highly ordered structure, in which integrated functions are demonstrated, supported by, in some cases, adaptable structrual changes (such as induced fit) becasuse of their structural flexibility. In addition, correlation of several types of "molecular space," through cooperative structural changes in the peptide framework, is key for environmental responsive functions of biological systems. These features of peptides could be applied to artificial systems in the design of functional solid materials. Crystalline materials are one of the easiest approaches to construct materials consisting of several units which correlate each other due to the structural regularity. Thus, until now, various crystalline nano-cavities (as was "molecular space") consisting of short peptide moieties (hereafter "peptide crystalline nano-cavities") have been reported, showing their potential for the design of artificial crystalline nano-cavities. Various features of biological systems, such as design properties and adaptable properties (such as induced fit), have been successfully mimicked in reported systems. Recently, dynamic features such as cooperative bindings have also been demonstrated using peptide crystalline cavities. These features could lead to catalysis supported by selective or efficient molecular capture as was "molecular space" in enzyme in near future. In this review, recent developments of peptide crystalline nano-cavities are overviewed.

Keywords Crystalline cavities · Peptides · Molecular inclusions · Mimic of molecular space in biological systems

R. Miyake (✉)
Department of Chemistry and Biochemistry, Graduate School of Humanities and Sciences, Ochanomizu University, Tokyo, Japan
e-mail: miyake.ryosuke@ocha.ac.jp

S. Shirakawa (ed.), *Designed Molecular Space in Material Science and Catalysis*,
https://doi.org/10.1007/978-981-13-1256-4_11

11.1 Introduction

Biological systems use various types of "molecular space" as sites for accurate recognition of environmental factors and substrates and/or for catalytic reactions. In the molecular space, multi-interactions with various functional groups of peptides undertake accurate recognition of substrates. In some cases, the recognitions are supported by structural changes to fit the target molecules (i.e., induced fit). In addition, cooperative structural changes between molecular spaces give biological systems environmental-responsive functions. In such biological functions, the structural features of peptides, as well as their design feature, play significant roles. Thus, peptides should also be suitable for the construction of artificial nano-cavities with highly efficient and accurate switching functions. For use as artificial materials, however, proteins, which possess such molecular space, contain more than 50 amino-acid residues (e.g., in the case of insulin), and are not suitable due to their low stability and challenging artificial synthesis pathways. To overcome this problem, crystalline nano-cavities that consist of short peptides ("peptide crystalline nano-cavities") can be used to provide suitable molecular space: crystal packing structures should help to construct framework structure corresponding to the higher structures of peptides. In this chapter, peptide crystalline cavities are reviewed to discuss their role on designing properties of cavities. The reported functions are summarized, and the possibility to further develop sophisticated functions, such as accurate and auto-active functions including catalytic reactions in artificial materials, is explored.

11.2 Crystalline Cavities Consist of Oligopeptide and Their Inclusion Properties

Crystalline peptide nano-cavities were originally founded in the molecular crystals of pure oligopeptides and may be subdivided into two groups, cyclic peptides and dipeptides. The pioneering development of crystalline cavities formed by cyclic peptides was reported by Gardhili and co-workers in 1993, as a part of a study of ion channel mimics using cyclic peptides (Fig. 11.1) [1]. They reported the synthesis of cyclic peptides with alternating sequences of D- and L-amino acids and found that the cyclic peptides form channels of 7–8 Å in diameter in the crystalline phase [1–3]. Generally, cyclization reactions are in low yield and need long reaction times due to the low concentration of the reaction mixture used to avoid intermolecular coupling reactions instead of intramolecular coupling reactions. To improve this problem, Granja and co-workers succeeded in the synthesis of cyclic peptides in good yield (50–75%) under normal conditions for the coupling reaction, by restricting the bond angle between the amino group and the carboxylic group using artificial amino acids (*cis*-3-aminocyclohexanecarboxylic acids) (Fig. 11.2) [4–7]. These cyclic peptides have been studied for their application as ion channels

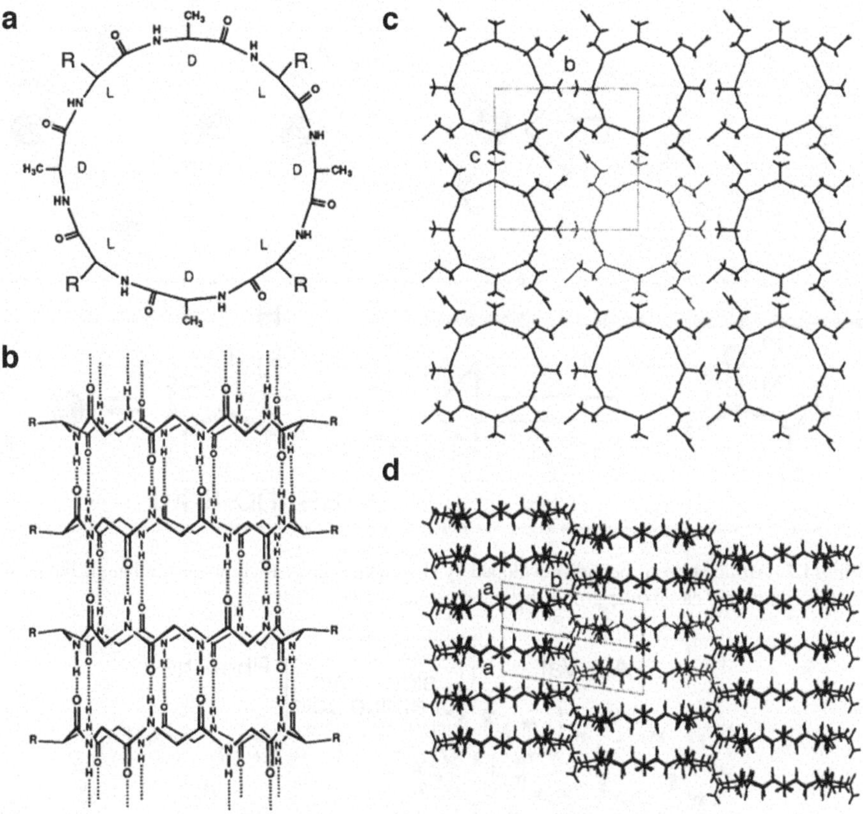

Fig. 11.1 Cyclic peptide reported by Gardhili. Molecular design (**a**) and its assembled tube structure through hydrogen bonds (**b**) and crystal structure from top view (**c**) and side view (**d**). (Reprinted by permission from Macmillan Publishers Ltds: Ref. 1, copyright 1993)

in membranes, but studies on their crystalline cavities were not a major focus. However, they did suggest the possibility of using crystalline cavities for various applications.

The other group consists of molecular crystals of dipeptides (Fig. 11.3): dipeptides form crystalline nano-cavities suitable for molecular uptake with diameters of 4–10 Å (dipeptides consists of hydrophobic amino acids, such as Leu, Phe, Vl, Ile, Ala, etc.). Görbitz reviewed crystalline nano-cavities of dipeptides and summarized the relationship between the size and properties of crystalline cavities and amino-acid sequence [8]. All reported dipeptides [8–11] that form crystalline peptide nano-cavities are listed in Table 11.1. The properties of crystalline cavities (hydrophobic or hydrophilic) are roughly determined by the location of the side chain of the dipeptides. Dipeptides that possess Leu or Phe in their side chain tend to form a hydrophilic cavity with the side chain located on the outside of the cavity. In contrast, dipeptides including Ala, Val, and Ile often form a hydrophobic cavity

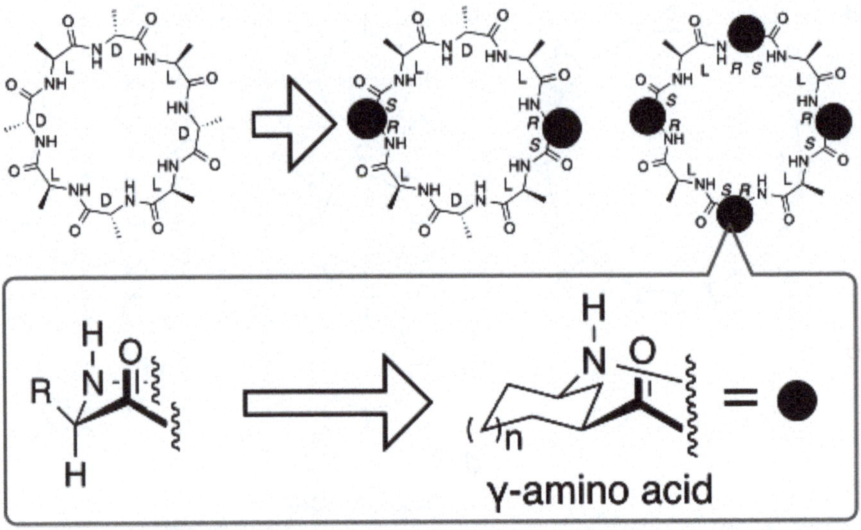

Fig. 11.2 Artificial cyclic peptide developed by Granja through restricting the bond angle between amino groups and carboxylic groups by using cyclic γ-amino acids

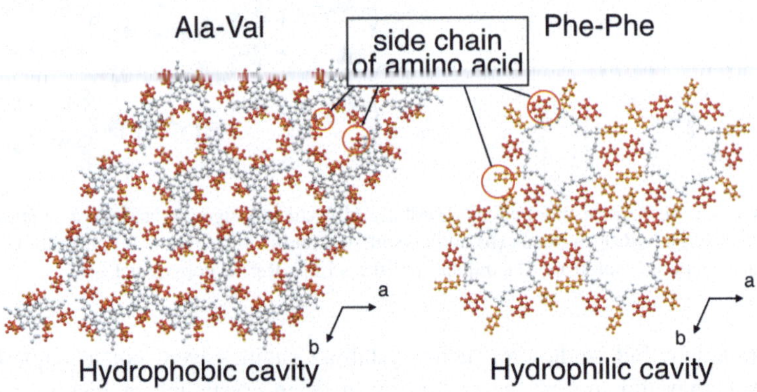

Fig. 11.3 Examples of crystalline cavities in molecular crystals of dipeptides possessing hydrophobic cavity (*a*: Ala-Val) and hydrophilic cavity (*b*: Phe-Phe). The functional groups in the side chain of the dipeptides are shown in red and orange. Crystalline solvent molecules are omitted for clarity

with their side chains located on the inside of the cavity. In most cases, the diameter of the cavity is 3–5 Å. However, Phe-Phe and Ile-Ile form cavities whose diameters are larger than 10 Å. Although the prediction of crystal structure is still difficult, minor modifications of dippetides are one of the useful approaches in the design of peptide crystalline cavities. Görbitz and Comotti reported the tuning of cavity size (2–6 Å) depended on the number of carbon atoms in the side

Table 11.1 Summary of reported crystalline peptide nano-cavities in molecular crystals of dipeptides [8]

Hydrophobic cavity		Crystalline solvent	Diameter of cavity/Å
Val-Ser	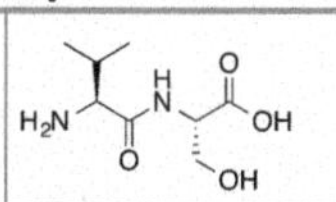	$0.23CF_3CH_2OH$	4.6
Val-Ala	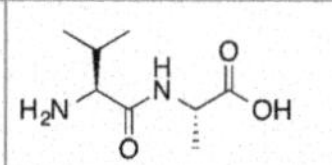	$0.33CH_3CN{\cdot}0.29H_2O$	4.2
		$0.12CH_3CN$	4.7
Ala-Ile	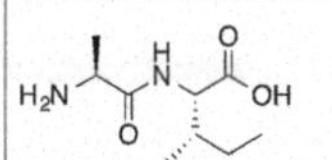	H_2O	4.7
Ile-Ala	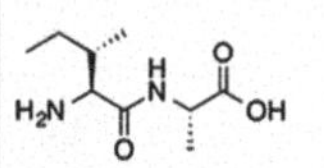	–	3.7
Val-Val	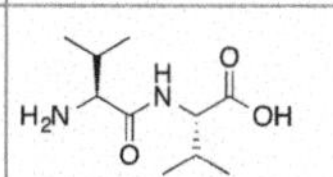	H_2O	4.4
Val-Ile	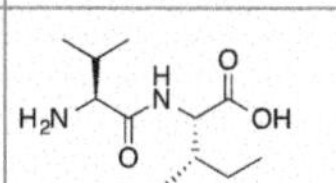	$0.22H_2O$	3.7
Ile-Val	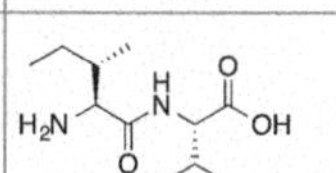	$0.22H_2O$	3.9
Ala-Val	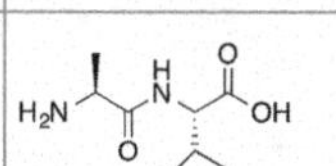	$0.35CH_3CN$	5.0
		$0.25C_3H_7OH{\cdot}0.22H_2O$	5.0, 5.2
Ile-Leu	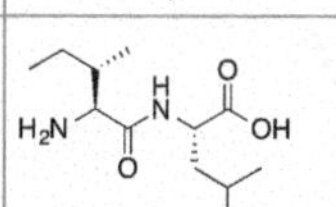	$0.91H_2O$	3.2
Leu-Leu	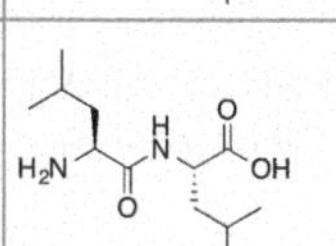	$0.87H_2O$	3.2
Leu-Phe	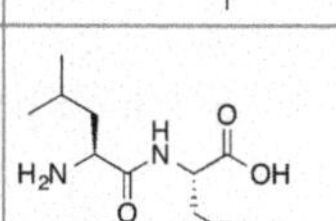	$0.86H_2O$	3.2
Phe-Leu	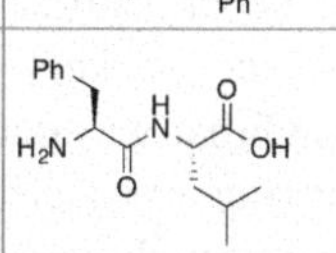	$1.26H_2O$	4.2

(continued)

Table 11.1 (continued)

Hydrophobic cavity		Crystalline solvent	Diameter of cavity/Å
Phe-Phe	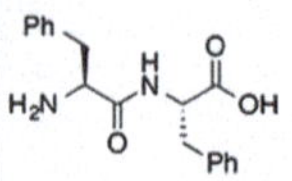	$2.47H_2O$	9.2
Phe-Trp	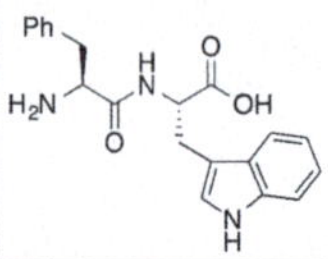	$0.75H_2O$	2.8
Trp-Gly	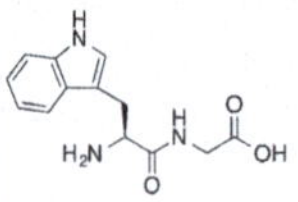	H_2O	4.7
Leu-Ser	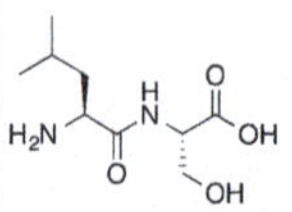	–	4.9

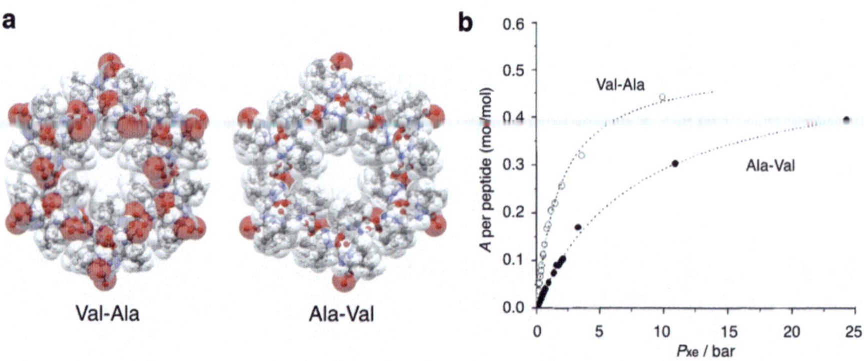

Fig. 11.4 Comparison of cavities in crystalline dipeptides (Val-Ala and Ala-Val) (**a**) and their Xe adsorption isotherm curves at 298 K (**b**). (Reproduced from ref. 10 by permission of John Wiley & Sons Ltd.)

chain of artificial peptides including artificial amino acids which possess various alkyl group (C_nH_{2n+1}: n = 2-4) as the side chain [12].

Several applications were reported for these peptide crystalline nano-cavities consisting of dipeptides. Ripemeester, Soldatov, and co-workers reported high affinity for adsorption of Xe gas (a chemically inert gas) that was demonstrated for crystalline Vla-Ala and Ala-Val, with sorption occurring most efficiently in the Val-Ala channels, which are slightly smaller than the channels of Ala-Val (difference of diameters ca. 0.2 Å) [10, 11] (Fig. 11.4). Also, Val-Ala shows high selectivity for CO_2 gas at ambient temperature [9].

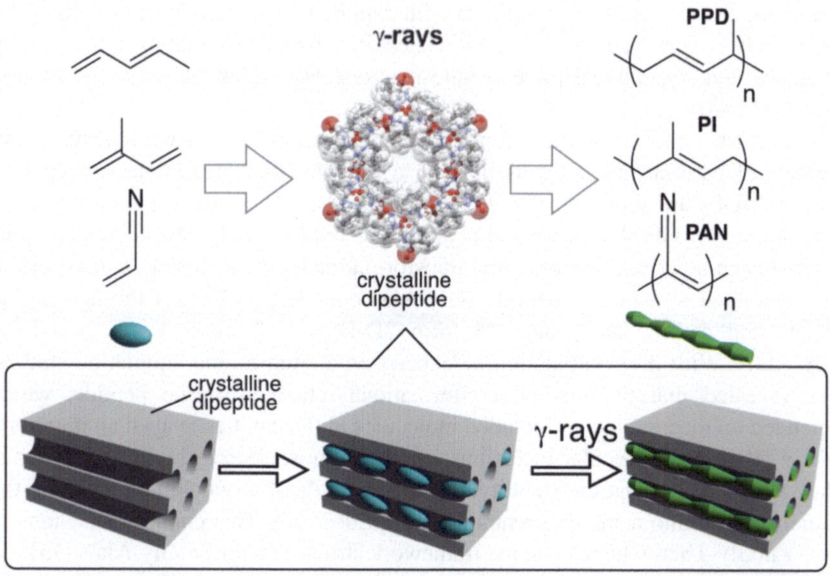

Fig. 11.5 Schematic figure showing the formation of a polymer by using crystalline peptide nano-cavities

Sozzani and co-workers reported that crystalline peptide cavities are useful nano-reactors for various polymers [13] (Fig. 11.5). These peptide cavities are ecologically benign and are easily decomposed without formation of harmful substrates. The results show that crystalline cavities consisting of pure peptides can provide ecologically friendly nano-reactors.

11.3 Crystalline Nano-cavities of Metal Complexes of Peptide Ligands

The overview of crystalline peptide nano-cavities formed from pure short peptide indicates their high potential for the design of crystalline cavities dependant on their amino-acid sequences. However, the design of cavities (size and properties) is still limited. For example, dipeptides that form crystalline peptide cavities mainly contain amino acids possessing hydrophilic functional groups at the side chain, except serine (−OH). On this point, crystalline metal complexes of peptide ligands are an important approach for the design of crystalline cavities since coordination bonds are appropriate for the buildup and design of cyclic structures and/or network structures. Peptides possess various functional groups on their amino-acid residues (amino groups, carboxylic groups, etc.) that could work as metal coordinating sites. In addition, there are a lot of artificial amino acids or peptides that possess various functional groups that could also function as metal coordinating sites. Up to now, crystalline cavities of metal complexes with peptide ligands have been developed by three kinds of approaches (peptide MOFs, cavities formed by folding and assembly strategies, and crystalline peptide metallo-macrocycles (which I described in the next section (11.4))).

Rosseinsky and co-workers reported a metal-organic framework (MOF) that consists of natural dipeptides as ligands, a peptide MOF, (examples of reported peptide MOFs are shown in Fig. 11.6) in 2010 [14]. They first reported that a Zn (II) complex of Gly-Ala possessed crystalline cavities which showed switching of gas inclusion by crystalline structural transformation by the inclusion of gas itself. In the crystalline structure, dipeptide (Gly-Ala) connects Zn^{2+} ions through amino groups and carboxylate groups to form 2D-layered framework compound Zn (Gly-Ala)$_2$. With the combination of experimental studies and simulation studies, they revealed that the gradual conformational changes of the peptide, which depended on the amounts of included molecules in the cavities, played an important role in adaptable molecular inclusions similar to those seen in biological systems. They also revealed that the stability and structural flexibility could be tuned by the sequence of amino acids. For example, dipeptides (Gly-Thr) connect Zn^{2+} ions to form Zn(Gly-Thr)$_2$ which has same framework structure with Zn(Gly-Ala)$_2$ [15]. Zn (Gly-Ala)$_2$ forms a flexible cavity which displays adaptable shape that evolves continuously from an open to a partially disordered closed structure in response to the level of guest loading. In contrast, Zn(Gly-Thr)$_2$ forms a rigid cavity and

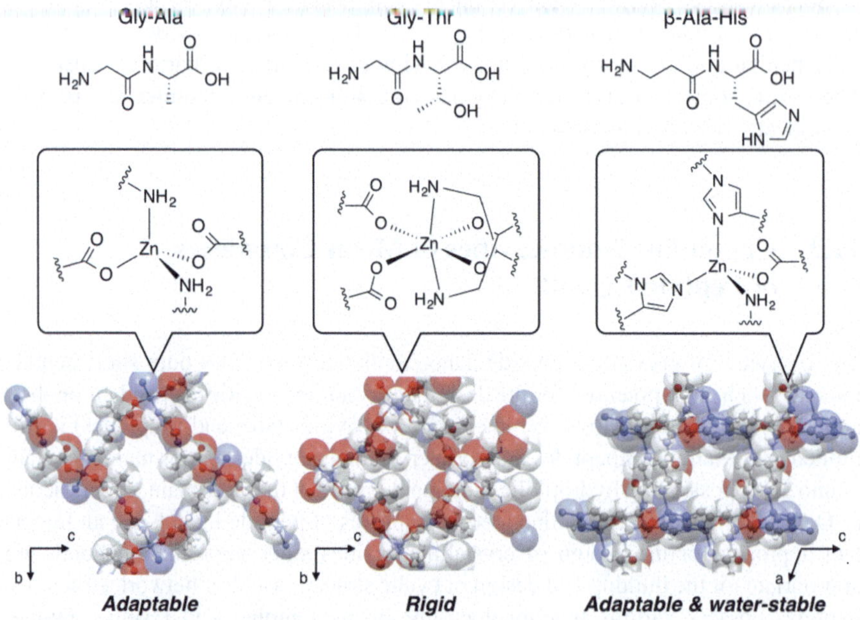

Fig. 11.6 Crystal structures of peptide MOFs as Zn complex of dipeptide (*a*: Gly-Ala, *b*: Gly-Thr, *c*: β-Ala-His). The minimum units of square frameworks in the crystal structures were shown in the bottom of the figure

maintains a framework structure after the removal of included molecules. Such structural properties can be tuned nonlinearly by mixing the ratio of dipeptide ligands [16]. This phenomenon was similar to that observed in single-point mutations of proteins, in which exchange of single amino acids can radically alter structure and functions. In addition, they reported that dipeptide βAla-His could form crystalline cavities in a 3D framework that is water-stable but has adaptable cavities: a similar 2D layer framework connects through coordination bonds with imidazolyl groups to form a 3D framework [17]. The series of peptide MOFs successfully demonstrated the design of molecular space that could be tuned by changing one of amino acids in the dipeptide ligand. Recently, they also succeeded to enlarge the size of crystalline cavities by using tripeptides. The Cu(II) complex of tripeptide (Gly-His-Gly or Gly-His-Lys) forms crystalline cavities that are much larger than that of the peptide MOF consisting of dipeptides [18]. By using these enlarged crystalline cavities, they demonstrated enantioselective separation of the chiral drug (+)-ephedrine that can be separated from its racemic mixture >50% [19].

In addition to the designability and structural flexibility of peptides, the other factor useful for the construction of molecular space in biological systems is the folding structure of peptides: folding structures are important especially for constructing large cavities. Crystalline cavities surrounded by folded short peptides were originally demonstrated by Fujita and Sawada in 2014 [20] (Fig. 11.7). In general, short peptides don't form stable folding structures. Thus, to form stable folding structures with short peptides, they applied a trimer peptide including proline, which forms a relatively rigid folding structure because the side chain

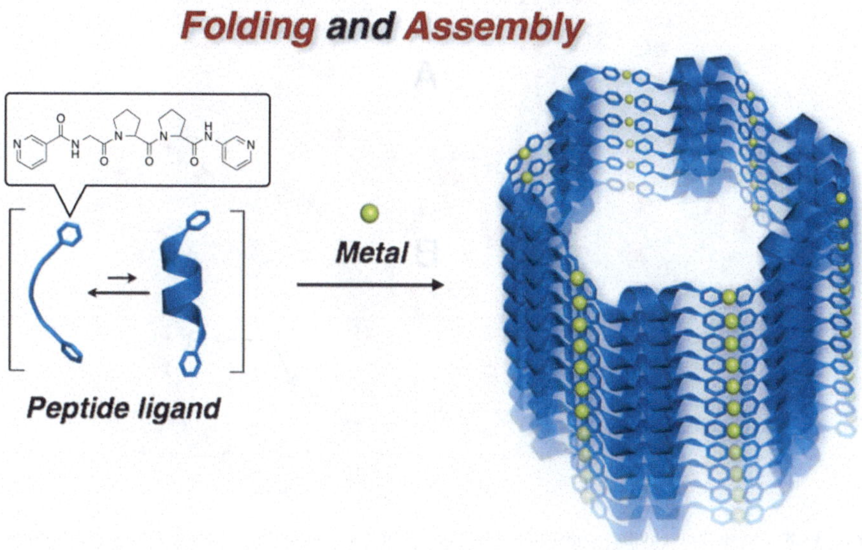

Fig. 11.7 Schematic figure showing the folding and assembly method to construct large peptide crystalline cavities. (Reproduced from Ref. 20 by permission of John Wiley & Sons Ltd.)

directly connects to amino groups and restricts the rotation of the main chain of the peptides. To use metal coordination bondsfor assembly process, they incorporated pyridine groups at both N-and C-termini as metal-binding sites into a native tripeptide (Gly-Pro-Pro). The tripeptide is the shortest residue to form a helical structure. They found that the Ag complex of the tripeptide ligand forms crystalline cavities. Base on the dihedral angle, the conformation of tripeptide could be assigned as the P_{II} helix conformation. As a result, large cavities (diameter is 20 Å) were successfully formed (Fig. 11.8). This pore is the largest cavity reported for peptide-based porous materials, which could include biological molecules such as oligosaccharide (maltopentaose) [20]. The results demonstrated the merit of using a folding and assembly approach to build up large crystalline cavities. Due to the chirality on the peptide framework, this crystalline cavity can be used for chiral recognition of 1,1′-bi-2-naphtol [20]. They also reported that the cavity size changes by modification of the tetrapeptide ligand (replacement of the 5-member ring of proline to 6-member ring or introducing functional groups at the meta position of pyridine) [21]. The properties of the cavities can be tuned by minor changes to the peptide sequence, as has been observed for other peptide crystalline cavities.

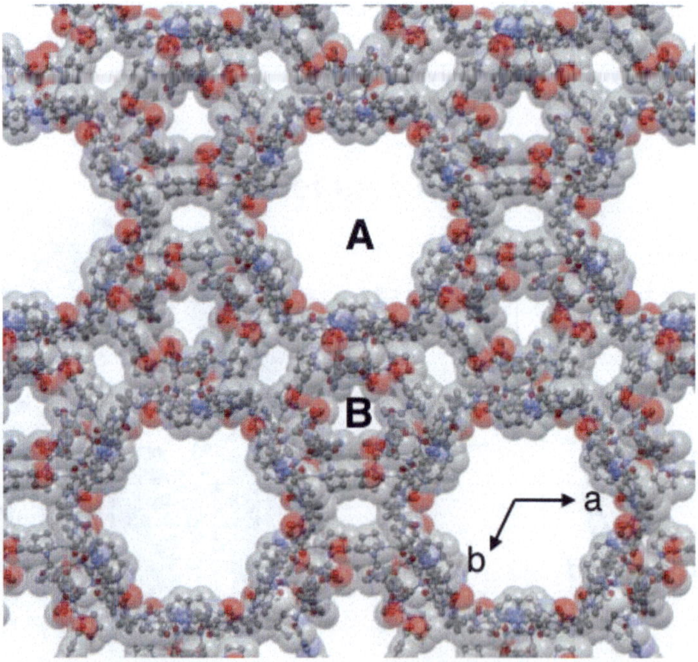

Fig. 11.8 Crystalline cavities constructed by the folding and assembly method using tetrapeptide ligands (Gly-Pro-Pro) possessing pyridine at the N- and C-termini as metal coordinating sites. Two kinds of crystalline cavities (*A* and *B*) are formed surrounded by helical assembly of the peptide ligands. Hydrogen atoms, counteranions, and molecules in the cavities are omitted for clarity

11.4 Switchable Heterogeneous Peptide Nano-cavities

Through peptide MOFs and crystalline cavities obtained using the folding and assembly strategy, the design feature of biological molecular space was principally achieved in artificial systems: peptide frameworks are suitable motifs for the design of crystalline cavities in which the arrangement of various functional groups and their structural flexibility was reflected (on design). On the other hand, reports of switching properties by cooperative structural changes have been limited. Adaptable behavior for guest molecules and sigmoidal behavior of guest inclusion by cooperative structural changes of peptide framework supported by the switching of a hydrogen bond network has been reported for a peptide MOF. However, mimicking of cooperative structural transformations between several cavities has not been reported for these peptide cavities.

Recently, such switching of heterogeneous cavities was demonstrated by using crystals of cyclic metal complexes of peptide ligands (crystalline peptide metallomacrocycles). Molecular crystals of cyclic complexes are suitable for stimuli-responsive structural transformation driven by ligand exchange at the metal center, since compared with crystalline network complexes, such as MOFs, several covalent bonds are replaced by weaker molecular interactions (Fig. 11.9a). In addition, they

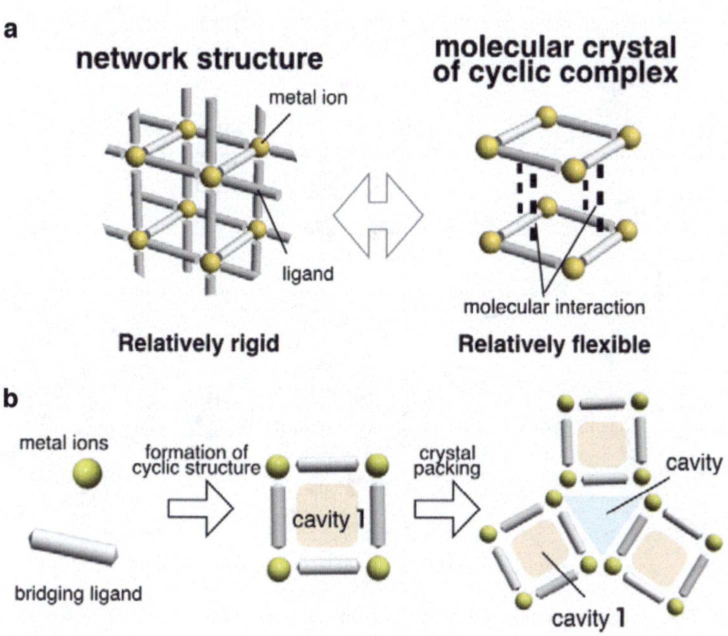

Fig. 11.9 Conceptual figure showing the comparison of structural feature between molecular crystal of cyclic complexes and network complex such as MOF (**a**) and showing the strategy to form heterogeneous cavity by using cyclic complexes (**b**)

are appropriate for designing heterogeneous cavities, both inside and outside of cyclic structures in the crystalline state (Fig. 11.9b). In 2008, we reported a cyclic peptide metal complex by using an artificial β-dipeptide **1** possessing two tridentate-binding sites on its side chain that act as a bridging ligand (refer to Fig. 11.10a) [22]. This artificial β-dipeptide **1** formed a cyclic Ni(II) complex $[Ni_4\mathbf{1}_4]^{8+}$ by connecting four Ni(II) ions and four β-dipeptide ligands in a head-to-tail manner in its crystal structure. The crystal packing structure changes depended on the kinds of counteranions. In the case of NO_3 salt, the peptide Ni(II)-macrocycles possess different crystalline cavities, inside and outside of the macrocycles, which show different affinity for water uptake and release. By using this NO_3 salt, we succeeded to demonstrate reversible structural transformation accompanying cooperative opening and closing of heterogeneous cavities in this crystalline Ni(II)-macrocycle [23] (Fig. 11.10, also refer to Fig. 11.12). This structural transformation occurs with ligand exchange at the Ni(II) center between coordinating sites of the dipeptide

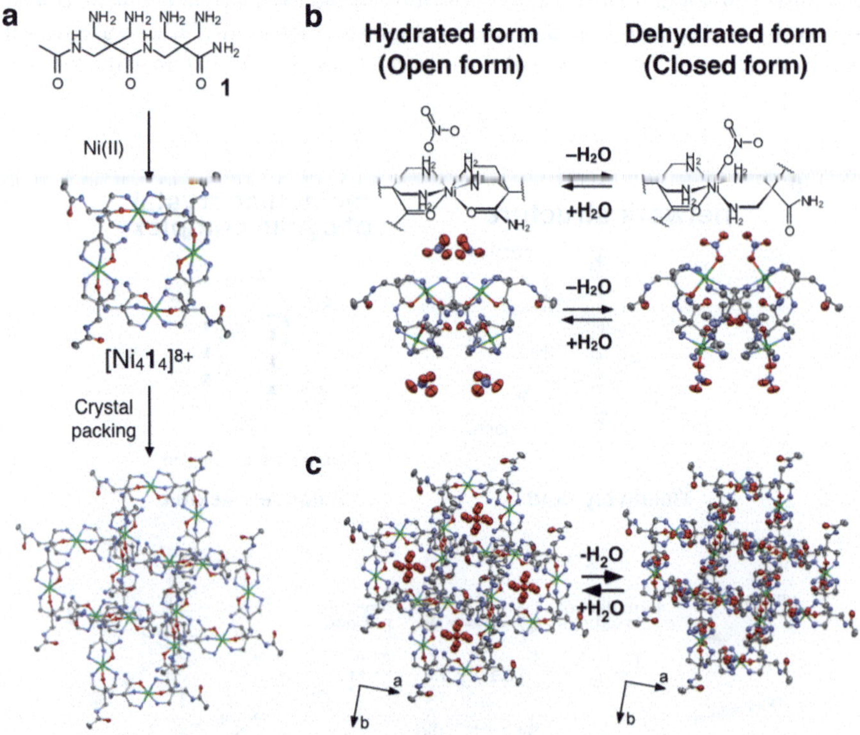

Fig. 11.10 Macrocyclic structure and crystal packing structure of the NO_3 salt of peptide Ni(II)-macrocycle $[Ni_4\mathbf{1}_4]^{8+}$ consisting of an artificial β-dipeptide **1** (**a**) and its structural transformation between an open form and a closed form by water uptake/release in macrocyclic units (**b**) and crystal packing structures (**c**). Hydrogen atoms, some crystalline water molecules, and some counteranions are omitted for clarification. (Reproduced from Ref. 23 (Royal Chemical Society))

ligand and nitrate ion by the recognition (inclusion) of water molecules in one of the cavities, supported by switching of the hydrogen bond network. The other feature of this motif is the existence of counterions in crystalline cavities: counteranions located inside the cavities since it is necessary to overcome the positive charge at the Ni(II) center. X-ray structural studies revealed that the counteranion (NO_3^- in this case) changes its location and supports this structural transformation by filling the gaps. Although various ionic interactions, between NH_4^+ and COO^- on the side chain, are also important for folding structures and could affect their structural transformations in biological systems, such ionic functional groups are difficult to incorporate into crystalline cavities of peptide metal complexes due to their metal coordinating properties. The use of counteranions would be one of the alternative approaches for mimicking electric interactions in the crystalline cavities.

This structural change can be tuned by the kinds of counteranions: the NO_3 salt of the Ni(II)-macrocycle undertakes the structural transformation mentioned above, while the BF_4 salt, which has almost the same structure in hydrated form as the NO_3 salt, does not undertake structural transformation to a closed form and maintains an open form after the removal of water molecules [24]. The crystalline cavities of the BF_4 salt are flexible and adsorbed various gaseous molecules whose diameter is larger than that of the crystalline cavities. The BF_4 salt of Ni(II)-macrocycles preferentially adsorbed CO_2 gas over CH_4 gas (Fig. 11.11) [25]. The ratio of the amount of adsorption is 22:1 at 100 kPa (293 K), which is relatively high among those reported for MOF compounds. Since a high adsorption enthalpy was observed for CO_2 gas adsorption, interaction between amino groups and CO_2 gas could be one of the reasons for the highly selective capture of CO_2 gas versus CH_4 gas.

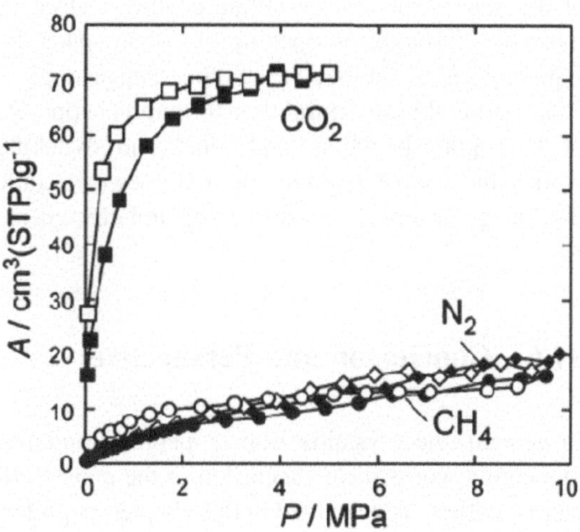

Fig. 11.11 Adsorption isotherm curves of various gases (CO_2, N_2, CH_4 gas) for the BF_4 salt at 293 K. (Reproduced from Ref. 24 (Royal Chemical Society))

11.5 Mimicking Cooperative Binding Systems in Artificial Systems

The cooperativity of several kinds of molecular space is an important element for the demonstration of accurate environmental-responsive functions in biological systems. In particular, heterotropic cooperative binding systems are desired as mimics for artificial systems since they could provide efficient molecular transfer or storage as well as accurate environmental sensing as is observed in the Bohr effect of hemoglobin and kinase in biological systems. Mimicking such cooperative in artificial systems would lead to sophisticated environmental-responsive (auto-active) functions.

In 2018, we succeeded to mimic such heterotropic cooperative binding systems by using the structural switch of heterogeneous crystalline cavities in a peptide Ni(II)-macrocycle which was described in the previous section (Fig. 11.12). In this crystal, upon cooperative structural transformation between an "open form" and a "closed form" induced by water uptake, the water uptake into one of the cavities opens/closes the other cavities cooperatively as is observed in heterotropic cooperative binding systems in biological systems [26]. As shown in Fig. 11.12b, the crystalline peptide Ni(II)-macrocycle adsorbs CO_2 gas up to 34 g·mol^{-1} by mixing appropriate water vapor pressure to transform into the "open form," while it adsorbs a negligible amount of CO_2 gas in the "closed form" without water vapor. The adsorption of CO_2 gas was supported by gas composition analysis by using gas chromatography. The reversible switching of CO_2 adsorption responding to surrounding humidity was clearly observed as IR spectrum changes of the macrocycles under CO_2 gas flow by changing the humidity in the gas. Structural study of its CO_2 including peptide Ni(II)-macrocycle revealed that cooperative opening and closing of the heterogeneous crystalline cavities caused the inclusion and release of CO_2 molecules through the opening of vacant space. In addition, inclusion of CO_2 and H_2O molecules stabilized each other, suggesting a cooperative effect between them. Considering this structural transformation supported by the conformational changes of the peptide backbone and concurrent switching of the hydrogen-bonding networks, the use of peptides as a framework could be a significant approach for designing various cooperative structural changes.

11.6 Conclusion and Perspective

In general, the crystallization of peptide moieties is not easy. However, various pioneering works have succeeded in the construction of several crystalline peptide nano-cavities. As described in this chapter, peptides provide a powerful motif for the design of functional crystalline cavities that can be tuned by the sequence of amino acids and other factors, such as counteranions. As a result, various crystalline cavities have been demonstrated: the diametr of reported cavities varies from 1 Å

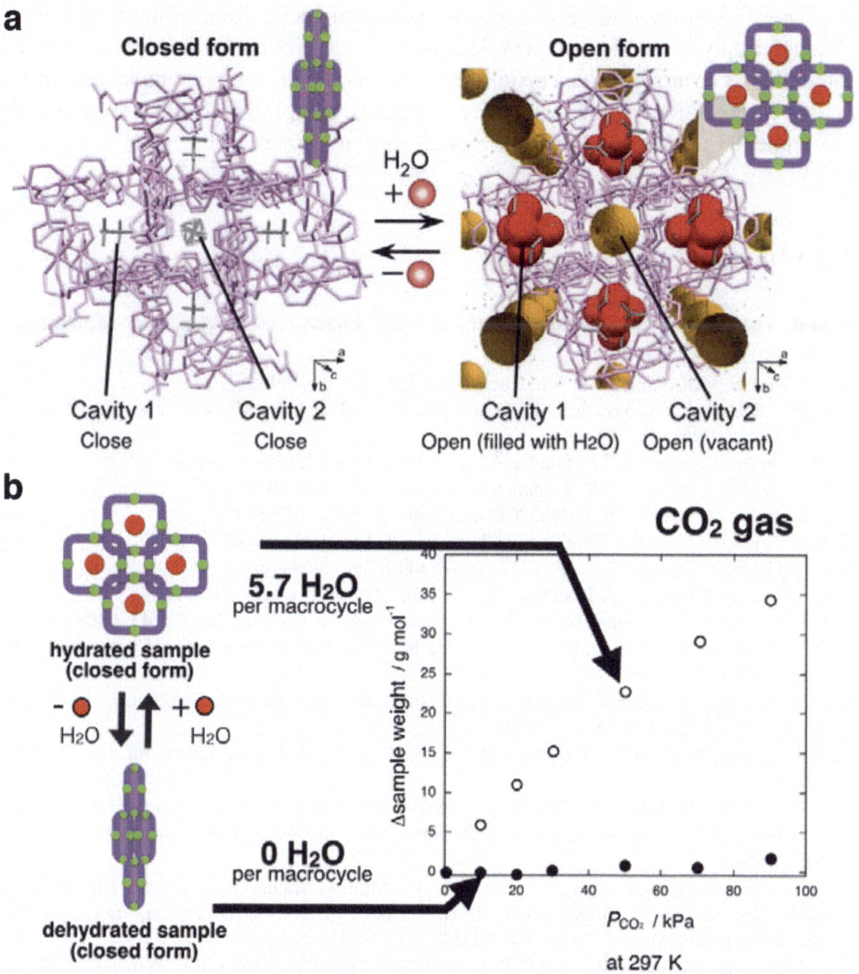

Fig. 11.12 Cooperative opening and closing of heterogeneous cavities in a peptide Ni(II)-macrocycle (**a**) and switching of CO_2 gas inclusion at 297 K by using the cooperative structural transformation (**b**). Estimated vacant space in the open form is colored ocher. (Reproduced from ref. 26 by permission of John Wiley & Sons Ltd.)

to 20 Å. However as seen in chiral recognition using such large cavities, selectivity for uptake is not so high and further improvement is needed. One of the possible solutions for enhancing selectivity is the use of cooperative binding, analogous to the effect known as allosterism in biological systems. Recently, cooperative binding systems have been demonstrated principally by using crystalline peptide metallo-macrocycles. Since the mimicking the main features of peptides in molecular space in biological systems were achieved by using peptide crystalline cavities, peptide crystalline cavities are expected to provide sophisticated combined systems showing integrated functions, leading to accurate molecular recognition in the near future. Although catalytic reactions in peptide crystalline cavities are limited to the report of Comotti and Görbitz, peptide crystalline cavities could also provide molecular space for catalytic reactions in the near future, including complex systems combined with environmental-responsive functions as was seen in kinase.

References

1. M.R. Ghadiri, J.R. Granja, R.A. Milligan, D.E. McRee, N. Khazanovich, Nature **366**, 324 (1993)
2. J.R. Granja, M.R. Ghadiri, J. Am. Chem. Soc. **116**, 10785 (1994)
3. M.R. Ghadiri, K. Kobayashi, J.R. Granja, R.K. Chadha, D.E. McRee, Angew. Chem. Int. Ed. **34**, 93 (1995)
4. M. Amorín, L. Castedo, J.R. Granja, J. Am. Chem. Soc. **125**, 2844 (2003)
5. M. Amorín, L. Castedo, J.R. Granja, Chem. Eur. J. **11**, 6539 (2005)
6. R.J. Brea, L. Castedo, J.R. Granja, Chem. Commun. **31**, 3267 (2007)
7. M. Amorín, L. Castedo, J.R. Granja, Chem. Eur. J. **14**, 2100 (2008)
8. C.H. Görbitz, Chem. Eur. J. **13**, 1022 (2007) and reference therein
9. A. Comotti, S. Bracco, G. Distefano, P. Sozzani, Chem. Commun., 284 (2009)
10. D.V. Soldatov, I.L. Moudrakovski, J.A. Ripmeester, Angew. Chem. Int. Ed. **43**, 6308 (2004)
11. D.V. Soldatov, I.L. Moudrakovski, E.V. Grachev, J.A. Ripmeester, J. Am. Chem. Soc. **128**, 6737 (2006)
12. V.N. Yadav, A. Comotti, P. Sozzani, S. Bracco, T. Bonge-Hansen, M. Hennum, C.H. Görbitz, Angew. Chem. Int. Ed. **54**, 15684 (2015)
13. G. Distefano, A. Comotti, S. Bracco, M. Beretta, P. Sozzani, Angew. Chem. Int. Ed. **51**, 9258 (2012)
14. J. Rabone, Y.-F. Yue, S.Y. Chong, K.C. Stylianou, J. Bacsa, D. Bradshaw, G.R. Darling, N.G. Berry, Y.Z. Khimyak, A.Y. Ganin, P. Wiper, J.B. Claridge, M.J. Rosseinsky, Science **329**, 1053 (2010)
15. A.P. Katsoulidis, K.S. Park, D. Antypov, C. Martí-Gastaldo, G.J. Miller, J.E. Warren, C.M. Robertson, F. Blanc, G.R. Darling, N.G. Berry, J.A. Purton, D.J. Adams, M.J. Rosseinsky, Angew. Chem. Int. Ed. **53**, 193 (2014)
16. C. Martí-Gastaldo, D. Antypov, J.E. Warren, M.E. Briggs, P.A. Chater, P.V. Wiper, G.J. Miller, Y.Z. Khimyak, G.R. Darling, N.G. Berry, M.J. Rosseinsky, Nat. Chem. **6**, 343 (2014)
17. C. Martí-Gastaldo, J.E. Warren, K.C. Stylianou, N.L. Flack, M.J. Rosseinsky, Angew. Chem. Int. Ed. **51**, 11044 (2012)
18. C. Martí-Gastaldo, J.E. Warren, M.E. Briggs, J.A. Armstrong, K.M. Thomas, M.J. Rosseinsky, Chem. Eur. J. **21**, 16027 (2015)
19. J. Navarro-Sánchez, A.I. Argente-García, Y. Moliner-Martínez, D. Roca-Sanjuán, D. Antypov, P. Campíns-Falcó, M.J. Rosseinsky, C. Martí-Gastaldo, J. Am. Chem. Soc. **139**, 4294 (2017)
20. T. Sawada, A. Matsumoto, M. Fujita, Angew. Chem. Int. Ed. **53**, 7228 (2014)
21. T. Sawada, M. Yamagami, S. Akinaga, T. Miyaji, M. Fujita, Chem. Asian J. **12**, 1715 (2017)
22. R. Miyake, S. Tashiro, M. Shiro, K. Tanaka, M. Shionoya, J. Am. Chem. Soc. **130**, 5646 (2008)
23. R. Miyake, M. Shionoya, Chem. Commun. **48**, 7553 (2012)
24. R. Miyake, M. Shionoya, Inorg. Chem. **53**, 5717 (2014)
25. R. Miyake, C. Kuwata, Y. Masumoto, Dalton Trans. **44**, 2993 (2015)
26. R. Miyake, C. Kuwata, M. Ueno, T. Yamada, Chem. Eur. J. **24**, 793 (2018)